ELECTROKINETIC PARTICLE TRANSPORT IN MICRO-/NANOFLUIDICS

Direct Numerical Simulation Analysis

SURFACTANT SCIENCE SERIES

FOUNDING EDITOR

MARTIN J. SCHICK
1918–1998

SERIES EDITOR

ARTHUR T. HUBBARD
Santa Barbara Science Project
Santa Barbara, California

1. Nonionic Surfactants, *edited by Martin J. Schick* (see also Volumes 19, 23, and 60)
2. Solvent Properties of Surfactant Solutions, *edited by Kozo Shinoda* (see Volume 55)
3. Surfactant Biodegradation, *R. D. Swisher* (see Volume 18)
4. Cationic Surfactants, *edited by Eric Jungermann* (see also Volumes 34, 37, and 53)
5. Detergency: Theory and Test Methods (in three parts), *edited by W. G. Cutler and R. C. Davis* (see also Volume 20)
6. Emulsions and Emulsion Technology (in three parts), *edited by Kenneth J. Lissant*
7. Anionic Surfactants (in two parts), *edited by Warner M. Linfield* (see Volume 56)
8. Anionic Surfactants: Chemical Analysis, *edited by John Cross*
9. Stabilization of Colloidal Dispersions by Polymer Adsorption, *Tatsuo Sato and Richard Ruch*
10. Anionic Surfactants: Biochemistry, Toxicology, Dermatology, *edited by Christian Gloxhuber* (see Volume 43)
11. Anionic Surfactants: Physical Chemistry of Surfactant Action, *edited by E. H. Lucassen-Reynders*
12. Amphoteric Surfactants, *edited by B. R. Bluestein and Clifford L. Hilton* (see Volume 59)
13. Demulsification: Industrial Applications, *Kenneth J. Lissant*
14. Surfactants in Textile Processing, *Arved Datyner*
15. Electrical Phenomena at Interfaces: Fundamentals, Measurements, and Applications, *edited by Ayao Kitahara and Akira Watanabe*
16. Surfactants in Cosmetics, edited by Martin M. Rieger (see Volume 68)
17. Interfacial Phenomena: Equilibrium and Dynamic Effects, *Clarence A. Miller and P. Neogi*
18. Surfactant Biodegradation: Second Edition, Revised and Expanded, *R. D. Swisher*
19. Nonionic Surfactants: Chemical Analysis, *edited by John Cross*
20. Detergency: Theory and Technology, *edited by W. Gale Cutler and Erik Kissa*
21. Interfacial Phenomena in Apolar Media, *edited by Hans-Friedrich Eicke and Geoffrey D. Parfitt*

ELECTROKINETIC PARTICLE TRANSPORT IN MICRO-/NANOFLUIDICS

Direct Numerical Simulation Analysis

Shizhi Qian

Old Dominion University
Norfolk, Virginia, U.S.A.

Ye Ai

Old Dominion University
Norfolk, Virginia, U.S.A.

CRC Press
Taylor & Francis Group
Boca Raton   London   New York

CRC Press is an imprint of the
Taylor & Francis Group, an **informa** business

CRC Press
Taylor & Francis Group
6000 Broken Sound Parkway NW, Suite 300
Boca Raton, FL 33487-2742

© 2012 by Taylor & Francis Group, LLC
CRC Press is an imprint of Taylor & Francis Group, an Informa business

No claim to original U.S. Government works

Printed in the United States of America on acid-free paper
Version Date: 20120409

International Standard Book Number: 978-1-4398-5438-9 (Hardback)

Library of Congress Cataloging-in-Publication Data

Qian, Shizhi.
  Electrokinetic particle transport in micro-/nanofluidics : direct numerical simulation analysis / Shizhi Qian, Ye Ai.
    p. cm. -- (Surfactant science)
  Includes bibliographical references and index.
  ISBN 978-1-4398-5438-9 (hardback)
    1. Electrophoresis. 2. Electrokinetics. 3. Nanofluids. 4. Microfluidics. 5. Molecular biotechnology. I. Ai, Ye, 1983- II. Title.

QP519.9.E434Q24 2012
541'.372--dc23   2011049955

Visit the Taylor & Francis Web site at
http://www.taylorandfrancis.com

and the CRC Press Web site at
http://www.crcpress.com

Contents

Preface

This book focuses on direct numerical simulations (DNSs) of electrokinetic particle transport in micro-/nanofluidics under a direct current (DC) electric field using continuum-based mathematical models and a commercial finite element package, COMSOL Multiphysics (http://www.comsol.com). The emphasis is placed on the DNS of DC electrokinetic phenomena, including electroosmosis, electrophoresis, dielectrophoresis, and induced-charge electrokinetics (ICEK), which have been widely used to manipulate fluids and particles in micro-/nanofluidic devices for various applications. The mathematical models for the DC electrokinetic phenomena and their implementations in COMSOL are detailed. A few results of representative examples that may help readers gain further insight into the complex but intriguing phenomena are discussed.

The text includes a brief introduction to micro-/nanofluidics and a minireview of the basic theories of DC electrokinetics (Chapter 1), numerical simulations of the electrical double layer (EDL) near a charged planar surface and electroosmotic flow in a charged nanopore (Chapter 2), DNS of electrokinetic particle transport in microfluidics using the thin EDL approximation (Chapters 3–6), and DNS of electrokinetic particle transport in nanofluidics (Chapters 7–10), which incorporates the finite EDL effect. Note that this text is *not* a summary of current research in the field of electrokinetics-based micro-/nanofluidics and omits any discussion of alternating current (AC) electrokinetics and detailed micro-/nanofabrication techniques for device fabrication.

We hope that this text may serve as a useful reference for researchers and graduate students in the micro-/nanofluidics community who might have limited experience or time in developing their own DNS codes to explore the active, growing, and interdisciplinary field of research. They could modify and extend the mathematical models, COMSOL files, and MATLAB® codes in this book for their own research. This text can also be used for a graduate course and has been used for a semester-long graduate course at Old Dominion University (Norfolk, VA) as well as Yeungnam University in South Korea.

This text primarily grew from some of our journal articles; therefore, we would like to acknowledge the following coauthors of some journal articles used as chapter material in this text: Professor Ali Beskok, Old Dominion University; Professor Xiangchun Xuan, Clemson University; Professor Sang W. Joo, Yeungnam University; Professor Ashutosh Sharma, Indian Institute of Technology at Kanpur; Professor Benjamin Mauroy, Université Paris 7; and Professor Jing Liu, Huazhong University of Science and Technology. Professor Howard H. Hu from the University of Pennsylvania also offered useful suggestions on Chapter 5. We also would like to acknowledge Barbara Glunn and Kathryn Younce from CRC Press for their support.

Shizhi Qian and Ye Ai
Norfolk, Virginia

MATLAB® is a trademark of The MathWorks, Inc. and is used with permission. The MathWorks does not warrant the accuracy of the text or exercises in this book. This book's use or discussion of MATLAB® software or related products does not constitute endorsement or sponsorship by The MathWorks of a particular pedagogical approach or particular use of the MATLAB® software.

For MATLAB® and Simulink® product information, please contact:

The MathWorks, Inc.
3 Apple Hill Drive
Natick, MA, 01760-2098 USA
Tel: 508-647-7000
Fax: 508-647-7001
E-mail: info@mathworks.com
Web: www.mathworks.com

1 Basics of Electrokinetics in Micro-/Nanofluidics

In the first part of this chapter, we briefly introduce the origin, development, and applications of microfluidics and the evolution from microfluidics to nanofluidics. Numerous applications of micro-/nanofluidics are related to the particle transport confined in micro-/nanoscale channels. Electrokinetics has been one of the most promising tools to manipulate particles in micro-/nanofluidics. Therefore, a comprehensive understanding of the electrokinetic particle transport in micro-/nanoscale channels is crucial to the development of micro-/nanofluidic devices. The second part of this chapter briefly summarizes the basics of electrokinetics under direct current (DC) electric fields, including electrical double layer (EDL), electroosmosis, electrophoresis, dielectrophoresis, and induced-charge electrokinetics (ICEK). Finally, the organization of this text is elaborated.

1.1 INTRODUCTION TO MICRO-/NANOFLUIDICS

Microfluidics refers to a set of technologies that handle fluids (including droplets and suspended particles) geometrically constrained to a small scale, typically larger than 1 μm but less than 1 mm. The original motivation for the development of the microfluidic systems came with the demand of microanalytical tools for biological and chemical applications, especially regarding the explosion of genomics in the 1980s (Whitesides 2006). Meanwhile, the significant advances in microfabrication technology, successfully utilized in microelectronics, also boost the development of microfluidics (Verpoorte and De Rooij 2003; Whitesides 2006). Analogous to the significant impact of integrated electronic circuits on computation and automation, microfluidics holds a similar promise of revolutionizing various biological and chemical applications. The tiny dimension of microchannels and highly integrated channel network fulfill the demands of many biochemical applications for small sample volume, low cost, rapid response, massive parallel and automatic analysis, great sensitivity and portability, and minimal cross contamination.

Microfluidic systems have extensive potential applications, including biodetection, chemical and biological reactors, medicine synthesis, clinical diagnostics, and environmental monitoring (Dittrich and Manz 2006; Melin and Quake 2007; Gomez 2008; Teh et al. 2008; Ahmed et al. 2010; Lombardi and Dittrich 2010; Wang and Wong 2010). A recent market research report from BCC

Research (2010) showed that the global market value of microfluidic products (also called lab-on-a-chip devices) was estimated as $2.6 billion in 2009, which was predicted to increase to nearly $6 billion in 2014, with a compound annual growth rate of 17.7%.

In recent years, there has been growing interest in nanofluidics-based sensing at the level of a single molecule, which requires that at least one characteristic dimension of the confined channel is below 100 nm. The evolution from microfluidics to nanofluidics is accompanied with new emerging physical phenomena (Sparreboom, van den Berg, and Eijkel 2009; Schoch, Han, and Renaud 2008; Daiguji 2010). For example, ion transport in nanofluidics is surface charge governed and independent of the bulk ionic concentration owing to the increasing surface-to-volume ratio (Schoch, van Lintel, and Renaud 2005; Stein, Kruithof, and Dekker 2004; Karnik et al. 2005; Nam et al. 2009; Cheng and Guo 2010; Daiguji 2010; Daiguji, Yang, and Majumdar 2003; Joshi et al. 2010). This unique phenomenon offers a probability for selective control of ion transport through nanopores for various applications (Baker, Choi, and Martin 2006; Garcia-Gimenez et al. 2009; Schoch, Han, and Renaud 2008; Vlassiouk, Smirnov, and Siwy 2008). The charge selectivity becomes more significant when the characteristic length of the nanofluidic system becomes comparable to the Debye screening length (Schoch, Han, and Renaud 2008).

Significant advances in nanofabrication technology also enable the study and application of nanofluidics (Kim et al. 2007; Lathrop et al. 2010; Jung et al. 2009; Kalman et al. 2009; Nam et al. 2009; Zhang, Wood, and Lee 2009; Joshi et al. 2010). A diode-like current-voltage behavior through an asymmetric nanopore, referred to as the ionic current rectification phenomenon, shows a potential application in nanofluidic logic circuits (Cheng and Guo 2009, 2010; Cruz-Chu, Aksimentiev, and Schulten 2009; Howorka and Siwy 2009; Vlassiouk, Kozel, and Siwy 2009; Yameen et al. 2009; Ai et al. 2010; Guo et al. 2010; Yusko, An, and Mayer 2010). Analogous to the metal oxide semiconductor field effect transistors (MOSFETs) in microelectronics, the surface charge of the nanopore can be controlled by an electrically addressable gate electrode fabricated along the outer surface of the nanopore. This kind of concept, referred to as a nanofluidic field effect transistor (FET), opens an opportunity to build large-scale integrated ionic circuits for complex biochemical analysis and computation. The translocation of DNA molecules through a nanopore can be utilized to interrogate the order of nucleotide bases in a single DNA molecule. Nanopore-based DNA sequencing has emerged as one of the most promising approaches to achieve high-throughput and affordable DNA sequencing (Storm et al. 2005; Rhee and Burns 2006; Healy, Schiedt, and Morrison 2007; Dekker 2007; Griffiths 2008; Howorka and Siwy 2009; Gupta 2008; Mukhopadhyay 2009; Derrington et al. 2010; Lathrop et al. 2010; McNally et al. 2010). Nanofluidics also provides potential applications in clean energy generation (van der Heyden et al. 2006; Pennathur, Eijkel, and van den Berg 2007; Xie et al. 2008; Wang and Kang 2010) and water purification and desalination (Kim et al. 2010; Shannon 2010).

1.2 PARTICLE TRANSPORT AND MANIPULATION IN MICRO-/NANOFLUIDICS

Transport and manipulation of micro-/nanoscale synthesized particles and bioparticles for biomedical applications have become critical issues in micro-/nanofluidics (Toner and Irimia 2005; Castillo, Dimaki, and Svendsen 2009; Kang and Li 2009). For example, blood is a complex mixture of various cells, including red blood cells, white blood cells, and platelets. If only red blood cells are required for analysis, they thus have to be separated from the others and concentrated and trapped in a microfluidic device prior to further genomic analysis or clinical diagnostics (Toner and Irimia 2005). Traditional manipulation techniques for macroscopic objects are not efficient for micro-/nanoscale particles due to the size effect. As a result, various techniques have been developed to manipulate particles in micro-/nanofluidics.

Electrokinetics refers to the use of electric fields to exert electrostatic forces on charged or polarizable fluids and suspended particles, which in turn induces the motions of fluids and particles. With significant advancement in micro-/nanofabrication technology, electric fields can be easily scaled down to the micro-/nanoscale, in which electrokinetics becomes one of the most dominant effects. As a result, electrokinetics has emerged as one of the most promising techniques to transport and manipulate particles in micro-/nanofluidics using only electric fields with no moving parts (Ramos et al. 1998; Hughes 2000; Wong et al. 2004; Kang and Li 2009; Karniadakis, Beskok, and Aluru 2005). Therefore, electrokinetics-based microfluidic devices have much better reliability than mechanical microfluidic devices. In addition, the fabrication of tiny electrodes is much easier than that of moving components in the micro-/nanoscale. Last, electrokinetics-based microfluidic devices are more compatible with the external microelectronic devices.

Based on the type of the applied electric field, electrokinetic phenomena can be categorized into DC and alternating current (AC) electrokinetics. In this book, we focus on the DC electrokinetics, in particular electroosmosis, electrophoresis, dielectrophoresis, and ICEK.

1.3 BASICS OF ELECTROKINETICS

There are many valuable books regarding the theory of electrokinetics and colloidal science, such as the books by Hunter (2001), Li (2004), and Masliyah and Bhattacharjee (2006). In this section, we briefly summarize the basics of DC electrokinetics, which will help in understanding subsequent chapters. We start with the EDL, which plays a crucial role in various electrokinetic phenomena.

1.3.1 ELECTRICAL DOUBLE LAYER

In general, most solid surfaces tend to gain surface charges when they are brought into contact with ionic aqueous solutions. The origin of the surface charge mainly arises from the adsorption or dissociation of chemical groups (Hunter 2001; Li 2004). The electrostatic interaction between the charged surface and the surrounding ions attracts counterions and repels co-ions from the charged surface. As a result, a thin layer predominantly occupied with more counterions is formed in the vicinity of the charged surface, referred to as the EDL. This layer consists of two layers, the stern and diffuse layers, as shown in Figure 1.1.

Ions within the stern layer are immobilized due to a very strong electrostatic force; ions within the diffuse layer are free to move. As a result, we mainly focus on the diffuse layer. The electric potential arising from the net charge within the diffuse layer obeys the classical Poisson equation:

$$-\varepsilon_0 \varepsilon_f \nabla^2 \phi = \sum_{i=1}^{n} F z_i c_i, \tag{1.1}$$

where ε_0 and ε_f are, respectively, the absolute permittivity of vacuum and the relative permittivity of the fluid; ϕ is the electric potential within the fluid; F is the Faraday constant; z_i is the valence of the ith ionic species; c_i is the molar concentration of the ith ionic species; and n is the total number of the ionic species.

The ionic fluxes, including the diffusive, electromigrative, and convective flux densities, are written as

$$\mathbf{N}_i = -D_i \nabla c_i - z_i \frac{D_i}{RT} F c_i \nabla \phi + \mathbf{u} c_i. \tag{1.2}$$

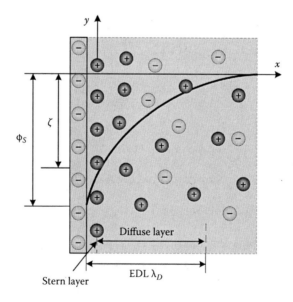

FIGURE 1.1   Schematics of an EDL formed adjacent to a negatively charged planar surface.

In the absence of fluid motion and at steady state, the ionic fluxes satisfy a simplified Nernst–Planck equation:

$$\nabla \bullet \mathbf{N}_i = \nabla \bullet (-D_i \nabla c_i - z_i \frac{D_i}{RT} F c_i \nabla \phi) = 0. \tag{1.3}$$

In this equation, D_i is the diffusivity of the ith ionic species; R is the universal gas constant; and T is the absolute temperature of the electrolyte solution. Equation (1.3) leads to an analytical solution of the ionic concentration, given as

$$c_i = C_{i0} \exp(-z_i \frac{F\phi}{RT}), \tag{1.4}$$

where C_{i0} is the bulk concentration of the ith species. Equation (1.4) is known as the famous Boltzmann distribution. By substituting Equation (1.4) into Equation (1.1), the Poisson–Boltzmann equation is obtained by assuming a binary symmetric ionic solution in a one-dimensional space,

$$\nabla^2 \frac{zF\phi}{RT} = \frac{1}{\lambda_D^2} \sinh(\frac{zF\phi}{RT}). \tag{1.5}$$

Here, $z = |z_i|$ and $\lambda_D = \sqrt{\varepsilon_0 \varepsilon_f RT \Big/ \sum_{i=1}^{2} F^2 z_i^2 C_{i0}}$ is the Debye length, characterizing the EDL thickness. It is shown that the Debye length depends on the bulk concentration of the ionic solution. For example, the Debye length of a charged surface immersed in a 100 mM KCl solution at room temperature (25°C) is about 1 nm. The use of the Poisson–Boltzmann equation implies that the EDL is at its equilibrium state in the absence of any disturbance from the external flow field and electric field. To satisfy the Boltzmann distribution, a far field is also required so that the EDL cannot interact with the other nearby EDLs.

When $\phi \ll \frac{RT}{zF}$, Equation (1.5) can be linearized using the Debye–Hückel approximation (Masliyah and Bhattacharjee 2006):

$$\nabla^2 \frac{zF\phi}{RT} = \frac{1}{\lambda_D^2} \frac{zF\phi}{RT}. \tag{1.6}$$

As a result, the distribution of the electric potential is derived as

$$\phi = \zeta \exp(-\frac{x}{\lambda_D}), \tag{1.7}$$

where ζ is the zeta potential at the shear plane, defined as the interface between the stern layer and the diffuse layer; x is the distance from the shear plane. It must be noted that Equation (1.7) is valid when the zeta potential is relatively small. If $\phi \gg \dfrac{RT}{zF}$, Equation (1.5) can be further derived as the Gouy–Chapman distribution (Masliyah and Bhattacharjee 2006):

$$\phi = 4\frac{RT}{zF}\operatorname{atanh}\left[\tanh(\frac{zF\zeta}{4RT})\exp(-\frac{x}{\lambda_D})\right]. \tag{1.8}$$

1.3.2 ELECTROOSMOSIS

When an external electric field is applied parallel to a stationary charged surface, the excessive counterions within the EDL of the charged surface migrate toward the oppositely charged electrode, dragging the viscous fluid with them. The induced flow motion arising from the electrostatic interaction between the net charge within the EDL and the applied electric field is called electroosmosis, also called electroosmotic flow (EOF), as shown in Figure 1.2.

The electrokinetic force acting on the liquid is written as

$$\mathbf{F} = \mathbf{E}\sum_{i=1}^{n} Fz_i c_i = -\varepsilon_0\varepsilon_f\nabla^2\phi\mathbf{E}, \tag{1.9}$$

where $\mathbf{E}$ is the externally applied electric field. Therefore, the fluid motion is governed by the modified Navier–Stokes (NS) equation,

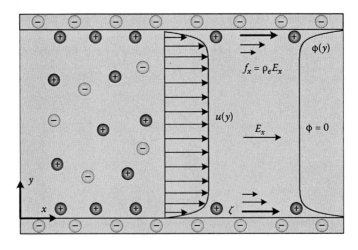

FIGURE 1.2 Schematics of EOF in a slit channel bearing a uniform negative surface charge.

$$\rho(\frac{\partial \mathbf{u}}{\partial t} + \mathbf{u} \bullet \nabla \mathbf{u}) = -\nabla p + \mu \nabla^2 \mathbf{u} - \varepsilon_0 \varepsilon_f \nabla^2 \phi \mathbf{E}, \tag{1.10}$$

and the continuity equation

$$\nabla \bullet \mathbf{u} = 0, \tag{1.11}$$

where ρ is the fluid density; $\mathbf{u}$ is the fluid velocity; p is the pressure; and μ is the fluid dynamic viscosity.

Assuming the external electric field is relatively weak compared to that induced by the surface charge of the solid surface, the ionic concentrations near the charged surface are not affected by the external electric field and the induced EOF. If the EOF is fully developed and steady and there is no external pressure gradient across the charged surface, Equations (1.10) and (1.11) lead to a simplified equation:

$$\mu \frac{d^2 u}{dy^2} = \varepsilon_0 \varepsilon_f \frac{d^2 \phi}{dy^2} E_x, \tag{1.12}$$

where u and E_x are, respectively, the x component fluid velocity and electric field imposed. Using the boundary conditions $u(y=0)=0$, $\frac{du}{dy}(y \to \infty)=0$, $\phi(y=0)=\zeta$, and $\frac{d\phi}{dy}(y \to \infty)=0$, Equation (1.12) can be easily integrated to obtain

$$u = \frac{\varepsilon_0 \varepsilon_f E_x}{\mu} [\phi(y) - \zeta]. \tag{1.13}$$

In this equation, the exact solution of $\phi(y)$ is given in Equation (1.8). As the electric potential due to the surface charge decays to zero in the bulk region, the velocity in the bulk region remains a constant, $u = -\frac{\varepsilon_0 \varepsilon_f E_x \zeta}{\mu}$. The EDL thickness is on the order of nanometers, which is much smaller than the characteristic length of microfluidic devices. As a result, the EOF velocity profile in a microchannel is almost uniform, referring to a plug-like flow, as shown in Figure 1.2. Therefore, one can use the constant velocity to describe the EOF velocity outside the EDL, which is known as the famous Smoluchowski slip velocity.

EOF has been widely utilized to convey fluids in micro-/nanofluidic devices for various applications, including microelectronics cooling (Jiang et al. 2002; Berrouche et al. 2009); high-performance liquid chromatographic separations (Chen, Ma, and Guan 2003, 2004); drug delivery (Hirvonen and Guy 1997; Pikal 2001; Chen, Choot, and Yan 2007); water management in fuel cells (Buie et al. 2006, 2007); and microinjection systems (Pu and Liu 2004; Wang, Chen, and

Chang 2006; Nie, Macka, and Paull 2007; Gan et al. 2000). Due to the intrinsic plug-like flow profile, EOF transport of species samples can highly diminish the dispersion problem, which remains a big issue in pressure-driven flows.

1.3.3 ELECTROPHORESIS

Electrophoresis refers to the migration of charged particles suspended in an aqueous solution subjected to an external electric field, as shown in Figure 1.3. The charged surface is stationary in EOF; however, it becomes mobile in electrophoresis.

The particle's electrophoretic velocity can be written as

$$\mathbf{U}_p = \eta\mathbf{E}, \tag{1.14}$$

where η is the particle's electrophoretic mobility. The governing equations for the steady fluid motion, the electric potential, and the ionic transport are as follows:

$$-\nabla p + \mu\nabla^2\mathbf{u} - \nabla\phi\sum_{i=1}^{n} F z_i c_i = 0, \tag{1.15}$$

$$\nabla\bullet\mathbf{u} = 0, \tag{1.16}$$

$$-\varepsilon_0\varepsilon_f\nabla^2\phi = \sum_{i=1}^{n} F z_i c_i, \tag{1.17}$$

$$\nabla\bullet(-D_i\nabla c_i - z_i\frac{D_i}{RT}Fc_i\nabla\phi + \mathbf{u}c_i) = 0. \tag{1.18}$$

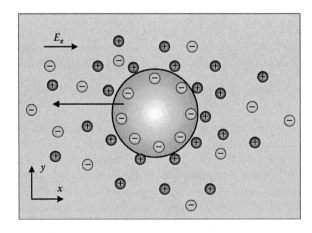

FIGURE 1.3 Schematics of electrophoretic motion of a negatively charged particle under an imposed electric field.

The inertial terms in the NS equations are neglected owing to the low Reynolds number. To determine the particle's steady electrophoretic velocity, one has to balance the hydrodynamic force acting on the particle by the electrostatic force acting on the particle. However, the strongly coupled Equations (1.15)–(1.18) do not lead to a simple analytical solution of the particle's electrophoretic velocity.

Equations (1.17) and (1.18) could be further simplified to a Poisson–Boltzmann equation as described in Equation (1.5) under appropriate conditions discussed in Section 1.4.1. When the zeta potential of the particle is relatively small ($\zeta < \dfrac{RT}{zF}$), the surface conduction within the EDL is negligible. Under a thin EDL ($\lambda_D \ll a$, where a is the characteristic size of the particle), the mobility of a particle suspended in an unbounded medium is described as $\dfrac{\varepsilon_0 \varepsilon_f \zeta}{\mu}$, which is known as the Helmholtz-Smoluchowski law (Masliyah and Bhattacharjee 2006). Under a thick EDL ($\lambda_D \gg a$), Hückel derived the particle mobility as $\dfrac{2\varepsilon_0 \varepsilon_f \zeta}{3\mu}$ (Masliyah and Bhattacharjee 2006). Later, Henry derived the famous Henry's function to account for the effect of finite EDL with an arbitrary thickness on electrophoresis of a sphere in an unbounded medium (Henry 1931). All these analytical solutions are based on equilibrium EDLs and low zeta potentials. In addition, the boundary effect is not considered.

In microfluidic devices, particles are usually confined in a microchannel with a comparable length scale to the particle size. As a result, the boundary effect plays an important role in particle electrophoresis in a confined channel. Keh and Anderson (1985) derived the velocities of a nonconducting rigid sphere near a single flat wall, within a slit channel and a long circular tube under the thin EDL assumption. As discussed previously, the finite EDL effect on particle electrophoresis must be taken into account when the characteristic length of the channel or the particle becomes comparable to the Debye length, which usually happens in nanofluidics. With the consideration of EDL, Ennis and Anderson (1997) derived the analytical approximation solutions for the velocity of a charged sphere near a single flat wall, within a slit channel and a cylindrical tube when the zeta potentials and the applied electric fields are relatively weak and EDLs of the particle and the charged boundary are not overlapped.

The developed analytical solutions are of great help in characterizing electrophoresis of spheres in simple micro-/nanochannels. However, many existing particles are not spherical. In addition, the channel geometries in real micro-/nanofluidic devices are usually complicated. As a result, one needs to determine particle electrokinetic motion numerically in complex micro-/nanochannels. In the numerical study of particle electrophoresis in complex microchannels, the EDL is incorporated with the charged surface as a single entity, referred to as the thin EDL approximation. The Smoluchowski slip velocity is used to describe the EOF near the charged surface. A mathematical model has been developed to track the particle motion dynamically under electrophoresis and EOF (Ye and Li 2004a, 2004b).

As many particles of interest in microfluidic applications, such as biological entities (Gomez 2008) and synthetic nanorods (Appell 2002; Patolsky, Zheng, and Lieber 2006), are nonspherical, increasing attention has been given to the electrophoresis of nonspherical particles in microchannels. Davison and Sharp implemented a transient model to predict the electrokinetic motion of a cylindrical particle through a slit channel (Davison and Sharp 2006, 2007) and an L-shaped microchannel (Davison and Sharp 2008). It was predicted that a cylindrical particle could experience an oscillatory motion in a straight channel (Davison and Sharp 2007), and an L-shaped channel could be used to control the orientation of cylindrical particles (Davison and Sharp 2008). However, the aforementioned numerical studies did not take into account the dielectrophoretic (DEP) effect in numerical modeling, which could play an important role in particle transport in complex microchannels. Investigation of the DEP effect on the electrokinetic particle transport in micro-/nanofluidics is one of the most important objectives in this book.

In the numerical study of particle electrophoresis in nanochannels, the finite EDL effect on particle transport must be considered. A quasistatic method, assuming all the physical fields at their equilibrium states for each particle position, was proposed to predict the particle's translational velocity (Liu, Bau, and Hu 2004; Liu, Qian, and Bau 2007; Hsu and Kuo 2006; Hsu, Kuo, and Ku 2006, 2008; Hsu, Yeh, and Ku 2006; Hsu et al. 2008). In particular, it has been found that the Poisson–Nernst–Planck plus Navier–Stokes (PNP-NS) model is valid for arbitrary EDL thickness; the Poisson–Boltzmann plus Navier–Stokes (PB-NS) model is not valid under EDL overlapping (Liu, Qian, and Bau 2007).

Electrophoresis has been widely used to pump, separate, and characterize colloidal particles and biological materials in microfluidics (Hunter 2001; Li 2004; Kang and Li 2009). In the recent nanopore-based biosensing technique, nanoparticles are also electrophoretically driven through a nanopore and give rise to a detectable change in the ionic current through the nanopore. This technique has been further developed to achieve affordable and high-throughput nanopore-based DNA sequencing (Storm et al. 2005; Rhee and Burns 2006; Healy, Schiedt, and Morrison 2007; Dekker 2007; Griffiths 2008; Howorka and Siwy 2009; Gupta 2008; Mukhopadhyay 2009; Derrington et al. 2010; Lathrop et al. 2010; McNally et al. 2010).

1.3.4 DIELECTROPHORESIS

Dielectrophoresis refers to the motion of polarizable particles immersed in an aqueous solution subjected to a spatially nonuniform electric field (Pohl 1978), as shown in Figure 1.4. The ratio of the polarizability of particles to that of the electrolyte solution determines the direction of the DEP force. A positive (negative) dielectrophoresis refers to the DEP force directed toward (away from) the region with a higher electric field. The DEP force is proportional to the square of the

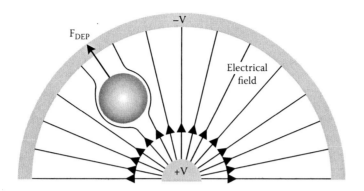

FIGURE 1.4 Schematics of negative dielectrophoresis of an uncharged particle subjected to a spatially nonuniform electric field.

electric field, indicating nonlinear electrokinetics. In addition, the DEP force is proportional to the third power of the particle size.

The time-averaged AC DEP force acting on a spherical particle of radius r obtained by a point dipole method is expressed as (Karniadakis, Beskok, and Aluru 2005)

$$\mathbf{F}_{DEP} = 2\pi r^3 \varepsilon_0 \varepsilon_f \, \mathrm{Re}\big[K(\omega)\big] \nabla |E_{rms}|^2, \tag{1.19}$$

where ω is the frequency of the AC electric field, and E_{rms} is the root mean square electric field strength. The applied electric field $\mathbf{E} = -\nabla\phi$ is related to the electric potential, which satisfies Laplace's equation:

$$\nabla \bullet (\overline{\varepsilon} \nabla \phi) = 0. \tag{1.20}$$

$\mathrm{Re}[K(\omega)]$ represents the real part of the Clausius–Mossotti factor, which is given by

$$K(\omega) = \frac{\overline{\varepsilon}_p - \overline{\varepsilon}_f}{\overline{\varepsilon}_p + 2\overline{\varepsilon}_f}. \tag{1.21}$$

In this equation, $\overline{\varepsilon}_k = \varepsilon_0 \varepsilon_k - i\dfrac{\delta_k}{\omega}$ is the complex permittivity, with δ_k denoting the corresponding conductivity. The point dipole method for the DEP force calculation is only valid when the particle size is much smaller than the characteristic length of the system and the presence of the particle does not significantly affect the electric field. However, the characteristic length of micro-/nanofluidic devices becomes comparable to the particle size, which renders the point dipole method inaccurate for DEP force calculation.

Previous studies have demonstrated that the most rigorous approach for DEP force calculation is direct integration of the Maxwell stress tensor (MST) over the particle surface (Al-Jarro et al. 2007; Rosales and Lim 2005; Wang, Wang, and Gascoyne 1997), which is written as

$$\mathbf{F}_{DEP} = \int \mathbf{T}^E \bullet \mathbf{n} d\Gamma = \int \overline{\varepsilon} \left[\mathbf{EE} - \frac{1}{2}(\mathbf{E} \bullet \mathbf{E})\mathbf{I} \right] \bullet \mathbf{n} d\Gamma. \qquad (1.22)$$

where $\mathbf{T}^E$ is the MST, and Γ denotes the surface of the particle. Wang, Wang, and Gascoyne (1997) revealed that the DEP force obtained by the point dipole method is only the first-order DEP force derived from the MST method.

Numerous experimental studies have implemented AC dielectrophoresis to manipulate colloidal particles and biological cells (Pethig 1996; Zhou, White, and Tilton 2005; Lewpiriyawong, Yang, and Lam 2010; Zhang and Zhu 2010; Koklu et al. 2010; Sabuncu et al. 2010; Park, Koklu, and Beskok 2009) and precisely deposit synthesized nanowires on electrodes (Krupke et al. 2003; Li et al. 2004, 2005; Kumar et al. 2010; Maruyama and Nakayama 2008; Monica et al. 2008; Chang and Hong 2009; Raychaudhuri et al. 2009). In addition, DEP particle-particle interaction arising from AC electric fields has been widely utilized to assemble biological cells and synthesized nanowires into functional structures (Hoffman, Sarangapani, and Zhu 2008; Seo et al. 2005; Tang et al. 2003; Wang et al. 2007; Gangwal, Cayre, and Velev 2008; Velev, Gangwal, and Petsev 2009).

In AC dielectrophoresis, electrodes are usually used to generate nonuniform electric fields in microfluidic devices and in turn induce dielectrophoresis of particles near the electrodes. In DC electrophoresis, usually DC dielectrophoresis is neglected. Recently, it has been found that DC dielectrophoresis also plays an important role in DC electrophoresis under certain conditions, which has been successfully implemented for particle separation (Barbulovic-Nad et al. 2006; Kang, Kang, et al. 2006; Kang, Xuan, et al. 2006; Hawkins et al. 2007; Li et al. 2007; Kang et al. 2008; Lewpiriyawong, Yang, and Lam 2008; Ozuna-Chacon et al. 2008; Parikesit et al. 2008) and particle focusing (Xuan, Raghibizadeh, and Li 2006; Thwar, Linderman, and Burns 2007; Sabounchi et al. 2008; Zhu et al. 2009; Zhu and Xuan 2009a, 2009b) in a continuous flow confined in a microfluidic device. However, the existing numerical models neglect DC dielectrophoresis in electrokinetic particle transport in micro-/nanofluidics, which could lead to inaccurate predictions.

1.3.5 INDUCED-CHARGE ELECTROKINETICS

Electrokinetic flows arising from the interaction between applied electric fields and ideally polarizable channels and particles (i.e., conducting channels and particles), referring to ICEK, have attracted much attention in the micro-/nanofluidics community (Bazant and Squires 2004, 2010; Squires and Bazant 2004). The main difference between conventional electrokinetics and ICEK is the origin of the surface charges. In conventional electrokinetics, the surface charge is gained due to the adsorption or dissociation of specific chemical groups. However, the surface charge in ICEK arises from the polarization of materials. The induced surface charge of an

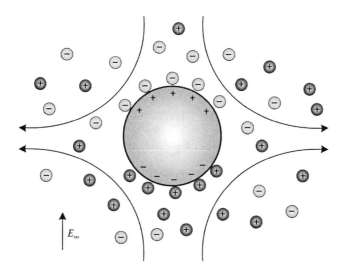

FIGURE 1.5   Schematics of induced-charge electroosmosis around a conducting particle.

ideally polarizable material is generally dipolar, as shown in Figure 1.5, in which the negative surface charge is induced near the anode side; the positive surface charge is generated near the side of the cathode. However, the net induced surface charge is zero, and the electric potential of the conducting material is a constant. The characteristic time scale of the induced surface charging is on the order of $a\lambda_D / D$, where a is the characteristic length of the ideally polarizable material, and D is the ionic diffusivity. Figure 1.5 shows that the flow field around a circular conducting particle is a quadrupolar EOF, which moves toward the particle along the direction of the electric field imposed and leaves the particle radically. The induced zeta potential is not a constant, however, and varies along the surface with an order of aE_∞. Therefore, the electroosmotic slip velocity of ICEK is proportional to the square of the electric field strength, $u \propto -\varepsilon_0\varepsilon_f aE_\infty^2/\mu$. Obviously, ICEK is a nonlinear electrokinetic phenomenon. As the induced zeta potential is tunable through the externally applied electric field and the geometry, ICEK-based microfluidics holds more versatile and sophisticated manipulations of fluids and suspended particles.

So far, ICEK has been successfully utilized to generate circulating flows for fluid stirring and mixing in microfluidics (Zhao and Bau 2007; Wu and Li 2008b, 2008a; Jain, Yeung, and Nandakumar 2009). Particle enrichment and trapping have also been experimentally demonstrated using the ICEK technique (Dhopeshwarkar et al. 2008; Yalcin et al. 2010, 2011).

1.4   ORGANIZATION OF THIS BOOK

This book provides comprehensive direct numerical simulations (DNSs) of DC electrokinetic particle transport in micro-/nanofluidics. The emphasis is placed on the DNS of DC electroosmosis, electrophoresis, dielectrophoresis, and

ICEK in micro-/nanofluidics. A commercial finite element package, COMSOL Multiphysics® 3.5a (http://www.comsol.com), was chosen to conduct all numerical simulations presented in this book. Detailed implementations and corresponding COMSOL files are included.

This chapter briefly discusses the origin, development, and applications of micro-/nanofluidics and summarizes the basic theories of DC electrokinetics and their applications in micro-/nanofluidics. Chapter 2 introduces how to model EDL near a planar surface and EOF in a nanopore using COMSOL Multiphysics 3.5a. The rest of this book can be divided into two parts, electrokinetic particle transport in microfluidics (Chapters 3–6) and electrokinetic particle transport in nanofluidics (Chapters 7–10), which mainly depends on the treatment of the EDL in the numerical modeling.

Chapter 3 presents the DNS of DC electrokinetic transport of circular particles in various microchannels, with several supportive experimental studies. Chapter 4 numerically and experimentally studies the electrokinetic transport of cylindrical-shaped algal cells in a straight microchannel. Chapter 5 numerically studies shear-induced and electrokinetics-induced particle deformation in a straight microchannel. Chapter 6 numerically investigates DEP particle–particle interaction and its relative motions. Chapters 7 and 8 discuss, respectively, the DNS of electrokinetic particle translocation through a nanopore using the PB-NS-based model and the PNP-NS-based model. Chapter 9 numerically demonstrates the feasibility of active regulation of DNA translocation through a nanopore using the FET control. Chapter 10 numerically studies the ICEK effect on electrokinetic particle translocation through a nanopore containing a floating electrode.

REFERENCES

Ahmed, T., T. S. Shimizu, and R. Stocker. 2010. Microfluidics for bacterial chemotaxis. *Integrative Biology* 2 (11-12):604–629.

Ai, Y., M. Zhang, S. W. Joo, M. A. Cheney, and S. Qian. 2010. Effects of electroosmotic flow on ionic current rectification in conical nanopores. *Journal of Physical Chemistry C* 114 (9):3883–3890.

Al-Jarro, A., J. Paul, D. W. P. Thomas, J. Crowe, N. Sawyer, F. R. A. Rose, and K. M. Shakesheff. 2007. Direct calculation of Maxwell stress tensor for accurate trajectory prediction during DEP for 2D and 3D structures. *Journal of Physics D-Applied Physics* 40 (1):71–77.

Appell, D. 2002. Nanotechnology: Wired for success. *Nature* 419 (6907):553–555.

Baker, L. A., Y. S. Choi, and C. R. Martin. 2006. Nanopore membranes for biomaterials synthesis, biosensing and bioseparations. *Current Nanoscience* 2 (3):243–255.

Barbulovic-Nad, I., X. Xuan, J. S. H. Lee, and D. Li. 2006. DC-dielectrophoretic separation of microparticles using an oil droplet obstacle. *Lab on a Chip* 6 (2):274–279.

Bazant, M. Z., and T. M. Squires. 2004. Induced-charge electrokinetic phenomena: Theory and microfluidic applications. *Physical Review Letters* 92 (6):066101.

Bazant, M. Z., and T. M. Squires. 2010. Induced-charge electrokinetic phenomena. *Current Opinion in Colloid and Interface Science* 15 (3):203–213.

BCC Research. 2010. Global biochips market: Microarrays and lab-on-a-chip to be worth $6 billion in 2014. June 6. http://bccresearch.blogspot.com/2010/01/global-biochips-market-microarrays-and.html.

Berrouche, Y., Y. Avenas, C. Schaeffer, H. C. Chang, and P. Wang. 2009. Design of a porous electroosmotic pump used in power electronic cooling. *IEEE Transactions on Industry Applications* 45 (6):2073–2079.

Buie, C. R., D. Kim, S. Litster, and J. G. Santiago. 2007. An electro-osmotic fuel pump for direct methanol fuel cells. *Electrochemical and Solid State Letters* 10 (11):B196–B200.

Buie, C. R., J. D. Posner, T. Fabian, C. A. Suk-Won, D. Kim, F. B. Prinz, J. K. Eaton, and J. G. Santiago. 2006. Water management in proton exchange membrane fuel cells using integrated electroosmotic pumping. *Journal of Power Sources* 161 (1):191–202.

Castillo, J., M. Dimaki, and W. E. Svendsen. 2009. Manipulation of biological samples using micro and nano techniques. *Integrative Biology* 1 (1):30–42.

Chang, Y. K., and F. C. N. Hong. 2009. The fabrication of ZnO nanowire field-effect transistors combining dielectrophoresis and hot-pressing. *Nanotechnology* 20 (23):235202.

Chen, L. X., J. Choot, and B. Yan. 2007. The microfabricated electrokinetic pump: A potential promising drug delivery technique. *Expert Opinion on Drug Delivery* 4 (2):119–129.

Chen, L. X., J. P. Ma, and Y. F. Guan. 2003. An electroosmotic pump for packed capillary liquid chromatography. *Microchemical Journal* 75 (1):15–21.

Chen, L. X., J. P. Ma, and Y. F. Guan. 2004. Study of an electroosmotic pump for liquid delivery and its application in capillary column liquid chromatography. *Journal of Chromatography A* 1028 (2):219–226.

Cheng, L. J., and L. J. Guo. 2009. Ionic current rectification, breakdown, and switching in heterogeneous oxide nanofluidic devices. *ACS Nano* 3 (3):575–584.

Cheng, L. J., and L. J. Guo. 2010. Nanofluidic diodes. *Chemical Society Reviews* 39 (3):923–938.

Cruz-Chu, E. R., A. Aksimentiev, and K. Schulten. 2009. Ionic current rectification through silica nanopores. *Journal of Physical Chemistry C* 113 (5):1850–1862.

Daiguji, H. 2010. Ion transport in nanofluidic channels. *Chemical Society Reviews* 39 (3):901–911.

Daiguji, H., P. Yang, and A. Majumdar. 2003. Ion transport in nanofluidic channels. *Nano Letters* 4 (1):137–142.

Davison, S. M., and K. V. Sharp. 2006. Boundary effects on the electrophoretic motion of cylindrical particles: Concentrically and eccentrically-positioned particles in a capillary. *Journal of Colloid and Interface Science* 303 (1):288–297.

Davison, S. M., and K. V. Sharp. 2007. Transient electrophoretic motion of cylindrical particles in capillaries. *Nanoscale and Microscale Thermophysical Engineering* 11 (1–2):71–83.

Davison, S. M., and K. V. Sharp. 2008. Transient simulations of the electrophoretic motion of a cylindrical particle through a 90 degrees corner. *Microfluidics and Nanofluidics* 4 (5):409–418.

Dekker, C. 2007. Solid-state nanopores. *Nature Nanotechnology* 2 (4):209–215.

Derrington, I. M., T. Z. Butler, M. D. Collins, E. Manrao, M. Pavlenok, M. Niederweis, and J. H. Gundlach. 2010. Nanopore DNA sequencing with MspA. *Proceedings of the National Academy of Sciences of the United States of America* 107 (37):16060–16065.

Dhopeshwarkar, R., D. Hlushkou, M. Nguyen, U. Tallarek, and R. M. Crooks. 2008. Electrokinetics in microfluidic channels containing a floating electrode. *Journal of the American Chemical Society* 130 (32):10480–10481.

Dittrich, P. S., and A. Manz. 2006. Lab-on-a-chip: microfluidics in drug discovery. *Nature Reviews Drug Discovery* 5 (3):210–218.

Ennis, J., and J. L. Anderson. 1997. Boundary effects on electrophoretic motion of spherical particles for thick double layers and low zeta potential. *Journal of Colloid and Interface Science* 185 (2):497–514.

Gan, W. E., L. Yang, Y. Z. He, R. H. Zeng, M. L. Cervera, and M. de la Guardia. 2000. Mechanism of porous core electroosmotic pump flow injection system and its application to determination of chromium(VI) in waste-water. *Talanta* 51 (4):667–675.

Gangwal, S., O. J. Cayre, and O. D. Velev. 2008. Dielectrophoretic assembly of metallodielectric janus particles in AC electric fields. *Langmuir* 24 (23):13312–13320.

Garcia-Gimenez, E., A. Alcaraz, V. M. Aguilella, and P. Ramirez. 2009. Directional ion selectivity in a biological nanopore with bipolar structure. *Journal of Membrane Science* 331 (1–2):137–142.

Gomez, F. A. 2008. *Biological Applications of Microfluidics.* New York: Wiley Interscience.

Griffiths, J. 2008. The realm of the nanopore. *Analytical Chemistry* 80 (1):23–27.

Guo, W., H. Xia, F. Xia, X. Hou, L. Cao, L. Wang, J. Xue, G. Zhang, Y. Song, D. Zhu, Y. Wang, and L. Jiang. 2010. Current rectification in temperature-responsive single nanopores. *ChemPhysChem* 11 (4):859–864.

Gupta, P. K. 2008. Single-molecule DNA sequencing technologies for future genomics research. *Trends in Biotechnology* 26 (11):602–611.

Hawkins, B. G., A. E. Smith, Y. A. Syed, and B. J. Kirby. 2007. Continuous-flow particle separation by 3D insulative dielectrophoresis using coherently shaped, dc-biased, ac electric fields. *Analytical Chemistry* 79 (19):7291–7300.

Healy, K., B. Schiedt, and A. P. Morrison. 2007. Solid-state nanopore technologies for nanopore-based DNA analysis. *Nanomedicine* 2 (6):875–897.

Henry, D. C. 1931. The cataphoresis of suspended particles. Part I. The equation of cataphoresis. *Proceedings of the Royal Society of London Series A* 133:106–129.

Hirvonen, J., and R. H. Guy. 1997. Iontophoretic delivery across the skin: Electroosmosis and its modulation by drug substances. *Pharmaceutical Research* 14 (9):1258–1263.

Hoffman, P. D., P. S. Sarangapani, and Y. X. Zhu. 2008. Dielectrophoresis and AC-induced assembly in binary colloidal suspensions. *Langmuir* 24 (21):12164–12171.

Howorka, S., and Z. Siwy. 2009. Nanopore analytics: Sensing of single molecules. *Chemical Society Reviews* 38 (8):2360–2384.

Hsu, J. P., Z. S. Chen, D. J. Lee, S. Tseng, and A. Su. 2008. Effects of double-layer polarization and electroosmotic flow on the electrophoresis of a finite cylinder along the axis of a cylindrical pore. *Chemical Engineering Science* 63 (18):4561–4569.

Hsu, J. P., and C. C. Kuo. 2006. Electrophoresis of a finite cylinder positioned eccentrically along the axis of a long cylindrical pore. *Journal of Physical Chemistry B* 110 (35):17607–17615.

Hsu, J. P., C. C. Kuo, and M. H. Ku. 2006. Electrophoresis of a toroid along the axis of a cylindrical pore. *Electrophoresis* 27 (16):3155–3165.

Hsu, J. P., C. C. Kuo, and M. H. Ku. 2008. Electrophoresis of a charge-regulated toroid normal to a large disk. *Electrophoresis* 29 (2):348–357.

Hsu, J. P., L. H. Yeh, and M. H. Ku. 2006. Electrophoresis of a spherical particle along the axis of a cylindrical pore filled with a Carreau fluid. *Colloid and Polymer Science* 284 (8):886–892.

Hughes, M. P. 2000. AC electrokinetics: applications for nanotechnology. *Nanotechnology* 11 (2):124–132.

Hunter, R. J. 2001. *Foundations of Colloid Science,* 2nd ed. Oxford, UK: Oxford University Press.

Jain, M., A. Yeung, and K. Nandakumar. 2009. Efficient micromixing using induced-charge electroosmosis. *Journal of Microelectromechanical Systems* 18 (2):376–384.

Jiang, L. N., J. Mikkelsen, J. M. Koo, D. Huber, S. H. Yao, L. Zhang, P. Zhou, J. G. Maveety, R. Prasher, J. G. Santiago, T. W. Kenny, and K. E. Goodson. 2002. Closed-loop electroosmotic microchannel cooling system for VLSI circuits. *IEEE Transactions on Components and Packaging Technologies* 25 (3):347–355.

Joshi, P., A. Smolyanitsky, L. Petrossian, M. Goryll, M. Saraniti, and T. J. Thornton. 2010. Field effect modulation of ionic conductance of cylindrical silicon-on-insulator nanopore array. *Journal of Applied Physics* 107 (5):054701.

Jung, J. Y., P. Joshi, L. Petrossian, T. J. Thornton, and J. D. Posner. 2009. Electromigration current rectification in a cylindrical nanopore due to asymmetric concentration polarization. *Analytical Chemistry* 81 (8):3128–3133.

Kalman, E. B., O. Sudre, I. Vlassiouk, and Z. S. Siwy. 2009. Control of ionic transport through gated single conical nanopores. *Analytical and Bioanalytical Chemistry* 394 (2):413–419.

Kang, K. H., Y. J. Kang, X. C. Xuan, and D. Q. Li. 2006. Continuous separation of microparticles by size with direct current-dielectrophoresis. *Electrophoresis* 27 (3):694–702.

Kang, K. H., X. C. Xuan, Y. Kang, and D. Li. 2006. Effects of dc-dielectrophoretic force on particle trajectories in microchannels. *Journal of Applied Physics* 99 (6):064702.

Kang, Y. J., and D. Q. Li. 2009. Electrokinetic motion of particles and cells in microchannels. *Microfluidics and Nanofluidics* 6 (4):431–460.

Kang, Y. J., D. Q. Li, S. A. Kalams, and J. E. Eid. 2008. DC-dielectrophoretic separation of biological cells by size. *Biomedical Microdevices* 10 (2):243–249.

Karniadakis, G., A. Beskok, and N. Aluru. 2005. *Microflows and Nanoflows: Fundamentals and Simulation.* New York: Springer.

Karnik, R., R. Fan, M. Yue, D. Y. Li, P. D. Yang, and A. Majumdar. 2005. Electrostatic control of ions and molecules in nanofluidic transistors. *Nano Letters* 5 (5):943–948.

Keh, H. J., and J. L. Anderson. 1985. Boundary effects on electrophoretic motion of colloidal spheres. *Journal of Fluid Mechanics* 153:417–439.

Kim, S. J., S. H. Ko, K. H. Kang, and J. Han. 2010. Direct seawater desalination by ion concentration polarization. *Nature Nanotechnology* 5 (4):297–301.

Kim, Y. R., J. Min, I. H. Lee, S. Kim, A. G. Kim, K. Kim, K. Namkoong, and C. Ko. 2007. Nanopore sensor for fast label-free detection of short double-stranded DNAs. *Biosensors and Bioelectronics* 22 (12):2926–2931.

Koklu, M., S. Park, S. D. Pillai, and A. Beskok. 2010. Negative dielectrophoretic capture of bacterial spores in food matrices. *Biomicrofluidics* 4 (3):034107.

Krupke, R., F. Hennrich, H. B. Weber, M. M. Kappes, and H. von Lohneysen. 2003. Simultaneous deposition of metallic bundles of single-walled carbon nanotubes using ac-dielectrophoresis. *Nano Letters* 3 (8):1019–1023.

Kumar, S., Z. C. Peng, H. Shin, Z. L. Wang, and P. J. Hesketh. 2010. AC dielectrophoresis of tin oxide nanobelts suspended in ethanol: Manipulation and visualization. *Analytical Chemistry* 82 (6):2204–2212.

Lathrop, D. K., E. N. Ervin, G. A. Barrall, M. G. Keehan, R. Kawano, M. A. Krupka, H. S. White, and A. H. Hibbs. 2010. Monitoring the escape of DNA from a nanopore using an alternating current signal. *Journal of the American Chemical Society* 132 (6):1878–1885.

Lewpiriyawong, N., C. Yang, and Y. C. Lam. 2008. Dielectrophoretic manipulation of particles in a modified microfluidic H filter with multi-insulating blocks. *Biomicrofluidics* 2 (3):034105.

Lewpiriyawong, N., C. Yang, and Y. C. Lam. 2010. Continuous sorting and separation of microparticles by size using AC dielectrophoresis in a PDMS microfluidic device with 3-D conducting PDMS composite electrodes. *Electrophoresis* 31 (15):2622–2631.

Li, D. 2004. *Electrokinetics in Microfluidics*. New York: Elsevier Academic Press.

Li, J. Q., Q. Zhang, N. Peng, and Q. Zhu. 2005. Manipulation of carbon nanotubes using AC dielectrophoresis. *Applied Physics Letters* 86 (15):153116.

Li, J. Q., Q. Zhang, D. J. Yang, and J. Z. Tian. 2004. Fabrication of carbon nanotube field effect transistors by AC dielectrophoresis method. *Carbon* 42 (11):2263–2267.

Li, Y. L., C. Dalton, H. J. Crabtree, G. Nilsson, and K. Kaler. 2007. Continuous dielectrophoretic cell separation microfluidic device. *Lab on a Chip* 7 (2):239–248.

Liu, H., H. H. Bau, and H. H. Hu. 2004. Electrophoresis of concentrically and eccentrically positioned cylindrical particles in a long tube. *Langmuir* 20 (7):2628–2639.

Liu, H., S. Qian, and H. H. Bau. 2007. The effect of translocating cylindrical particles on the ionic current through a nanopore. *Biophysical Journal* 92 (4):1164–1177.

Lombardi, D., and P. S. Dittrich. 2010. Advances in microfluidics for drug discovery. *Expert Opinion on Drug Discovery* 5 (11):1081–1094.

Maruyama, H., and Y. Nakayama. 2008. Trapping protein molecules at a carbon nanotube tip using dielectrophoresis. *Applied Physics Express* 1 (12):124001.

Masliyah, J. H., and S. Bhattacharjee. 2006. *Electrokinetic and Colloid Transport Phenomena*. New York: Wiley.

McNally, B., A. Singer, Z. L. Yu, Y. J. Sun, Z. P. Weng, and A. Meller. 2010. Optical recognition of converted DNA nucleotides for single-molecule DNA sequencing using nanopore arrays. *Nano Letters* 10 (6):2237–2244.

Melin, J., and S. R. Quake. 2007. Microfluidic large-scale integration: The evolution of design rules for biological automation. *Annual Review of Biophysics and Biomolecular Structure* 36:213–231.

Monica, A. H., S. J. Papadakis, R. Osiander, and M. Paranjape. 2008. Wafer-level assembly of carbon nanotube networks using dielectrophoresis. *Nanotechnology* 19 (8):085303.

Mukhopadhyay, R. 2009. DNA sequencers: The next generation. *Analytical Chemistry* 81 (5):1736–1740.

Nam, S. W., M. J. Rooks, K. B. Kim, and S. M. Rossnagel. 2009. Ionic field effect transistors with sub-10 nm multiple nanopores. *Nano Letters* 9 (5):2044–2048.

Nie, F. Q., M. Macka, and B. Paull. 2007. Micro-flow injection analysis system: On-chip sample preconcentration, injection and delivery using coupled monolithic electroosmotic pumps. *Lab on a Chip* 7:1597–1599.

Ozuna-Chacon, S., B. H. Lapizco-Encinas, M. Rito-Palomares, S. O. Martinez-Chapa, and C. Reyes-Betanzo. 2008. Performance characterization of an insulator-based dielectrophoretic microdevice. *Electrophoresis* 29 (15):3115–3122.

Parikesit, G. O. F., A. P. Markesteijn, O. M. Piciu, A. Bossche, J. Westerweel, I. T. Young, and Y. Garini. 2008. Size-dependent trajectories of DNA macromolecules due to insulative dielectrophoresis in submicrometer-deep fluidic channels. *Biomicrofluidics* 2 (2):024103.

Park, S., M. Koklu, and A. Beskok. 2009. Particle trapping in high-conductivity media with electrothermally enhanced negative dielectrophoresis. *Analytical Chemistry* 81 (6):2303–2310.

Patolsky, F., G. F. Zheng, and C. M. Lieber. 2006. Nanowire-based biosensors. *Analytical Chemistry* 78 (13):4260–4269.

Pennathur, S., J. C. T. Eijkel, and A. van den Berg. 2007. Energy conversion in microsystems: Is there a role for micro/nanofluidics? *Lab on a Chip* 7 (10):1234–1237.

Pethig, R. 1996. Dielectrophoresis: Using inhomogeneous AC electrical fields to separate and manipulate cells. *Critical Reviews in Biotechnology* 16 (4):331–348.

Pikal, M. J. 2001. The role of electroosmotic flow in transdermal iontophoresis. *Advanced Drug Delivery Reviews* 46 (1–3):281–305.

Pohl, H. A. 1978. *Dielectrophresis*. Cambridge, UK: Cambridge University Press.

Pu, Q. S., and S. R. Liu. 2004. Microfabricated electroosmotic pump for capillary-based sequential injection analysis. *Analytica Chimica Acta* 511 (1):105–112.

Ramos, A., H. Morgan, N. G. Green, and A. Castellanos. 1998. AC electrokinetics: A review of forces in microelectrode structures. *Journal of Physics D–Applied Physics* 31 (18):2338–2353.

Raychaudhuri, S., S. A. Dayeh, D. L. Wang, and E. T. Yu. 2009. Precise semiconductor nanowire placement through dielectrophoresis. *Nano Letters* 9 (6):2260–2266.

Rhee, M., and M. A. Burns. 2006. Nanopore sequencing technology: Research trends and applications. *Trends in Biotechnology* 24:580–586.

Rosales, C., and K. M. Lim. 2005. Numerical comparison between Maxwell stress method and equivalent multipole approach for calculation of the dielectrophoretic force in single-cell traps. *Electrophoresis* 26 (11):2057–2065.

Sabounchi, P., A. M. Morales, P. Ponce, L. P. Lee, B. A. Simmons, and R. V. Davalos. 2008. Sample concentration and impedance detection on a microfluidic polymer chip. *Biomedical Microdevices* 10 (5):661–670.

Sabuncu, A. C., J. A. Liu, S. J. Beebe, and A. Beskok. 2010. Dielectrophoretic separation of mouse melanoma clones. *Biomicrofluidics* 4 (2):021101.

Schoch, R. B., J. Y. Han, and P. Renaud. 2008. Transport phenomena in nanofluidics. *Reviews of Modern Physics* 80 (3):839–883.

Schoch, R. B., H. van Lintel, and P. Renaud. 2005. Effect of the surface charge on ion transport through nanoslits. *Physics of Fluids* 17 (10):100604.

Seo, H. W., C. S. Han, D. G. Choi, K. S. Kim, and Y. H. Lee. 2005. Controlled assembly of single SWNTs bundle using dielectrophoresis. *Microelectronic Engineering* 81 (1):83–89.

Shannon, M. A. 2010. Water desalination fresh for less. *Nature Nanotechnology* 5 (4):248–250.

Sparreboom, W., A. van den Berg, and J. C. T. Eijkel. 2009. Principles and applications of nanofluidic transport. *Nature Nanotechnology* 4 (11):713–720.

Squires, T. M., and M. Z. Bazant. 2004. Induced-charge electro-osmosis. *Journal of Fluid Mechanics* 509:217–252.

Stein, D., M. Kruithof, and C. Dekker. 2004. Surface-charge-governed ion transport in nanofluidic channels. *Physical Review Letters* 93 (3):035901.

Storm, A. J., C. Storm, J. H. Chen, H. Zandbergen, J. F. Joanny, and C. Dekker. 2005. Fast DNA translocation through a solid-state nanopore. *Nano Letters* 5 (7):1193–1197.

Tang, J., B. Gao, H. Z. Geng, O. D. Velev, L. C. Qin, and O. Zhou. 2003. Assembly of ID nanostructures into sub-micrometer diameter fibrils with controlled and variable length by dielectrophoresis. *Advanced Materials* 15 (16):1352–1354.

Teh, S. Y., R. Lin, L. H. Hung, and A. P. Lee. 2008. Droplet microfluidics. *Lab on a Chip* 8 (2):198–220.

Thwar, P. K., J. J. Linderman, and M. A. Burns. 2007. Electrodeless direct current dielectrophoresis using reconfigurable field-shaping oil barriers. *Electrophoresis* 28 (24):4572–4581.

Toner, M., and D. Irimia. 2005. Blood-on-a-chip. *Annual Review of Biomedical Engineering* 7:77–103.

van der Heyden, F. H. J., D. J. Bonthuis, D. Stein, C. Meyer, and C. Dekker. 2006. Electrokinetic energy conversion efficiency in nanofluidic channels. *Nano Letters* 6 (10):2232–2237.

Velev, O. D., S. Gangwal, and D. N. Petsev. 2009. Particle-localized AC and DC manipulation and electrokinetics. *Annual Reports Section "C" (Physical Chemistry)* 105:213–246.

Verpoorte, E., and N. F. De Rooij. 2003. Microfluidics meets MEMS. *Proceedings of the IEEE* 91 (6):930–953.

Vlassiouk, I., T. R. Kozel, and Z. S. Siwy. 2009. Biosensing with nanofluidic diodes. *Journal of the American Chemical Society* 131 (23):8211–8220.

Vlassiouk, I., S. Smirnov, and Z. Siwy. 2008. Ionic selectivity of single nanochannels. *Nano Letters* 8 (7):1978–1985.

Wang, D. Q., R. Zhu, Z. Y. Zhou, and X. Y. Ye. 2007. Controlled assembly of zinc oxide nanowires using dielectrophoresis. *Applied Physics Letters* 90 (10):103110.

Wang, T. H., and P. K. Wong. 2010. Transforming Microfluidics into Laboratory Automation. *Jala* 15 (3):A15–A16.

Wang, M. R., and Q. J. Kang. 2010. Electrochemomechanical energy conversion efficiency in silica nanochannels. *Microfluidics and Nanofluidics* 9 (2–3):181–190.

Wang, P., Z. Chen, and H. C. Chang. 2006. An integrated micropump and electrospray emitter system based on porous silica monoliths. *Electrophoresis* 27 (20):3964–3970.

Wang, X. J., X. B. Wang, and P. R. C. Gascoyne. 1997. General expressions for dielectrophoretic force and electrorotational torque derived using the Maxwell stress tensor method. *Journal of Electrostatics* 39 (4):277–295.

Whitesides, G. M. 2006. The origins and the future of microfluidics. *Nature* 442 (7101):368–373.

Wong, P. K., T. H. Wang, J. H. Deval, and C. M. Ho. 2004. Electrokinetics in micro devices for biotechnology applications. *IEEE-ASME Transactions on Mechatronics* 9 (2):366–376.

Wu, Z. M., and D. Q. Li. 2008a. Micromixing using induced-charge electrokinetic flow. *Electrochimica Acta* 53 (19):5827–5835.

Wu, Z. M., and D. Q. Li. 2008b. Mixing and flow regulating by induced-charge electrokinetic flow in a microchannel with a pair of conducting triangle hurdles. *Microfluidics and Nanofluidics* 5 (1):65–76.

Xie, Y. B., X. W. Wang, J. M. Xue, K. Jin, L. Chen, and Y. G. Wang. 2008. Electric energy generation in single track-etched nanopores. *Applied Physics Letters* 93 (16):163116.

Xuan, X. C., R. Raghibizadeh, and D. Li. 2006. Wall effects on electrophoretic motion of spherical polystyrene particles in a rectangular poly(dimethylsiloxane) microchannel. *Journal of Colloid and Interface Science* 296:743–748.

Yalcin, S. E., A. Sharma, S. Qian, S. W. Joo, and O. Baysal. 2010. Manipulating particles in microfluidics by floating electrodes. *Electrophoresis* 31 (22):3711–3718.

Yalcin, S. E., A. Sharma, S. Qian, S. W. Joo, and O. Baysal. 2011. On-demand particle enrichment in a microfluidic channel by a locally controlled floating electrode. *Sensors and Actuators B: Chemical* 153 (1):277–283.

Yameen, B., M. Ali, R. Neumann, W. Ensinger, W. Knoll, and O. Azzaroni. 2009. Single conical nanopores displaying pH-tunable rectifying characteristics. Manipulating ionic transport with zwitterionic polymer brushes. *Journal of the American Chemical Society* 131:2070–2071.

Ye, C. Z., and D. Q. Li. 2004a. Electrophoretic motion of two spherical particles in a rectangular microchannel. *Microfluidics and Nanofluidics* 1 (1):52–61.

Ye, C. Z., and D. Q. Li. 2004b. 3-D transient electrophoretic motion of a spherical particle in a T-shaped rectangular microchannel. *Journal of Colloid and Interface Science* 272 (2):480–488.

Yusko, E. C., R. An, and M. Mayer. 2010. Electroosmotic flow can generate ion current rectification in nano- and micropores. *ACS Nano* 4 (1):477–487.

Zhang, B., M. Wood, and H. Lee. 2009. A silica nanochannel and Its applications in sensing and molecular transport. *Analytical Chemistry* 81 (13):5541–5548.

Zhang, L., and Y. X. Zhu. 2010. Dielectrophoresis of Janus particles under high frequency ac-electric fields. *Applied Physics Letters* 96 (14):141902.

Zhao, H., and H. H. Bau. 2007. Microfluidic chaotic stirrer utilizing induced-charge electro-osmosis. *Physical Review E* 75 (6):066217.

Zhou, H., L. R. White, and R. D. Tilton. 2005. Lateral separation of colloids or cells by dielectrophoresis augmented by AC electroosmosis. *Journal of Colloid and Interface Science* 285 (1):179–191.

Zhu, J., T.-R. Tzeng, G. Hu, and X. Xuan. 2009. DC dielectrophoretic focusing of particles in a serpentine microchannel. *Microfluidics and Nanofluidics* 7 (6):751–756.

Zhu, J., and X. Xuan. 2009a. Dielectrophoretic focusing of particles in a microchannel constriction using DC-biased AC electric fields. *Electrophoresis* 30 (15):2668–2675.

Zhu, J., and X. Xuan. 2009b. Particle electrophoresis and dielectrophoresis in curved microchannels. *Journal of Colloid and Interface Science* 340 (2):285–290.

2 Numerical Simulations of Electrical Double Layer and Electroosmotic Flow in a Nanopore

In the first part of this chapter, numerical simulation of the electrical double layer (EDL) formed next to a charged planar surface is demonstrated using COMSOL Multiphysics®. The electrostatics and the ionic mass transport are governed by the Poisson–Nernst–Planck (PNP) equations without considering fluid motion. The numerical predictions of the electric field and ionic concentrations are in good agreement with the analytical solution. In the second part of this chapter, numerical simulation of an electroosmotic flow (EOF) in a nanopore is demonstrated. The model for EOF includes the modified Stokes equations for the flow field, the Poisson equation for the electrostatics, and the Nernst–Planck equations with the convective term for the ionic mass transport. The predicted fully developed EOF velocity quantitatively agrees with the analytical solution.

2.1 ELECTRICAL DOUBLE LAYER

The EDL formed in the vicinity of a negatively charged planar surface is shown in Figure 2.1. Since the EDL along the charged planar surface (y direction) is uniform, the electric potential and ionic concentrations only vary along the x direction, which is perpendicular to the charged surface. Therefore, a one-dimensional (1D) model is used in the present study. OP is the computational domain, and its length is much larger than the Debye length to ensure capturing the variations of the electric potential and ionic concentrations within the EDL. The electrolyte enclosed in the computational domain is an incompressible binary KCl solution.

As described in Chapter 1, the electric potential is governed by the classical Poisson's equation

$$-\varepsilon_0 \varepsilon_f \nabla^2 \phi = F(c_1 z_1 + c_2 z_2), \tag{2.1}$$

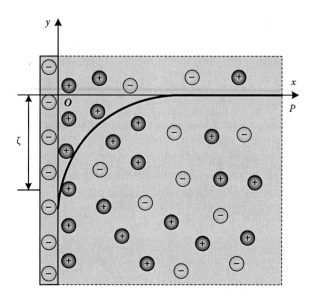

FIGURE 2.1   Schematics of an EDL formed adjacent to a negatively charged surface.

where ε_0 and ε_f are, respectively, the absolute permittivity of vacuum and the relative permittivity of the fluid; ϕ is the electric potential within the fluid; F is the Faraday constant; c_1 and c_2 are, respectively, the molar concentrations of the cations (K^+) and anions (Cl^-) in the electrolyte solution; z_1 and z_2 are, respectively, the valences of cations ($z_1 = 1$ for K^+) and anions ($z_2 = -1$ for Cl^-). The ionic concentrations are determined by the Nernst–Planck equations

$$\nabla \bullet \mathbf{N}_i = 0, \ i = 1 \text{ and } 2, \tag{2.2}$$

where $\mathbf{N}_i = \mathbf{u}c_i - D_i \nabla c_i - z_i \dfrac{D_i}{RT} Fc_i \nabla \phi$ is the ionic flux density of the ith ionic species, in which $\mathbf{u}$ is the fluid velocity; D_i is the diffusivity of the ith ionic species; R is the universal gas constant; and T is the absolute temperature of the electrolyte solution. In the absence of fluid motion (i.e., $\mathbf{u} = \mathbf{0}$), the convective term, $\mathbf{u}c_i$, in the ionic flux density is zero.

Appropriate boundary conditions are required to solve the PNP equations. The electric potential at the point O is the zeta potential of the charged surface, $\phi = \zeta$. The electric potential decays to zero in the bulk region. Therefore, the electric potential at point P is zero. The point O represents the interface between the stern layer and the diffuse layer of the EDL, which is impermeable to ions. Therefore, an insulation boundary condition, $\mathbf{n} \bullet \mathbf{N}_i = 0$, $i = 1$ and 2, is applied at point O, where $\mathbf{n}$ is a normal unit vector pointed from the charged surface into the fluid. The ionic concentrations recover the bulk concentration at point P, which is far away from the EDL.

After launching COMSOL Multiphysics 3.5a, the model navigator is prompted first, as shown in Figure 2.2. The **New** tab lists all the predefined modules that are organized in a tree structure. One can choose and couple any predefined modules to build a numerical model, indicating a capability of multiphysics modeling. In the model navigator, we can set up the space dimension, name the dependent variables, and give the abbreviation of the selected module. The element type is also listed here, and the default element type is quadratic. The right-hand side of the navigator gives a brief introduction to the selected module, which can give you a clear idea if this module fits your numerical model. The **Model Library** tab in the model navigator lists COMSOL example problems, and they are useful to COMSOL beginners. It is strongly recommended to find out if there is an example in the model library similar to the problem that you are going to model using COMSOL. For example, Figure 2.3 shows a selected COMSOL example that studies the flow field past a cylinder. The detailed implementations of these examples are given in a COMSOL manual file named modlib.pdf.

Figure 2.4 shows the model navigator setting of a model for the EDL near a charged surface. The space dimension is 1D. To add more than one module to your numerical model, click **Multiphysics** on the lower right corner of the model navigator. "Nernst–Planck without Electroneutrality" is defined under **Chemical Engineering Module|Mass Transport**. The default dependent variable is c. However, we have two ionic species (K^+ and Cl^-) in the fluid. Therefore, revise the dependent variables to c1 and c2, which represent concentrations of K^+ and Cl^-, respectively. Poisson's equation under **COMSOL Multiphysics|PDE Modes|Classical PDEs** is then added into the multiphysics model. The default dependent variable u is revised to V as the electric potential.

FIGURE 2.2 COMSOL Multiphysics Model navigator. (Image courtesy of COMSOL, Inc.)

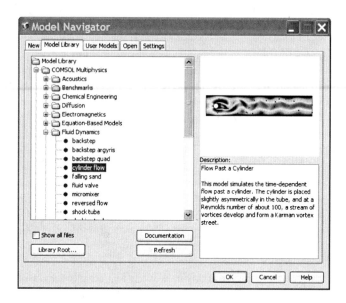

FIGURE 2.3   Model library in COMSOL Multiphysics. (Image courtesy of COMSOL, Inc.)

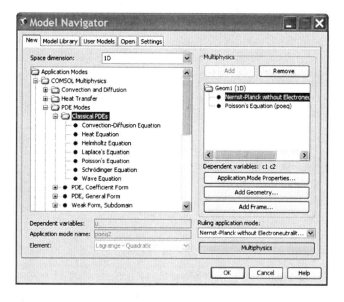

FIGURE 2.4   COMSOL Multiphysics Model navigator with Poisson–Nernst–Planck equations. (Image courtesy of COMSOL, Inc.)

Figure 2.5 shows the main graphical user interface (GUI) of COMSOL Multiphysics 3.5a. The top two rows are, respectively, the main menu and the shortcut icons. The left-hand side of the main GUI is the model tree, which is further split into two subwindows. The top subwindow contains the tree structure listing all the modules and related settings, while the bottom subwindow displays information about the selected setting in the top subwindow. For example, if **Geom1** in the top subwindow is selected, the bottom subwindow gives the information about the space dimension, reference coordinates, dependent variables, and base unit system. The model tree can be hidden by clicking the model tree icon in the row of shortcut icons. Most of the settings in the model tree can also be achieved in the main menu **Options** and **Physics**. The right-hand side of the main GUI is the area showing the geometry, discretized mesh, and postprocessing plots. Since the present model is 1D, we only draw one line in the x direction, as shown in Figure 2.5.

The default unit system is the International System of Units (SI), which can be modified in the main menu **Physics|Model Setting**, as shown in Figure 2.6. Here, we keep SI as the base unit system. The unchanged properties and parameters are usually defined in **Options|Constants**, which is also accessible in the model tree. Figure 2.7 lists all the constants used in the present case. If the unit system is activated in the numerical model, it is a good habit to write the unit in a bracket after the value. The subsequent expressions using the defined constants will automatically calculate the unit, which can be used to check if the expression is written correctly.

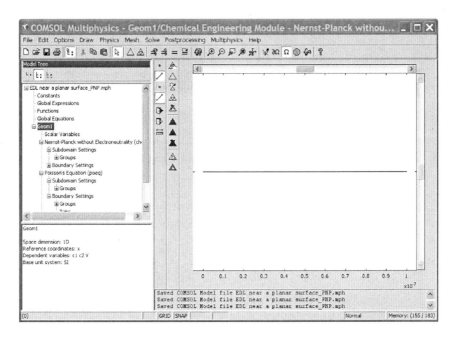

FIGURE 2.5 Main GUI of COMSOL Multiphysics 3.5a with a 1D geometry. (Image courtesy of COMSOL, Inc.)

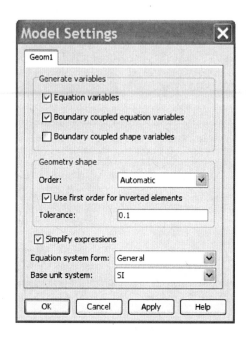

FIGURE 2.6   Model settings. (Image courtesy of COMSOL, Inc.)

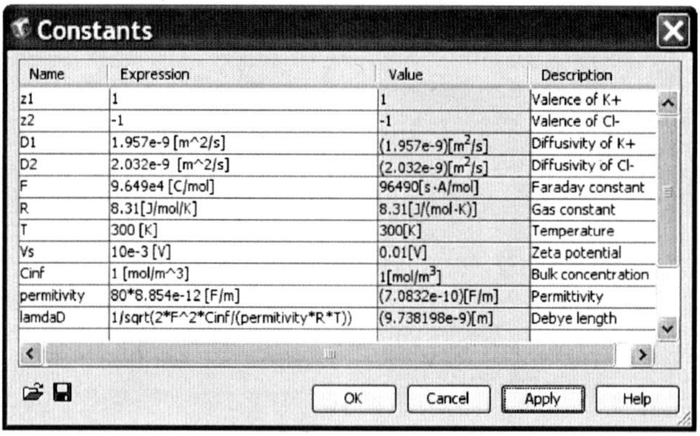

Name	Expression	Value	Description
z1	1	1	Valence of K+
z2	-1	-1	Valence of Cl-
D1	1.957e-9 [m^2/s]	(1.957e-9)[m²/s]	Diffusivity of K+
D2	2.032e-9 [m^2/s]	(2.032e-9)[m²/s]	Diffusivity of Cl-
F	9.649e4 [C/mol]	96490[s·A/mol]	Faraday constant
R	8.31[J/mol/K]	8.31[J/(mol·K)]	Gas constant
T	300 [K]	300[K]	Temperature
Vs	10e-3 [V]	0.01[V]	Zeta potential
Cinf	1 [mol/m^3]	1[mol/m³]	Bulk concentration
permitivity	80*8.854e-12 [F/m]	(7.0832e-10)[F/m]	Permittivity
lamdaD	1/sqrt(2*F^2*Cinf/(permitivity*R*T))	(9.738198e-9)[m]	Debye length

FIGURE 2.7   Constants for the model of EDL. (Image courtesy of COMSOL, Inc.)

To assign parameters for the governing equation and the initial guess values, one needs access to the subdomain settings. One can easily open the subdomain settings of each module in the model tree. If one prefers to hide the model tree, there is another way to access the subdomain settings for a specific module. First, open the **Multiphysics** menu in the main GUI of COMSOL Multiphysics 3.5a and select the module of interest. Subsequently, click **Physics|Subdomain Settings** to launch the window of subdomain settings. Figure 2.8 shows the window of subdomain settings for the Nernst–Planck without Electroneutrality module, abbreviated as chekf. We have two ionic species in the present study; thus, two dependent variables are defined previously in the model navigator. Figures 2.8 and 2.9 show, respectively, the subdomain settings for K$^+$ and Cl$^-$ ions. For the Potential in Figures 2.8 and 2.9, we put V, which is the defined dependent variable for the electric potential. As a result, Poisson's equation is coupled with the Nernst–Planck equations. In the bulk region, the ionic concentration is the bulk

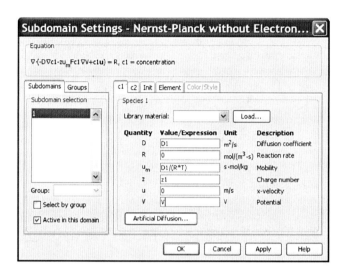

FIGURE 2.8 Subdomain settings of Nernst–Planck equation: c1. (Image courtesy of COMSOL, Inc.)

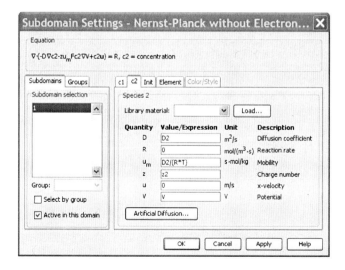

FIGURE 2.9 Subdomain settings of Nernst–Planck equation: c2. (Image courtesy of COMSOL, Inc.)

value. Therefore, we set the initial guess values of the two ionic species to the bulk concentration, as shown in Figure 2.10.

The access of the boundary settings of each module is similar to that of the subdomain settings. Figure 2.11 shows the boundary settings of point O, which is an insulation boundary and thus impermeable to ions. One can give a common name for the boundaries with the same boundary setting in the **Groups** tab in Figure 2.11. The boundary settings for both ionic species are the same. Point P at the other end of the computation domain is in the bulk region. Therefore, the ionic concentration is the bulk value, as shown in Figure 2.12. Again, the boundary settings for both ionic species at point P are the same.

Similarly, one can specify the subdomain settings for Poisson's equation as shown in Figure 2.13. The source term is the net charge in the fluid, as described in Equation (2.1). Therefore, the Nernst–Planck equations are also coupled to Poisson's equation, which refers to a two-way coupling. The electric potential in the bulk region is zero. Therefore, we can keep the default zero initial guess value. Figure 2.14 shows the boundary settings of point O, where the electric potential is the zeta potential. The electric potential at point P decays to zero in the bulk region, as shown in Figure 2.15.

Once the subdomain settings and boundary settings are complete, the computational domain has to be discretized into finite elements. The parameters for mesh generation can be assigned in the **Mesh|Free Mesh Parameters**. Figure 2.16 shows the assignment of the maximum element size for the computational domain. The number of elements can be checked in **Mesh|Mesh Statistics**, as shown in Figure 2.17.

Keep the default parameters for solver and click **Solve|Solve Problem** to run the simulation. COMSOL Multiphysics 3.5a also integrates very

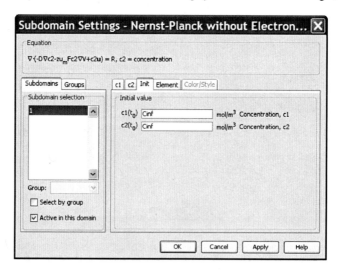

FIGURE 2.10   Subdomain settings of Nernst–Planck equation: initial guess values for c1 and c2. (Image courtesy of COMSOL, Inc.)

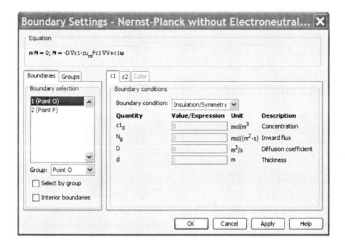

FIGURE 2.11   Boundary settings of Nernst–Planck equations at point O. (Image courtesy of COMSOL, Inc.)

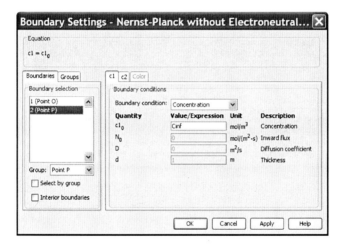

FIGURE 2.12   Boundary settings of Nernst–Planck equations at point P. (Image courtesy of COMSOL, Inc.)

powerful postprocessing tools in the **Postprocessing** menu. When clicking **Postprocessing|Plot Parameters**, we are able to obtain Figure 2.18. If one wants to plot the figure in a new window, change Main axes in the **Plot in** option to New figure. As we are plotting a line-type figure, go to the **Line** tab, as shown in Figure 2.19. Choose V in the **predefined quantities**, and the corresponding expression is given in the **Expression**. One can type any other expressions based on the defined variables to plot your results of interest. Figure 2.20 shows the variation of the electric potential along the *x* axis plotted in a new window. This figure can be edited by clicking the **Edit plot** icon on the top of this figure. If one wants

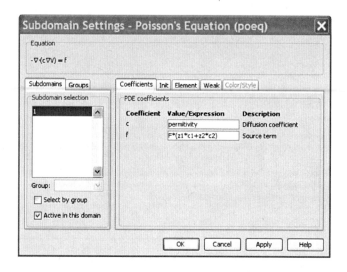

FIGURE 2.13   Subdomain settings of Poisson's equation. (Image courtesy of COMSOL, Inc.)

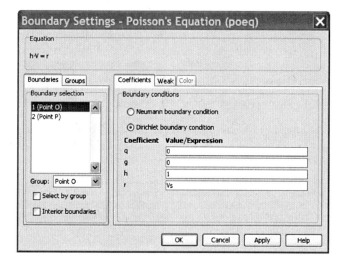

FIGURE 2.14   Boundary settings of Poisson's equation at point O. (Image courtesy of COMSOL, Inc.)

to export the data and postprocess them in some other software, click **Export Current Plot** icon on the left-hand side of the **Edit plot** icon. Detailed implementations are summarized in Table 2.1.

Figure 2.21 compares the numerical results of the electric potential near a charged surface submersed in 1 mM and 10 mM KCl solutions to the analytical solutions described in Equation (1.8). The electric potential is normalized by

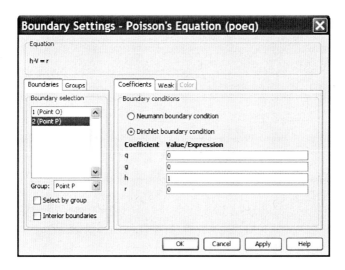

FIGURE 2.15 Boundary settings of Poisson's equation at point P. (Image courtesy of COMSOL, Inc.)

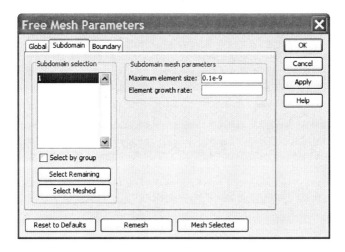

FIGURE 2.16 Free mesh parameters for subdomain 1. (Image courtesy of COMSOL, Inc.)

RT/F, and the x axis is normalized by the Debye length at 10 mM KCl solution. Obviously, the numerical results are in excellent agreement with the analytical solutions. Figure 2.22 also indicates excellent agreements between the numerically obtained ionic concentrations and the analytical solutions given in Equation (1.4). It is also revealed that the EDL thickness decreases as the bulk concentration increases.

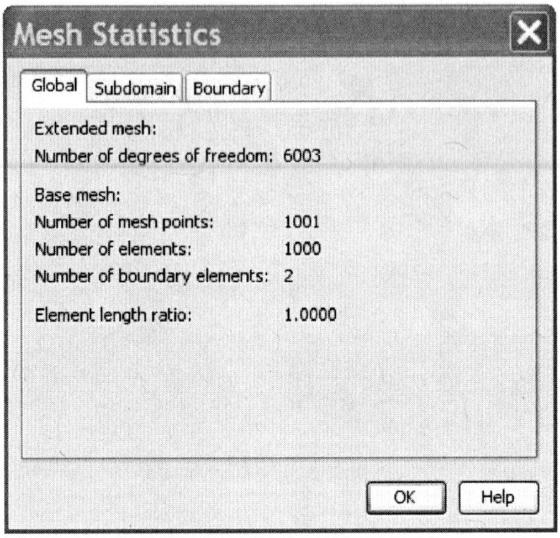

FIGURE 2.17 Mesh statistics. (Image courtesy of COMSOL, Inc.)

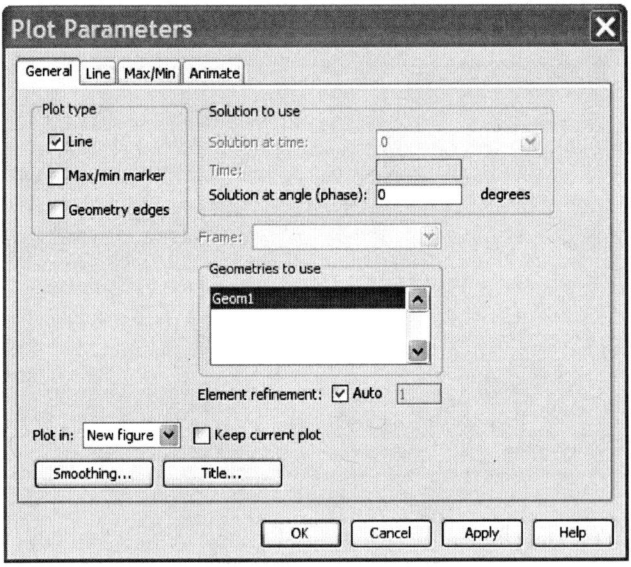

FIGURE 2.18 General settings in plot parameters. (Image courtesy of COMSOL, Inc.)

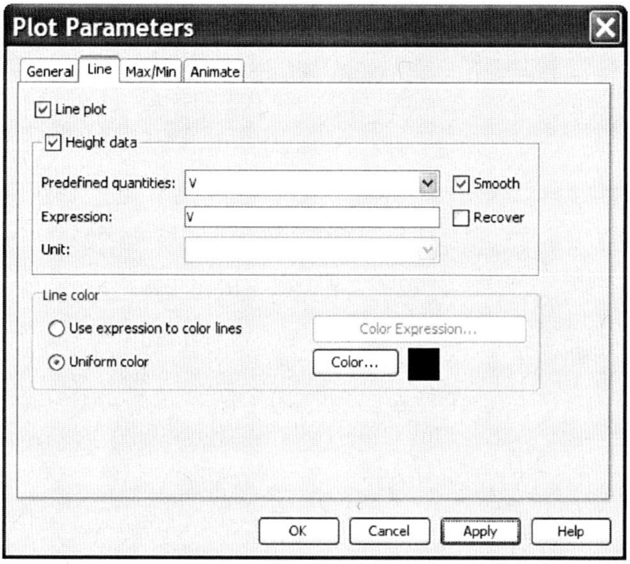

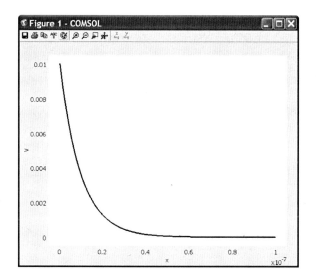

FIGURE 2.19 Line plot in plot parameters. (Image courtesy of COMSOL, Inc.)

FIGURE 2.20 Variation of electric potential along the x axis. (Image courtesy of COMSOL, Inc.)

TABLE 2.1

Model Setup in the GUI of COMSOL Multiphysics 3.5a

Model Navigator	Select **1D** in space dimension and click **Multiphysics** button
	Select **Chemical Engineering Module|Mass Transport| Nernst–Planck without Electroneutrality|Steady-state analysis**. Remove the predefined variables in the **Dependent variables** and enter c1 and c2. Click **Add** button
	Select **COMSOL Multiphysics|PDE Modes|Classical PDEs|Poisson's Equation**. Click **Add** button
Option Menu|**Constants**	Define variables in Figure 2.7
Physics Menu|**Subdomain Setting**	chekf mode
	Subdomain 1
	Tab c1
	D = D1; R = 0; u_m = D1/(R*T); z = z1; u = 0; V = V
	Tab c2
	D = D2; R = 0; u_m = D2/(R*T); z = z2; u = 0; V = V
	Tab Init
	c1(t_0) = Cinf; c2(t_0) = Cinf
	poeq mode
	Subdomain 1
	c = permittivity; ρ = F*(z1*c1+z2*c2)
Physics Menu|**Boundary Setting**	chekf mode
	Point 1(Point O): Insulation/Symmetry for c1 and c2
	Point 2(Point P): Concentration c1 = Cinf and c2 = Cinf
	poeq mode
	Point 1(Point O): q =0; g = 0; h = 1; r = Vs
	Point 2(Point P): q =0; g = 0; h = 1; r = 0
Mesh|**Free Mesh Parameters**	Tab Subdomain
	Subdomain 1
	Maximum element size = 1e–10
Solve Menu|	Click **= Solve Problem**
Postprocessing Menu|	Result check: V
	Postprocessing|Plot Parameters
	Tab **Line**
	Predefined quantities: V
	Click Apply

2.2 ELECTROOSMOTIC FLOW IN A NANOPORE

Next, we simulate the EOF in a charged cylindrical nanopore with a finite EDL, as shown in Figure 2.23. The governing equations include the PNP equations and the modified Stokes equations:

$$-\varepsilon_0\varepsilon_f\nabla^2\phi = F(c_1z_1 + c_2z_2), \qquad (2.3)$$

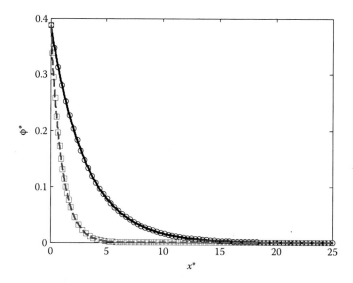

FIGURE 2.21   Distribution of electric potential near a charged planar surface under different bulk concentrations (solid line and circles, 1 mM; dashed line and squares, 10 mM). Lines and symbols represent, respectively, the analytical and numerical results.

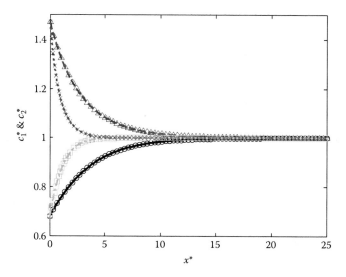

FIGURE 2.22   Distribution of ionic concentrations (c1, K$^+$; c2, Cl$^-$) near a charged planar surface under different bulk concentrations (solid line, dashed line, circles and triangles, 1 mM; dotted line, dash-dotted line, squares, and crosses, 10 mM). Lines and symbols represent, respectively, the analytical and numerical results. Solid line, dash-dotted line, circles, and squares represent the dimensionless concentration of K$^+$ normalized by the bulk concentration, while the others represent that of Cl$^-$.

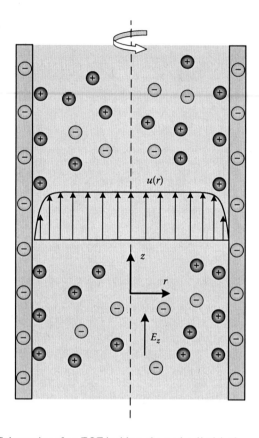

FIGURE 2.23   Schematics of an EOF inside a charged cylindrical nanopore.

$$\nabla \bullet \mathbf{N}_i = 0, \ i = 1 \text{ and } 2,\tag{2.4}$$

$$-\nabla p + \mu \nabla^2 \mathbf{u} - F(z_1 c_1 + z_2 c_2)\nabla \phi = 0,\tag{2.5}$$

$$\nabla \bullet \mathbf{u} = 0.\tag{2.6}$$

The inertial terms in the Navier–Stokes equations are neglected due to a very low Reynolds number. In these equations, p and $\mathbf{u}$ are, respectively, the pressure and the fluid velocity vector; μ is the dynamic viscosity of the fluid. Due to the axisymmetric nature of this problem, the axial symmetry (two-dimensional, 2D) model is selected in the Model Navigator, as shown in Figure 2.24. The **Chemical Engineering Module|Momentum Transport|Laminar Flow|Incompressible Navier–Stokes|Steady-state analysis, AC/DC Module|Statics, Electric| Electrostatics**, and **Chemical Engineering Module|Mass Transport|Nernst– Planck without Electroneutrality|Steady-state analysis** are added to the multiphysics model. The **Incompressible Navier–Stokes** and **Electrostatics** are also included in the basic **COMSOL Multiphysics Module**. However, the **Incompressible Navier–Stokes** defined in the **Chemical Engineering Module** and **Electrostatics** defined in the **AC/DC Module** offer more options, which might be useful for some cases.

The axisymmetric computational domain is shown in Figure 2.25, where $r = 0$ represents the axis of the cylindrical tube. In all the modules, the boundary condition along the axis is axial symmetry. In the **Incompressible Navier–Stokes**, the pressures on the upper and lower boundaries are both zero to eliminate the

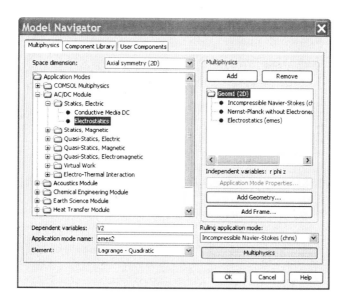

FIGURE 2.24 COMSOL Multiphysics Model navigator with Poisson–Nernst–Planck equations and incompressible Navier–Stokes equations. (Image courtesy of COMSOL, Inc.)

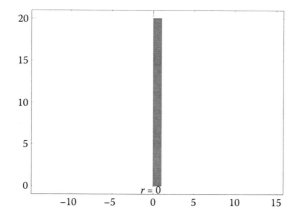

FIGURE 2.25 Geometry of the axial symmetry model of a cylindrical nanopore.

pressure-driven flow. On the wall of the nanopore, a no-slip boundary condition, that is, $\mathbf{u} = 0$, is used. In the **Nernst–Planck without Electroneutrality**, the ionic concentrations on the upper and lower boundaries for both ionic species are the bulk concentration. The wall of the nanopore is insulation. In the **Electrostatics**, a potential difference is applied between the upper and lower boundaries to generate an axial electric field along the nanopore. The wall of the nanopore is the surface charge boundary condition.

The bulk concentration C_0 as the ionic concentration scale, RT/F as the potential scale, the tube radius R_t as the length scale, $U_0 = \varepsilon_0 \varepsilon_f R^2 T^2 / (\mu R_t F^2)$ as the velocity scale, and $\mu U_0 / R_t$ as the pressure scale are selected to normalize the previous governing equations,

$$-\nabla^{*2} \phi^* = \frac{(\kappa R_t)^2}{2} (c_1^* z_1 + c_2^* z_2), \tag{2.7}$$

$$\nabla^* \bullet (\mathbf{u}^* c_i^* - D_i^* \nabla^* c_i^* - z_i D_i^* c_i^* \nabla^* \phi^*) = 0, \ i = 1 \text{ and } 2, \tag{2.8}$$

$$-\nabla^* p^* + \nabla^{*2} \mathbf{u}^* - \frac{(\kappa R_t)^2}{2} (z_1 c_1^* + z_2 c_2^*) \nabla^* \phi^* = 0, \tag{2.9}$$

$$\nabla^* \bullet \mathbf{u}^* = 0. \tag{2.10}$$

In these equations, $\kappa^{-1} = \lambda_D = \sqrt{\varepsilon_0 \varepsilon_f RT / \sum_{i=1}^{2} F^2 z_i^2 C_{i0}}$ is the Debye length. The surface charge density on the wall of the nanopore is accordingly normalized by $\varepsilon_0 \varepsilon_f RT / (FR_t)$.

Before specifying the subdomain and boundary settings, we first define the dimensional and dimensionless constants in **Options|Constants**, as shown in Figure 2.26. Figure 2.27 shows the subdomain settings of the **Incompressible Navier–Stokes module**. To solve the dimensionless equations in COMSOL Multiphysics 3.5a, one has to make the coefficients in the equations shown on the top of Figure 2.27 identical to normalized Equations (2.9) and (2.10). As a result, the density is zero as we neglect the inertial term, and the fluid viscosity is unity. F_r and F_z are, respectively, the r and z components of the electrostatic force arising from the interaction between the net charge inside the EDL and the applied electric field. Vr and Vz denote, respectively, the partial derivative of the electric potential with respect to r and z. As the density is set to be zero, the default stabilization option in the **Stabilization** tab should be turned off. Figure 2.28 shows the axial symmetry boundary condition on the axis of the computational domain. The lower boundary is the inlet, where pressure and normal viscous stress are zero, as shown in Figure 2.29. The upper boundary is the outlet, where the pressure

Constants

Name	Expression	Value	Description
rho	1000	1000	Fluid density
vis	1e-3	0.001	Fluid viscosity
sigma	-0.001	-0.001	Surface charge
D1	1.957e-9	1.957e-9	Diffusivity of K+
D2	2.032e-9	2.032e-9	Diffusivity of Cl-
Van	0.05	0.05	Applied potential
z1	1	1	Valence of K+
z2	-1	-1	Valence of Cl-
F0	96485.3415	96485.3415	Faraday constant
R	8.31	8.31	Gas constant
T	300	300	Temperature
C0	10	10	Bulk concentration
eps0	8.854187817e-12	8.854188e-12	Permittivity of vacuum
Rt	50e-9	5e-8	Length scale
lamda	(eps_r*eps0*R*T/(2*F0^2...	3.07967e-9	Debye length
ai	Rt/lamda	16.235507	Length scale vs Debye length
Van_dless	Van/(R*T/F0)	1.935125	Dimensionless potential
D1_dless	D1/(eps_r*eps0*R^2*T^2/...	4.138379	Dimensionless diffusivity of K+
D2_dless	D2/(eps_r*eps0*R^2*T^2/...	4.296978	Dimensionless diffusivity of Cl-
sigma_dless	sigma*F0*Rt/(eps_r*eps0*...	-2.731935	Dimensionless surface charge
eps_r	80	80	Relative permittivity of fluid

OK Cancel Apply Help

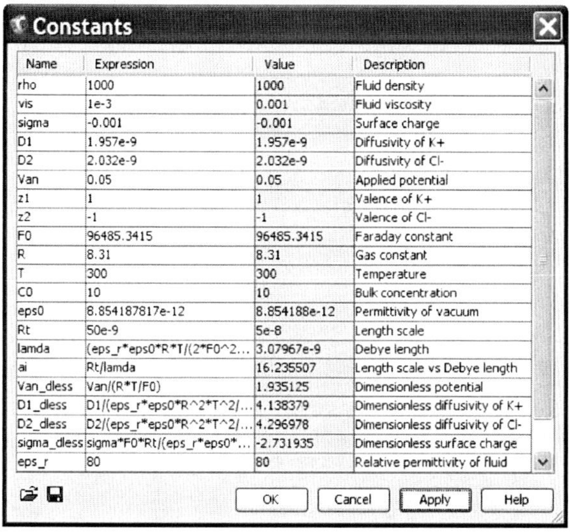

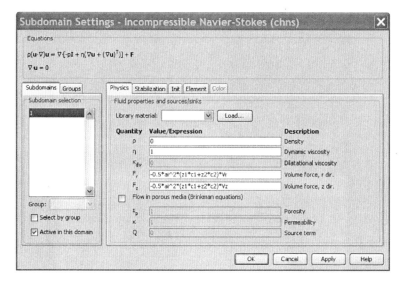

FIGURE 2.26 Constants for the model of EOF in a cylindrical nanopore. (Image courtesy of COMSOL, Inc.)

FIGURE 2.27 Subdomain settings of dimensionless incompressible Navier–Stokes equations. (Image courtesy of COMSOL, Inc.)

and normal viscous stress are also zero. Figure 2.30 shows the no-slip boundary condition on the wall of the nanopore.

Figure 2.31 shows the subdomain settings of the **Nernst–Planck without Electroneutrality module**. The coefficients are given based on the normalized Equation (2.8). The boundary settings are similar to those shown in Section 2.1.

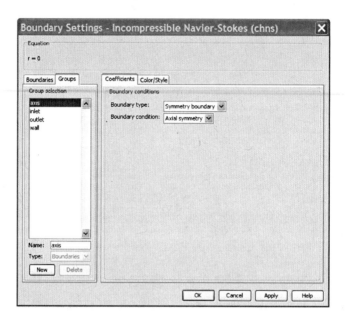

FIGURE 2.28   Boundary settings of the axis in the incompressible Navier–Stokes module. (Image courtesy of COMSOL, Inc.)

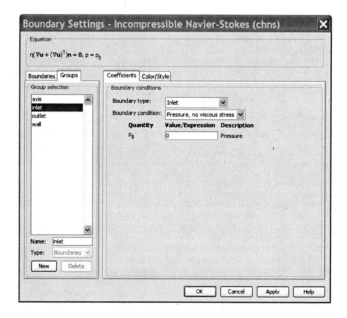

FIGURE 2.29   Boundary settings of the inlet in the incompressible Navier–Stokes module. (Image courtesy of COMSOL, Inc.)

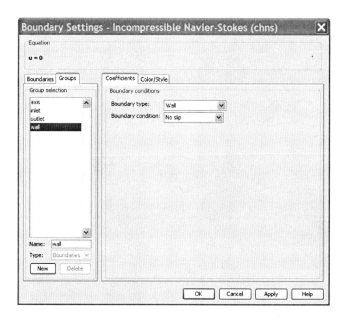

FIGURE 2.30 Boundary settings of the wall in the incompressible Navier–Stokes module. (Image courtesy of COMSOL, Inc.)

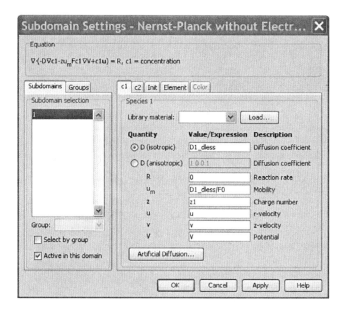

FIGURE 2.31 Subdomain settings of dimensionless Nernst–Planck equations. (Image courtesy of COMSOL, Inc.)

Figure 2.32 shows the normalized Poisson equation. Figure 2.33 shows the surface charge boundary condition on the wall of the nanopore, where sigma_dless is the normalized surface charge density.

As the computational domain is a rectangle, it is easy to apply structured elements. One can click **Mesh|Mapped Mesh Parameters** to specify the parameters for the generation of structured elements. In the **boundary** tab, one can specify the number of elements on the selected boundary, as shown

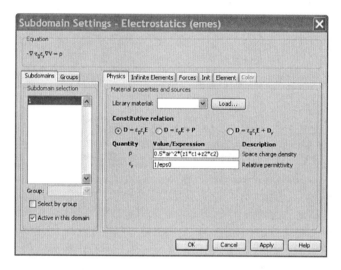

FIGURE 2.32   Subdomain settings of dimensionless Poisson's equation. (Image courtesy of COMSOL, Inc.)

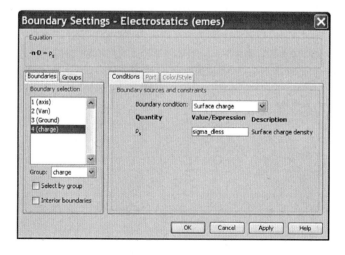

FIGURE 2.33   Boundary settings of the wall in the dimensionless Poisson's module. (Image courtesy of COMSOL, Inc.)

in Figure 2.34. The **element ratio** in the **Distribution** option can control the growth rate of the element along selected boundaries. For example, the ratio of 3 can ensure a denser mesh near the EDL and a sparser mesh in the bulk region, as shown in Figure 2.35. The nonuniform mesh laterally across the nanopore is shown in Figure 2.36.

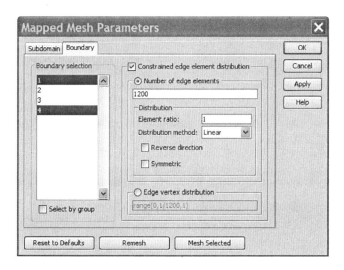

FIGURE 2.34 Mapped mesh parameters of the axis and the wall. (Image courtesy of COMSOL, Inc.)

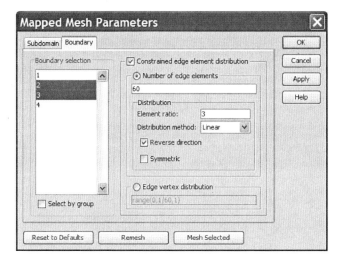

FIGURE 2.35 Mapped mesh parameters of the inlet and the outlet. (Image courtesy of COMSOL, Inc.)

Keep the default parameters and click **Solve|Solve Problem** to run the simulation. There are more options in the **Postprocessing|Plot Parameters** for the axial symmetry (2D) model than 1D model, as shown in Figure 2.37. In the **Plot type**, you can activate any plot types and have them in a single figure. Detailed implementations are summarized in Table 2.2.

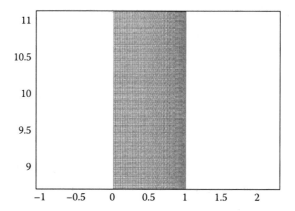

FIGURE 2.36 Generated structured mesh.

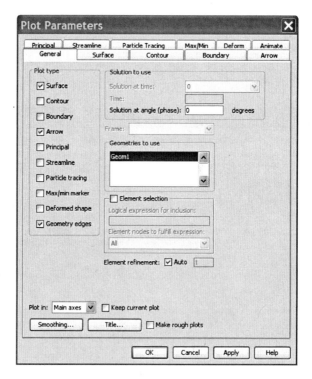

FIGURE 2.37 General setting in plot parameters. (Image courtesy of COMSOL, Inc.)

TABLE 2.2
Model Setup in the GUI of COMSOL Multiphysics 3.5a

Model Navigator	Select **Axial Symmetry (2D)** in space dimension and click **Multiphysics** button.
	Select **Chemical Engineering Module|Momentum Transport|Laminar Flow|Incompressible Navier–Stokes|Steady-state analysis**. Click **Add** button.
	Select **Chemical Engineering Module|Mass Transport|Nernst–Planck without Electroneutrality|Steady-state analysis**. Remove the predefined variables in the **Dependent variables** and enter c1 and c2. Click **Add** button.
	Select **AC/DC Module|Statics, Electric|Electrostatics**. Click **Add** button.
Option Menu|**Constants**	Define variables in Figure 2.26.
Physics Menu|	chns model
Subdomain Setting	Subdomain 1
	Tab Physics
	$\rho = 0$; $\eta = 1$; $F_r = -0.5*ai^2*(z1*c1+z2*c2)*Vr$;
	$F_z = -0.5*ai^2*(z1*c1+z2*c2)*Vz$
	Tab Stabilization
	Deactivate Streamline diffusion and Crosswind diffusion
	chekf mode
	Subdomain 1
	Tab c1
	$D = D1\_dless$; $R = 0$; $u_m = D1\_dless/F0$; $z = z1$; $u = u$; $v = v$; $V = V$
	Tab c2
	$D = D2\_dless$; $R = 0$; $u_m = D2\_dless/F0$; $z = z2$; $u = u$; $v = v$; $V = V$
	Tab Init
	$c1(t_0) = 1$; $c2(t_0) = 1$
	emes mode
	Subdomain 1
	$\rho = 0.5*(ai)^2*(z1*c1+z2*c2)$; $\varepsilon_r = 1/eps0$
Physics Menu|**Boundary**	chns mode
Setting	Axis: Symmetry boundary|Axial symmetry
	Lower boundary: Inlet|Pressure, no viscous stress $P_0 = 0$
	Upper boundary: Outlet|Pressure, no viscous stress $P_0 = 0$
	Nanopore wall: Wall|No slip
	chekf mode
	Axis: Axial symmetry
	Two ends of the nanopore: Concentration c1 = 1 and c2 = 1
	Nanopore wall: Insulation/Symmetry for c1 and c2
	emes mode
	Axis: Axial symmetry
	Lower boundary: $V_0 = Van\_dless$

continued

TABLE 2.2 (CONTINUED)
Model Setup in the GUI of COMSOL Multiphysics 3.5a

	Upper boundary: Ground
	Nanopore wall: Surface charge ρ_s = sigma_dless
Mesh|Mapped Mesh	Tab Boundary
Parameters	Axis and nanopore wall: 1, 4|Number of edge elements: 1,200
	Element ratio:1
	Inlet and outlet: 2, 3|Number of edge elements: 60 Element ratio:3
Solve Menu|	Click = Solve Problem
Postprocessing Menu|	Result check: c2 and flow field
	Postprocessing|Plot Parameters
	Tab Surface
	Predefined quantities: Concentration, c2
	Tab Arrow
	Predefined quantities: Velocity field
	Arrow positioning: r points: 15; z points: 20
	Click Apply
	Result check: fluid velocity along a cross section
	Postprocessing|Cross-Section Plot Parameters
	Tab Line/Extrusion
	Expressions: v*eps_r*eps0*R^2*T^2/vis/Rt/F0^2
	x-axis data: r
	Cross-section line data: r0 = 0; z0 = 10; r1 = 1; z1 = 10
	Click Apply

Figure 2.38 shows the surface plot of the ionic concentration of Cl⁻ and the arrow plot of the flow field in a single figure. It is shown that the EDL is very thin next to the charged surface. Because the nanopore is negatively charged, negative ions are repelled from the EDL. The fluid velocity in the bulk region is uniform. Thus, the EOF is usually called plug-like flow if the channel width is larger than the Debye length. One can also plot other variables in the computational domain in **Postprocessing\|Cross-Section Plot Parameters**. Figure 2.39 shows the parameters to plot the dimensional z-component fluid velocity in the middle cross section of the cylindrical tube. The analytical solution of the fully developed axial EOF velocity is available (Newman and Thomas-Alyae 2004; White and Bund 2008):

$$v_z(r) = -\frac{\lambda_D \sigma E_z}{\mu I_1(R_t / \lambda_D)}\left[I_0(R_t / \lambda_D) - I_0(r / \lambda_D)\right], \qquad (2.11)$$

where σ is the surface charge density of the nanopore; E_z is the imposed axial electric field; I_i is the modified Bessel functions of the first kind of order i; and R_t is the radius of the nanopore. Note that the closed-form solution was derived when the EDL was at equilibrium state, the EDLs of the opposing walls were

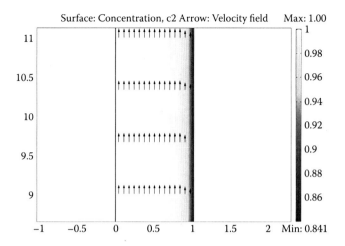

FIGURE 2.38   Surface plot of c2 and arrow plot of the flow field.

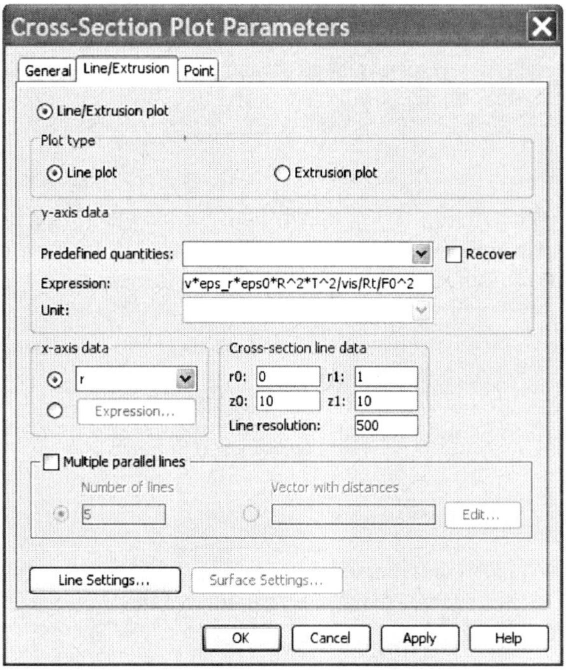

FIGURE 2.39   Cross-section plot parameters. (Image courtesy of COMSOL, Inc.)

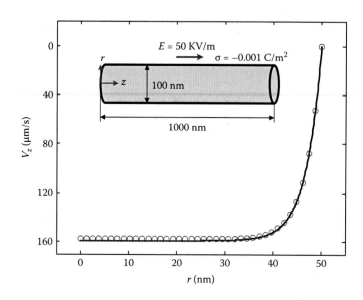

FIGURE 2.40 Comparison between the analytical (solid line) and numerical (circles) results of the axial EOF velocity in a cylindrical nanotube. The bulk electrolyte is 10 mM KCl solution, the surface charge density of the nanotube is $\sigma = -0.001$ C/m², and the externally imposed axial electric field is 50 KV/m. The inset shows a schematic view of the nanotube with dimensions. (From Ai, Y., M. Zhang, S. W. Joo, M. A. Cheney, and S. Qian. 2010. Effects of electroosmotic flow on ionic current rectification in conical nanopores. *Journal of Physical Chemistry C* 114:3883–3890 with permission of ACS.)

not overlapped (i.e., $R_t / \lambda_D \gg 1$), and the surface charge or zeta potential of the nanopore wall was relatively low.

We compare the numerically obtained fluid velocity across the nanopore to the analytical solution given in Equation (2.11). The surface charge and radius of the tube are, respectively, $\sigma = -0.001$ C/m² and $R_t = 50$ nm. Figure 2.40 shows good agreement between the numerical prediction and the analytical solution. The fluid velocity increases rapidly within the EDL and becomes a constant in the bulk region.

2.3   CONCLUDING REMARKS

This chapter introduced detailed implementations of COMSOL to solve the EDL near a charged planar surface using a 1D model and the EOF in a charged cylindrical nanopore using an axial symmetry 2D model. The electric potential exponentially decays within the EDL and recovers zero in the bulk region. Within the EDL, counterions dominate over co-ions due to the electrostatic interaction between the surface charge and ions of the fluid solution. The net charge is zero in the region outside the EDL. The EOF flow velocity becomes a constant in the region outside the EDL when the channel width is much larger than the Debye length. The numerical predictions were in good agreement with the existing analytical solutions, which validates the accuracy of the numerical predictions obtained from COMSOL.

REFERENCES

Ai, Y., M. Zhang, S. W. Joo, M. A. Cheney, and S. Qian. 2010. Effects of electroosmotic flow on ionic current rectification in conical nanopores. *Journal of Physical Chemistry C* 114:3883–3890.

Newman, J., and K. E. Thomas-Alyae. 2004. *Electrochemical Systems*, 3rd ed. Hoboken, NJ: Wiley.

White, H. S., and A. Bund. 2008. Ion current rectification at nanopores in glass membranes. *Langmuir* 24 (5):2212–2218.

3 Transient Electrokinetic Motion of a Circular Particle in a Microchannel

The use of electrokinetics for both biological and synthetic particle manipulations, including particle separation, assembly, sorting, focusing, and characterization in microfluidic devices, has recently gained significant attention. Microfluidic devices in general have complex geometries, giving rise to spatially nonuniform electric fields. Even in a straight uniform microchannel, finite-size particles distort the electric field, yielding spatially nonuniform electric fields around them. Arising from the spatially nonuniform electric fields and the surface charges on the particle surfaces and the microchannel walls, electroosmosis, electrophoresis, and dielectrophoresis typically coexist in the process of the electrokinetic particle transport in a microfluidic device. The particles experience both hydrodynamic and dielectrophoretic (DEP) forces and distort both electric and flow fields, which in turn affect the mentioned forces acting on the particles. Therefore, the full fluid-particle-electric field interactions as well as dielectrophoresis should be taken into account in the mathematical model to predict the electrokinetic particle transport in a microfluidic device accurately.

In this chapter, a transient multiphysics model taking these factors into account is developed to investigate electrokinetic particle transport in a microchannel. Under thin electrical double layer (EDL) approximation, the model, composed of the Stokes and continuity equations for the fluid flow, the Laplace equation for the electric field imposed, and Newton's equations for the particle motion, is simultaneously solved using the arbitrary Lagrangian–Eulerian (ALE) finite-element method. The induced DEP force is calculated by integrating the Maxwell stress tensor over the particle surface. Electrokinetic motions of a charged circular particle in a straight microchannel, a converging-diverging microchannel, a Y-shaped microchannel, and an L-shaped microchannel are investigated, respectively. The numerical predictions are in quantitative agreement with the existing experimental data. Due to the induced DEP force, lateral migration in particle electrophoresis is obtained. Particle focusing, trapping, and separation are obtained in a converging-diverging microchannel. A particle experiences both translation and rotation through an L-shaped microchannel. The induced DEP force affects both the particle's velocity and its trajectory and must be taken into account in the study of electrokinetic particle transport in microfluidic devices where spatially nonuniform electric fields are present.

3.1 INTRODUCTION

Electrokinetic particle transport offers a means to pump and manipulate particles using only electric fields with no moving parts. Other inherent advantages include nonintrusion, low cost, easy implementation, and favorable scaling with size. The electrokinetic particle motion is generally induced by both electrostatic and hydrodynamic forces acting on the particle. Electrokinetic transport of particles in microchannels with simple geometries such as parallel-plate (Keh and Anderson 1985; Unni, Keh, and Yang 2007), cuboid (Xuan, Raghibizadeh, and Li 2005; Ye and Li 2004b), and cylindrical tube (Keh and Anderson 1985; Qian and Joo 2008; Qian et al. 2008; Xuan, Ye, and Li 2005; Ye, Xuan, and Li 2005) with uniform electric fields has been extensively studied. It has also been applied to separate and characterize particles based on charges (Leopold, Dieter, and Ernst 2004; Rodriguez and Armstrong 2004; Dietrich et al. 2008; Gloria et al. 2008).

Microfluidic devices in general have complex geometries, such as the L-shaped, Y-shaped, and constricted microchannels, giving rise to spatially nonuniform electric fields. Even in a uniform microchannel, the presence of a particle with a size comparable to the channel cross section may significantly distort the electric field lines around the particle. In the presence of spatially nonuniform electric fields, the DEP effect arises along with the mentioned electrophoretic and electroosmotic effects due to the induced dipole moment on the particles. The induced DEP force and torque acting on the particle will affect particle motion, including both translation and rotation, which in turn affect the flow and electric fields.

Many investigators have utilized the resultant DEP forces under nonuniform direct current (DC) electric fields in microfluidic devices for various particle manipulations (Cummings and Singh 2003; Lapizco-Encinas et al. 2004a, 2004b; Ying et al. 2004; Lapizco-Encinas and Rito-Palomares 2007; Hawkins et al. 2007; Kang et al. 2008; Lewpiriyawong, Yang, and Lam 2008). For example, Kang, Xuan et al. (2006) experimentally demonstrated lateral particle migration in a constricted microchannel due to the DC DEP effect, which was then utilized for particle separation (Barbulovic-Nad et al. 2006; Hawkins et al. 2007; Kang et al. 2008, 2009; Kang, Kang, et al. 2006; Lewpiriyawong, Yang, and Lam 2008) and focusing (Thwar, Linderman, and Burns 2007; Zhu and Xuan 2009). Zhu et al. (2009) employed the DC DEP force generated in a serpentine microchannel to achieve particle focusing. Liang et al. (2010) achieved particle focusing in a straight uniform microchannel by the induced DEP force.

Despite many applications of DEP-based particle manipulations, a comprehensive analysis of electrokinetic particle transport under nonuniform DC electric fields is still limited. The majority of previous theoretical studies of particle electrokinetic transport in L-shaped (Davison and Sharp 2008), T-shaped (Ye and Li 2004b), and converging-diverging microchannels (Qian, Wang, and Afonien 2006) have neglected the DEP effect. A numerical model based on the Lagrangian tracking method has been developed to understand the DEP effects on particle electrokinetic motion in microchannels (Kang, Xuan, et al. 2006). However, the distortion of the flow and electric fields by the finite-size particles and the particle rotation were neglected in this model (Kang, Xuan, et al. 2006). Instead, a correction factor has to be introduced to account for the particle size effects on the DEP force and is determined by fitting the numerical predictions to the experimental data.

In this chapter, dynamic motion of a charged, circular particle driven by a DC electric field in a microchannel is numerically investigated with full consideration of the particle-fluid-electric field interactions as well as the induced DEP force obtained by directly integrating the Maxwell stress tensor over the particle surface under the assumption of a thin EDL. Section 3.2 introduces the mathematical model composed of the Stokes equations for flow field and the Laplace equation for electric field defined in the ALE kinematics. Section 3.3 describes the detailed numerical implementation in COMSOL Multiphysics® 3.5a. Section 3.4 introduces some representative results on the electrokinetic motion of a circular particle in a straight uniform channel, a converging-diverging microchannel, a Y-shaped microchannel, and an L-shaped microchannel, with the focus on the effect of the induced DEP effect. Concluding remarks are given in the final section.

3.2 MATHEMATICAL MODEL

We consider a charged circular particle of diameter d electrokinetically moving through a converging-diverging microchannel filled with an incompressible and Newtonian aqueous solution of density ρ, dynamic viscosity μ, and relative permittivity ε_f, as shown in Figure 3.1a. The channel shape and dimensions are taken from an existing fabricated device (Xuan, Xu, and Li 2005), as shown in Figure 3.1b. A two-dimensional (2D) Cartesian coordinate system (x, y) is used in the present study, with its origin fixed at the center of the throat. The computational domain Ω is surrounded by the inlet AJ, the outlet EF, the channel wall ABCDE and FGHIJ, and the particle surface Γ. An electric potential difference is applied between segments AJ and EF to impose an electric field across the microchannel. The particle bearing a uniform zeta potential ζ_p is initially located upstream with a center-to-center distance h away from the centerline of the microchannel. The channel wall also carries a uniform zeta potential ζ_w, which generates an electroosmotic flow (EOF) through the microchannel. The converging-diverging section is symmetric with respect to the throat ($L_b = L_c$), which is narrowest along the entire microchannel. The widths of the uniform section and the throat are, respectively, a and b.

Since the typical EDL thickness is about a few nanometers, which is much smaller than the particle size and the characteristic length of the microchannel, a thin EDL approximation is appropriate and adopted in this study. In the thin EDL approximation, the particle and its adjacent EDL are considered as a whole entity, and the EOF due to the net charge inside the EDL is described by the famous Smoluckowski electroosmotic slip velocity (Ye and Li 2004a; Davison

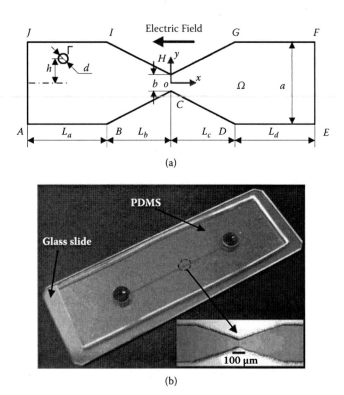

(a)

(b)

FIGURE 3.1 (a) A two-dimensional schematic view of a circular particle of diameter d and zeta potential ζ_p migrating in a converging-diverging microchannel. The zeta potential of the channel wall is ζ_w. An axial electric field E is externally applied between the outlet and inlet of the channel; (b) a converging-diverging microchannel fabricated with polydimethylsiloxane (PDMS). The inset shows the converging-diverging section of the microchannel. (From Ai, Y., S. W. Joo, Y. Jiang, X. Xuan, and S. Qian. 2009. Transient electrophoretic motion of a charged particle through a converging-diverging microchannel: Effect of direct current-dielectrophoretic force. *Electrophoresis* 30:2499–2506. With permission from Wiley-VCH.)

and Sharp 2008). The electric field outside the EDL is governed by the Laplace equation because the net charge outside the EDL is zero,

$$\nabla^2 \phi = 0 \qquad \text{in } \Omega, \tag{3.1}$$

where ϕ is the electric potential. The electric potentials applied on the inlet AJ and the outlet EF are, respectively,

$$\phi = 0 \qquad \text{on AJ}, \tag{3.2}$$

and

$$\phi = \phi_0 \qquad \text{on EF.} \qquad (3.3)$$

All the other rigid surfaces are electrically insulating,

$$\mathbf{n} \bullet \nabla\phi = 0, \qquad (3.4)$$

where $\mathbf{n}$ is the unit normal vector pointing from the boundary surface into the fluid.

In the thin EDL approximation, the net charge outside the EDL is zero, which leads to a zero electrostatic body force. In addition, the Reynolds number of the electrokinetic flow in the microchannel is very low (typically lower than 0.01). As a result, the fluid motion can be modeled by the Stokes equations without any electrostatic body force, described as

$$\rho\frac{\partial\mathbf{u}}{\partial t} - \mu\nabla^2\mathbf{u} + \nabla p = 0 \quad \text{in } \Omega, \qquad (3.5)$$

$$\nabla \bullet \mathbf{u} = 0 \quad \text{in } \Omega. \qquad (3.6)$$

In these equations, $\mathbf{u}$ is the fluid velocity vector, and p is the pressure. Both the fluid velocities and the pressure are initially zero in the computational domain. The pressure at the inlet and the outlet are both zero, and there is no pressure gradient imposed. The electrokinetic flow inside the EDL is incorporated into the Smoluckowski slip velocity. Therefore, the fluid velocity on the channel wall is

$$\mathbf{u} = \frac{\varepsilon_0\varepsilon_f\zeta_w}{\mu}(\mathbf{I} - \mathbf{nn}) \bullet \nabla\phi \quad \text{on ABCDE and FGHIJ,} \qquad (3.7)$$

where ε_0 is the permittivity of the vacuum. The quantity $(\mathbf{I} - \mathbf{nn}) \bullet \nabla\phi$ defines the electric field tangential to the charged surface, with $\mathbf{I}$ denoting the second-order unit tensor. As the particle translates and rotates, the fluid velocity on the particle surface includes the translational velocity $\mathbf{U}_p$, rotational velocity ω_p, and Smoluckowski slip velocity addressing the induced EOF:

$$\mathbf{u} = \frac{\varepsilon_0\varepsilon_f\zeta_p}{\mu}(\mathbf{I} - \mathbf{nn}) \bullet \nabla\phi + \mathbf{U}_p + \omega_p \times (\mathbf{x}_s - \mathbf{x}_p) \quad \text{on } \Gamma. \qquad (3.8)$$

In this equation, $\mathbf{x}_s$ and $\mathbf{x}_p$ are, respectively, the position vector of the surface and center of the particle.

The particle's translational velocity is governed by Newton's second law:

$$m_p\frac{d\mathbf{U}_p}{dt} = \mathbf{F}_{net}, \qquad (3.9)$$

where m_p is the mass of the particle, and $\mathbf{F}_{net}$ is the net force acting on it. The Coulomb force arising from the charges on the particle surface exactly cancels the hydrodynamic force due to the flow field within the EDL (Ye and Li 2004a). Therefore, the net force $\mathbf{F}_{net}$ includes the hydrodynamic force due to the flow field originating in the outer region of the EDL, and the DC DEP force $\mathbf{F}_{DEP}$ arising from the interaction between the dielectric particle and the spatially nonuniform electric field:

$$\mathbf{F}_H = \int \mathbf{T}^H \bullet \mathbf{n} d\Gamma = \int \left[-p\mathbf{I} + \mu \left(\nabla \mathbf{u} + \nabla \mathbf{u}^T \right) \right] \bullet \mathbf{n} d\Gamma, \tag{3.10}$$

$$\mathbf{F}_{DEP} = \int \mathbf{T}^E \bullet \mathbf{n} d\Gamma = \int \varepsilon_0 \varepsilon_f \left[\mathbf{E}\mathbf{E} - \frac{1}{2} (\mathbf{E} \bullet \mathbf{E}) \mathbf{I} \right] \bullet \mathbf{n} d\Gamma. \tag{3.11}$$

In these equations, $\mathbf{T}^H$ and $\mathbf{T}^E$ are, respectively, the hydrodynamic and Maxwell stress tensors. $\mathbf{E}$ is the electric field related to the electric potential by $\mathbf{E} = -\nabla \phi$. The integration of the first term of the integrand in the right-hand side of Equation (3.11) vanishes due to Equation (3.4), which renders the integration of the Maxwell stress tensor as the pure DEP force.

The rotational velocity of the particle is determined by

$$I_p \frac{d\omega_p}{dt} = \int (\mathbf{x}_s - \mathbf{x}_p) \times \left(\mathbf{T}^H \bullet \mathbf{n} + \mathbf{T}^E \bullet \mathbf{n} \right) d\Gamma, \tag{3.12}$$

where I_p is the moment of inertia of the particle, and the right-hand side of Equation (3.12) is the net torque exerted on the particle.

The center $\mathbf{x}_p$ and the orientation θ_p of the particle are expressed by

$$\mathbf{x}_p = \mathbf{x}_{p0} + \int_0^t \mathbf{U}_p \, dt. \tag{3.13}$$

and

$$\theta_p = \theta_{p0} + \int_0^t \omega_p \, dt. \tag{3.14}$$

where $\mathbf{x}_{p0}$ and θ_{p0} denote, respectively, the initial position and orientation of the particle.

The particle radius a, as the length scale; the electric potential on segment EF ϕ_0, as the potential scale; $U_0 = \varepsilon_0 \varepsilon_f \zeta_p \phi_0 / (\mu a)$ as the velocity scale; and $\mu U_0 / a$ as the pressure scale are selected to normalize all the previous governing equations:

$$\nabla^{*2} \phi^* = 0 \quad \text{in } \Omega, \tag{3.15}$$

$$\mathrm{Re}\frac{\partial \mathbf{u}^*}{\partial t^*} - \nabla^{*2}\mathbf{u}^* + \nabla^* p^* = 0 \quad \text{in } \Omega, \tag{3.16}$$

$$\nabla^* \bullet \mathbf{u}^* = 0 \quad \text{in } \Omega, \tag{3.17}$$

and also the corresponding boundary conditions,

$$\mathbf{n} \bullet \nabla^* \phi^* = 0 \quad \text{on ABCDE and FGHIJ}, \tag{3.18}$$

$$\phi^* = 1 \quad \text{on EF}, \tag{3.19}$$

$$\mathbf{u}^* = \gamma(\mathbf{I} - \mathbf{nn}) \bullet \nabla^* \phi^* \quad \text{on ABCDE and FGHIJ}, \tag{3.20}$$

$$\mathbf{u}^* = \mathbf{U}_p^* + \omega_p^* \times (\mathbf{x}_s^* - \mathbf{x}_p^*) + (\mathbf{I} - \mathbf{nn}) \bullet \nabla^* \phi^* \quad \text{on } \Gamma. \tag{3.21}$$

In these equations, $\mathrm{Re} = \rho U_0 a/\mu$, and $\gamma = \zeta_w/\zeta_p$ is the ratio of the zeta potential of the particle to that of the channel wall.

The force and torque are, respectively, normalized by μU_0 and $a\mu U_0$, yielding the dimensionless equations of particle motion:

$$m_p^* \frac{d\mathbf{U}_p^*}{dt^*} = \int (\mathbf{T}^{H*} \bullet \mathbf{n} + \mathbf{T}^{E*} \bullet \mathbf{n}) d\Gamma^*, \tag{3.22}$$

$$I_p^* \frac{d\omega_p^*}{dt^*} = \frac{\phi_0}{\zeta_p} \int (\mathbf{x}_s^* - \mathbf{x}_p^*) \times (\mathbf{T}^{H*} \bullet \mathbf{n} + \mathbf{T}^{E*} \bullet \mathbf{n}) d\Gamma^*, \tag{3.23}$$

where the mass and the moment of inertia are, respectively, normalized by $a\mu/U_0$ and $a^3\mu/U_0$,

$$\mathbf{T}^{H*} = -p^*\mathbf{I} + \left(\nabla^*\mathbf{u}^* + \nabla^*\mathbf{u}^{*T}\right),$$

and

$$\mathbf{T}^{E*} = \mathbf{E}^*\mathbf{E}^* - \frac{1}{2}\left(\mathbf{E}^* \bullet \mathbf{E}^*\right)\mathbf{I}.$$

3.3 NUMERICAL IMPLEMENTATION IN COMSOL

The coupled equations discussed are solved in COMSOL Multiphysics 3.5a using the predefined ALE method, which tracks the particle motion in a Lagrangian fashion and at the same time solves the fluid flow and the electric field in an Eulerian framework. The theoretical derivation of the ALE technique was originally introduced by Hughes, Lu, and Zimmerman (1981). Detailed numerical implementations of the ALE technique on the simulation of particle motion in viscous fluids was first introduced by Hu et al. (Hu, Joseph, and Crochet 1992; Hu, Patankar, and Zhu 2001).

Briefly, the ALE method updates the particle's location and orientation by deforming the mesh after each computational time step using Equations (3.22) and (3.23). As the particle translates and rotates, the mesh becomes highly deformed, which could affect the accuracy of the numerical results. Therefore, we usually assign a minimum mesh quality level below which the mesh deformation is forced to stop. A new geometry is then generated based on the deformed mesh. Subsequently, an undeformed mesh is generated on the new geometry to continue the computation until the next remeshing.

Here, we demonstrate how to set up the mathematical model in COMSOL Multiphysics 3.5a and solve it using the ALE method. First, open COMSOL Multiphysics 3.5a to get the Model Navigator and click **Multiphysics** on the lower right corner of the prompt window, as shown in Figure 3.2. The ALE module is defined in **COMSOL Multiphysics|Deformed Mesh|Moving Mesh (ALE)**. Because we are dealing with time-independent problems, select **Transient analysis** and then click **Add** in the **Multiphysics** tab. A spatial frame, called ale, is automatically generated. The Stokes flow neglecting the inertial terms is

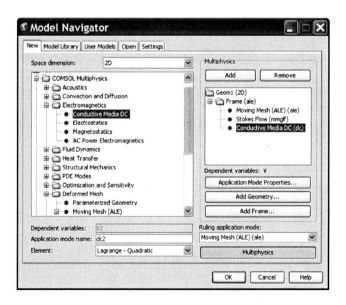

FIGURE 3.2   Multiphysics Model Navigator. (Image courtesy of COMSOL, Inc.)

predefined in **MEMS Module|Microfluidics|Stokes Flow**; add the **Transient analysis** into the **Multiphysics** tab. Finally, add **COMSOL Multiphysics|Elec tromagnetics|Conductive Media DC** and click OK to go to the main GUI of COMSOL Multiphysics 3.5a.

Unchanged parameters used in the simulation are listed in Table 3.1; for example, the fluid properties and the normalization scales can be defined in **Options|Constants**, as shown in Figure 3.3. It is a good habit to define constants at the beginning and then use them in the following settings. Because one parameter may be used in several settings, it is easier to modify the parameter in **Options|Constants** than to modify the number in all the related settings. In addition, it is also good to write proper descriptions of the defined constants, which is helpful for reminding yourself or others of the physical meaning of the constant in the future. Draw a dimensionless converging-diverging channel with a particle inside, as shown in Figure 3.4.

To modify the subdomain and boundary settings of each module, one first has to select the correct module. There are two ways to do this in COMSOL Multiphysics 3.5a. The first way is to open the **Multiphysics** menu in the main graphical user interface (GUI) of COMSOL Multiphysics 3.5a, in which all the modules used are listed and can be selected (Table 3.2). The subdomain and boundary settings of the selected module can be modified in **Physics|Subdomain Settings|Boundary Settings**, respectively. Another way is to set up each module in the **Model Tree** by double clicking the corresponding subdomain settings/boundary settings. The **Model Tree** can be activated or hidden by the icon on the right-hand side of the printer icon in the main GUI of COMSOL Multiphysics 3.5a. In addition, some other settings (e.g., constants, expressions, and global equations) can be modified

TABLE 3.1
Constant Table

Variable	Value or Expression	Description
eps_r	80	Relative permittivity of fluid
eps0	8.854187817e-12 [F/m]	Permittivity of vacuum
rho	1e3 [kg/m^3]	Particle/fluid density
Xp0	−70	Dimensionless initial x location of particle
Yp0	0	Dimensionless initial y location of particle
Masss	rho*pi*a^2	Dimensional particle mass
Mass	Masss*Uc/eta/a	Dimensionless particle mass
Ir	1/2*Masss*a^2*Uc/eta/a^3	Dimensionless moment of inertia
a	10E-06 [m]	Particle radius
zetar	2.5	Zeta potential ratio
ep	32e-3 [V]	Zeta potential of particle
eta	1e-3 [pa*s]	Fluid viscosity
Uc	eps_r*eps0*ep* v0 /(eta*a)	Velocity scale
v0	15 [V]	Applied voltage
Re	rho*Uc*a/eta	Reynolds number
depr	v0/ep	DEP force coefficient

Name	Expression	Value	Description
eps_r	80	80	Relative permittivity of fluid
eps0	8.854187817e-12	8.854188e-12	Permittivity of vacuum
rho	1e3	1000	Particle/fluid density
Xp0	70	70	Dimensionless initial x location of particle
Yp0	0	0	Dimensionless initial y location of particle
Mass	Masss*Uc/eta/a	1.068144	Dimensional particle mass
Ir	1/2*Masss*a^2*Uc/eta/a^3	0.534072	Dimensionless moment of inertia
a	10e-6 [m]	1e-5	Particle radius
zetar	2.5	2.5	Zeta potential ratio
ep	32e-3	0.032	Zeta potential of particle
eta	1.0e-3	0.001	Fluid viscosity
Uc	eps_r*eps0*ep*v0/(eta*a)	0.034	Velocity scale
Masss	rho*pi*a^2	3.141593e-7	Dimensionless particle mass
v0	15	15	Applied voltage
Re	rho*Uc*a/eta	0.340001	Reynolds number
depr	v0/ep	468.75	Dielectrophoretic force coefficient

FIGURE 3.3   Constants window. (Image courtesy of COMSOL, Inc.)

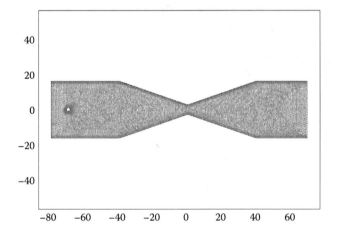

FIGURE 3.4   Geometry of the converging-diverging microchannel and the generated mesh.

in the **Model Tree**. Figure 3.5 shows the subdomain settings for the ale module. Subdomain 1 is free displacement while keeping all the other settings at the default values. Subsequently, we need to specify the boundary conditions for the ale module. The x and y component mesh velocity on the particle surface are, respectively, up+Ur and vp+Vr, as shown in Figure 3.6. These variables will be defined later in the chapter. The x and y component displacements on the other boundaries are zero, as shown in Figure 3.7. Right click the ale module in the **Model Tree**; the properties settings will be prompted, as shown in Figure 3.8. Change the default Laplace method to the Winslow method for smoothing and turn on the allow remeshing. Keep the other setting at the default value.

TABLE 3.2
Model Setup in the GUI of COMSOL Multiphysics 3.5a

Model Navigator	Select **2D** in space dimension and click **Multiphysics** button
	Select **COMSOL Multiphysics|Deformed Mesh|Moving Mesh (ALE)|Transient analysis**. Click **Add** button.
	Select **AC/DC Module|Statics, Electric|Conductive Media DC**. Click **Add** button
	Select **MEMS Module|Microfluidics|Stokes Flow|Transient analysis**. Click **Add** button.
Option Menu|**Constants**	Define constants in Table 3.1.
Physics Menu|**Subdomain Setting**	ale model
	Subdomain 1
	Free displacement
	emdc mode (external electric field)
	Subdomain 1
	$J_e = 0;\ Q_j = 0;\ d = 1;\ \sigma = 1$
	mmglf mode
	Subdomain 1
	Tab Physics
	$\rho = Re;\ \eta = 1;$ Thickness $= 1$
Physics Menu|**Boundary Setting**	ale mode
	Particle surface: Mesh velocity $vx = up+Ur$
	$vy = vp+Vr$
	Other boundaries: Mesh displacement $dx = 0$ $dy = 0$
	emdc mode
	Left boundary: Ground
	Right boundary: $V_0 = 1$
	Other boundaries: Electric insulation
	mmglf mode
	Left boundary: Inlet|Pressure, no viscous stress $P_0 = 0$
	Right boundary: Outlet|Pressure, no viscous stress $P_0 = 0$
	Particle surface: Wall|Moving/leaking wall $Uw = up+Ur+Ueofp$
	$Vw = vp+Vr+Veofp$
	Channel wall: Wall|Moving/leaking wall $Uw = Ueof$ $Vw = Veof$
Other Settings	Properties of ale model (**Physics** Menu|**Properties**)
	Smoothing method: Winslow
	Allow remeshing: On
	Properties of mmglf model (**Physics** Menu|**Properties**)
	Weak constraints: On
	Constraint type: Nonideal
	Deactivate the unit system (**Physics** Menu|**Model Setting**)
	Base unit system: None

continued

TABLE 3.2 (CONTINUED)
Model Setup in the GUI of COMSOL Multiphysics 3.5a

Options Menu|**Integration** **Coupling Variables|** **Boundary Variables**	Boundary selection: 11, 12, 13, 14 (Particle surface)
	Name: F_x Expression: -lm3; Integration order: 4; Frame: Frame(mesh)
	Name: F_y Expression: -lm4; Integration order: 4; Frame: Frame(mesh)
	Name: Fd_x Expression: depr*(tEx_emdc^2+tEy_emdc^2)*nx/2; Integration order: 4; Frame: Frame(mesh)
	Name: Fd_y Expression: depr*(tEx_emdc^2+tEy_emdc^2)*ny/2; Integration order: 4; Frame: Frame(mesh)
	Name: Tr Expression: (x-Xpar)*(depr*(tEx_emdc^2+tEy_ emdc^2)*ny/2-lm4)-(y-Ypar)*(depr*(tEx_emdc^2+tEy_ emdc^2)*nx/2-lm3); Integration order: 4; Frame: Frame(mesh)
Options Menu|**Expressions|** **Boundary Expressions**	Boundary selection: 11, 12, 13, 14 (Particle surface)
	Name: Ueofp Expression: -tEx_emdc
	Name: Veofp Expression: -tEy_emdc
	Name: Ur Expression: -omega*(y-Ypar)
	Name: Vr Expression: omega*(x-Xpar)
	Boundary selection: 2, 3, 4, 5, 6, 7, 8, 9 (Channel wall)
	Name: Ueof Expression: -zetar*tEx_emdc
	Name: Veof Expression: -zetar*tEy_emdc
Physics Menu| **Global Equations**	Name: up; Expression: Mass*upt-(F_x+Fd_x); Init (u): 0; Init (ut): 0
	Name: Xpar; Expression: Xpart-up; Init (u): Xp0; Init (ut): 0
	Name: vp; Expression: Mass*vpt-(F_y+Fd_y); Init (u): 0; Init (ut): 0
	Name: Ypar; Expression: Ypart-vp; Init (u): Yp0; Init (ut): 0
	Name: omega; Expression: Ir*omegat-Tr; Init (u): 0; Init (ut): 0
	Name: ang; Expression: angt-omega; Init (u): 0; Init (ut): 0
Mesh|Free Mesh **Parameters**	Tab subdomain
	Subdomain 1|Maximum element size: 1
	Tab boundary
	Channel wall: 2, 3, 4, 5, 6, 7, 8, 9 |**Maximum element size**: 0.4
	Particle surface: 11, 12, 13, 14 |**Maximum element size**: 0.05
Solve Menu|	**Solver Parameters**
	Tab General
	Times: Range(0, 2, 20000)
	Relative tolerance: 1e-4
	Absolute tolerance: 1e-6
	Tab Timing Stepping
	Activate Use stop condition
	minqual1_ale-0.6
	Click = **Solve Problem**
Postprocessing Menu|	Result check: up, Xpar, vp, Ypar, omega, ang
	Global Variables Plot|Select all the predefined quantities, click >>
	Click Apply

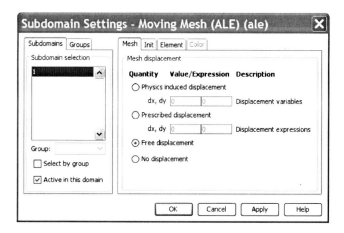

FIGURE 3.5   Subdomain settings for the ale module. (Image courtesy of COMSOL, Inc.)

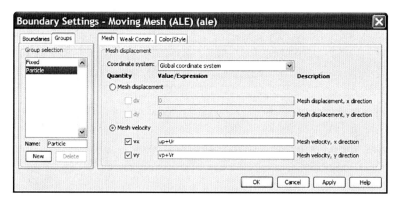

FIGURE 3.6   Boundary settings on the particle surface for the ale module. (Image courtesy of COMSOL, Inc.)

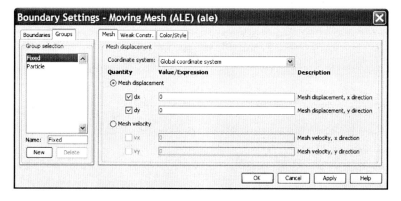

FIGURE 3.7   Boundary settings on the other boundaries for the ale module. (Image courtesy of COMSOL, Inc.)

Figure 3.9 shows the subdomain settings for the Stokes flow module, called mmglf. As we are solving dimensionless equations, the equations shown in Figure 3.9 should be identical to Equations (3.16) and (3.17). Therefore, the fluid density in Figure 3.9 is actually the dimensionless Reynolds number derived in Equation (3.16), and the dynamics viscosity is unity. The volumetric force in the x and y directions is zero for each because of a zero net charge outside

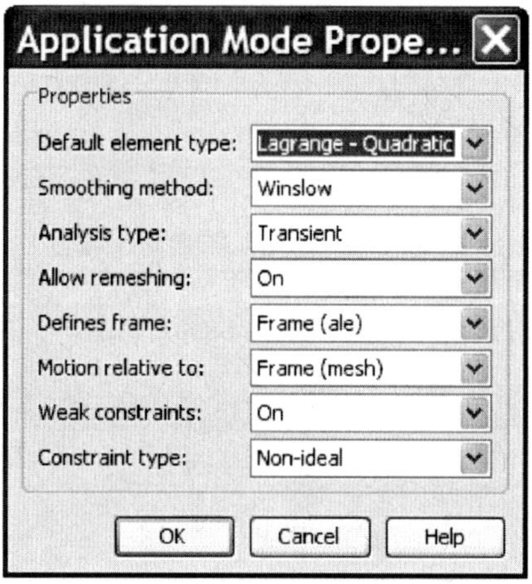

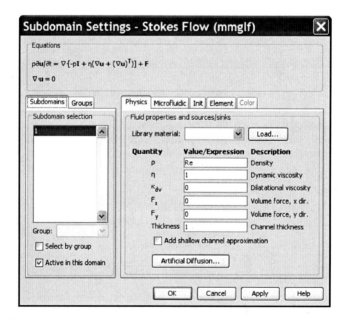

FIGURE 3.8 Properties of the ale module. (Image courtesy of COMSOL, Inc.)

FIGURE 3.9 Subdomain settings for the mmglf module. (Image courtesy of COMSOL, Inc.)

the EDL. The normal viscous stress and pressure on the inlet and outlet are both zero, as shown in Figures 3.10 and 3.11. The x and y component fluid velocities on the channel wall are, respectively, Ueof and Veof, as shown in Figure 3.12. These boundaries can be named in the group tab of the boundary selection in Figure 3.12. Ueof and Veof are defined further to address the EOF due to the surface charge on the channel wall. The fluid velocity on the particle surface includes three parts, as described in Equation (3.21), which are specified as shown in Figure 3.13. COMSOL Multiphysics 3.5a has predefined weak constraints to obtain a more accurate calculation of the hydrodynamic stress. However, it needs to be activated in the properties of the mmglf module, as shown in Figure 3.14. Basically, the weak constraints should be turned on, and the constraint type is nonideal. It is also seen that the inertial term is turned off in the present study, which could also be activated if the inertial effect is important.

Figure 3.15 shows the subdomain settings for the conductive media DC, called emdc. To be the same as dimensionless Equation (3.15), the thickness and the electric conductivity are both unity. The dimensionless electric potentials on the inlet and the outlet are, respectively, 0 and 1, as shown in Figures 3.16 and 3.17. The other boundaries are electric insulation, as shown in Figure 3.18.

Equations (3.22) and (3.23), which determine the particle's translational and rotational motions, are ordinary differential equations (ODEs). They can be defined in the **Physics|Global Equations** as shown in Figure 3.19 and then solved together with the other PDEs. Name (u) in Figure 3.19 defines the variable that needs to be solved. Expressions defined in equation f(u, ut, utt, t) are all zero. For

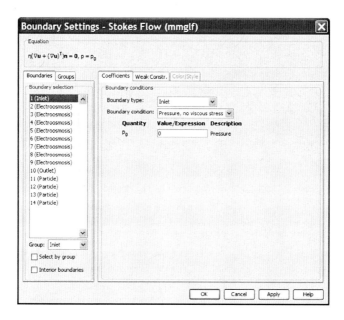

FIGURE 3.10   Boundary settings at the inlet for the mmglf module. (Image courtesy of COMSOL, Inc.)

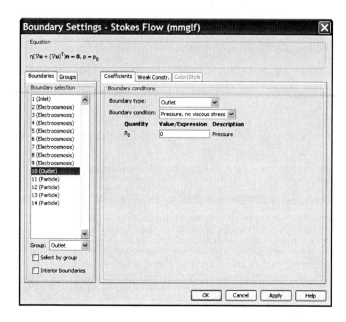

FIGURE 3.11 Boundary settings at the outlet for the mmglf module. (Image courtesy of COMSOL, Inc.)

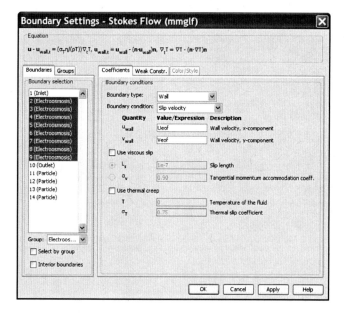

FIGURE 3.12 Boundary settings at the channel wall for the mmglf module. (Image courtesy of COMSOL, Inc.)

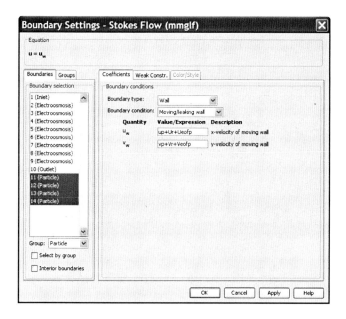

FIGURE 3.13   Boundary settings at the particle surface for the mmglf module. (Image courtesy of COMSOL, Inc.)

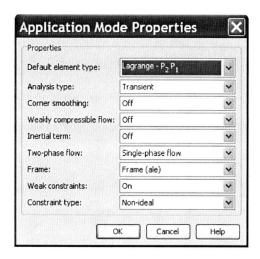

FIGURE 3.14   Properties of the mmglf module. (Image courtesy of COMSOL, Inc.)

example, Mass*upt-(F_x+Fd_x) refers to the equation Mass*upt-(F_x+Fd_x)=0, in which F_x and Fd_x are, respectively, the x component hydrodynamic force and the x component DEP force. Therefore, all the terms should be moved to the left-hand side when defining an equation. In COMSOL, the variable ut is the time derivative of the dependent variable u. For example, up is the x component particle velocity, which implies that upt is the x component acceleration. Init (u)

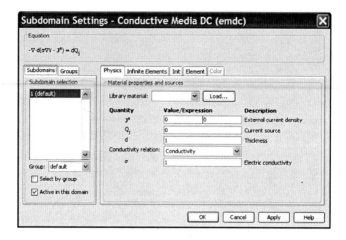

FIGURE 3.15   Subdomain settings for the emdc module. (Image courtesy of COMSOL, Inc.)

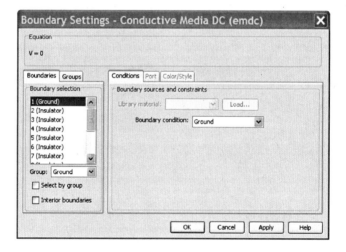

FIGURE 3.16   Boundary settings at the inlet for the emdc module. (Image courtesy of COMSOL, Inc.)

in Figure 3.19 is the initial value of the defined variable, and Init (ut) is the initial value of the time derivative of the defined variable. In Figure 3.19, up, Xpar, vp, Ypar, omega, and ang are, respectively, the x component particle velocity, x position of the particle's center of mass, y component particle velocity, y position of the particle's center of mass, rotational velocity, and angle. The initial location of the particle's center of mass is (Xp0, Yp0), and the initial angle is zero. The initial translational and rotational velocities are all zero.

To solve the ODEs defined in **Physics|Global Equations**, the force and torque acting on the particle should be known. From Equations (3.22) and (3.23),

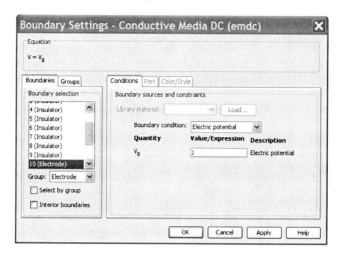

FIGURE 3.17   Boundary settings at the outlet for the emdc module. (Image courtesy of COMSOL, Inc.)

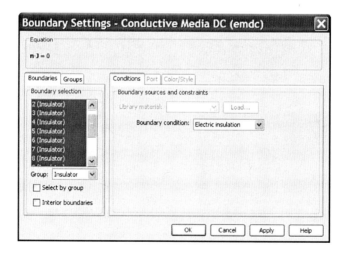

FIGURE 3.18   Boundary settings at the other boundaries for the emdc module. (Image courtesy of COMSOL, Inc.)

we can know that the calculations of force and torque are integrations over the particle surface. Figure 3.20 shows how to obtain the forces and torque acting on the particle in **Options|Integration Coupling Variables|Boundary Variables**. First select the boundaries for the integration. Name in Figure 3.20 defines the integral, and Expression defines the integrand. Integration order is 4, and Frame is mesh frame, which describes the configuration just after the latest remeshing. Here, lm3 and lm4 represent, respectively, the x and y component hydrodynamic stress (weak constraints) acting on the fluid predefined in COMSOL for a highly accurate calculation. The negative sign represents the hydrodynamic stress acting

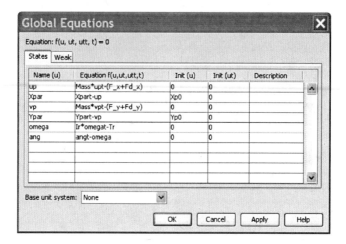

FIGURE 3.19   Global equations. (Image courtesy of COMSOL, Inc.)

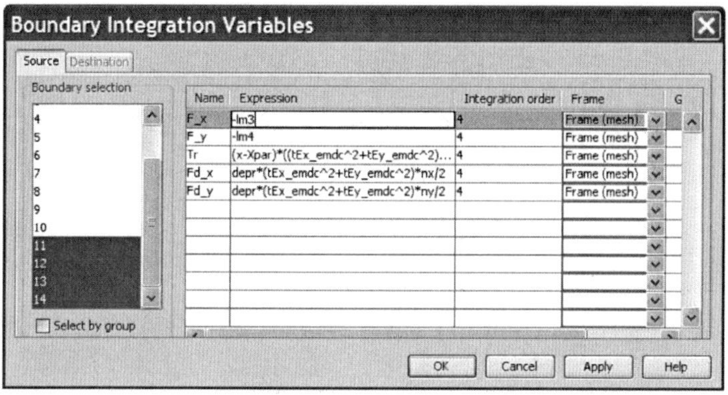

FIGURE 3.20   Boundary integration variables for force and torque calculation. (Image courtesy of COMSOL, Inc.)

on the particle but not the fluid. In the emdc module, Ex_emdc and Ey_emdc are, respectively, the x and y component electric field. tEx_emdc and tEy_emdc are, respectively, the predefined x- and y-component tangential electric field on the boundary.

The variables Ur and Ueofp used in the fluid boundary condition on the particle surface are defined in **Options|Expressions|Expressions**. First select the boundaries that you are going to use to define boundary expressions. Name in Figure 3.21 defines the variable representing the boundary expression. Omega is the rotational velocity and is positive when the particle rotates counterclockwise. The expression -omega*(y-Ypar) transfers the rotational velocity to the linear velocity. Because the expression is defined on the particle surface, the y variable is the y coordinate on the particle surface. The dimensionless Smoluckowski slip

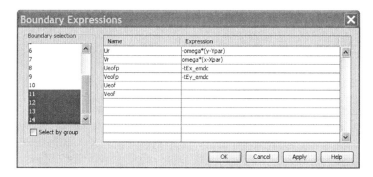

FIGURE 3.21 Boundary expressions on the particle surface. (Image courtesy of COMSOL, Inc.)

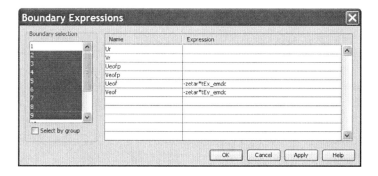

FIGURE 3.22 Boundary expressions on the channel wall. (Image courtesy of COMSOL, Inc.)

velocity is the dimensionless tangential electric field, as shown in Equation (3.21). Figure 3.22 shows how to set up the dimensionless Smoluckowski slip velocity on the channel wall.

When all these settings are complete, we generate mesh in the computational domain. As we are using the ALE method, triangular mesh has to be used. The parameters that control the mesh generation can be specified in **Mesh|Free Mesh Parameters**. To control the overall mesh in the computational domain, we usually specify the maximum element size of subdomains, as shown in Figure 3.23. To generate fine mesh locally, we can also specify the maximum element size on certain boundaries. For example, the mesh on the particle surface should be very dense to capture the flow field and electric field around the particle. Therefore, we set the maximum element size on the particle surface to 0.05, as shown in Figure 3.24. The maximum element size on the channel wall is specified as 0.4, as shown in Figure 3.25. The mesh based on these parameters is shown in Figure 3.4.

The **Solve|Solver Manager** can specify the initial value and the solution components to solve. Figure 3.26 shows all the variables for solving in the **Solve for**

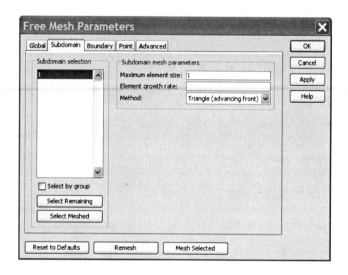

FIGURE 3.23   Free-mesh parameters for the subdomain. (Image courtesy of COMSOL, Inc.)

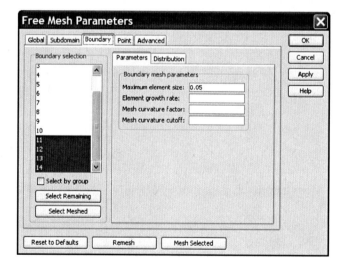

FIGURE 3.24   Free-mesh parameters for the particle surface. (Image courtesy of COMSOL, Inc.)

tab. One can determine which dependent variables to solve for by clicking the variables. Use Ctrl+click for multiple selections. From Figure 3.26, we can also see the predefined weak constraints in the mmglf module. The two variables following p, u, and v are, respectively, the x and y component hydrodynamic stress. Every time before we set up the **Options|Integration Coupling Variables|Boundary Variables**, we need to check which weak constraints represent the x- and y-component hydrodynamic stress in the **Solve for** tab.

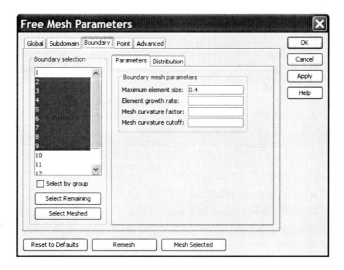

FIGURE 3.25   Free-mesh parameters for the channel wall. (Image courtesy of COMSOL, Inc.)

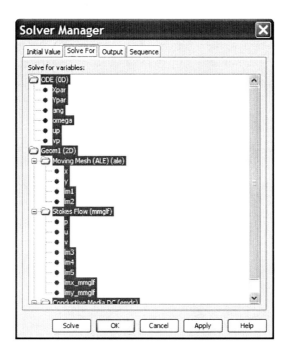

FIGURE 3.26   Solver manager. (Image courtesy of COMSOL, Inc.)

The **Solve|Solver Parameters** can control the solver settings, as shown in Figure 3.27. We can see the analysis type of ale and mmglf modules are transient. As a result, the solver that we use for this study is the time-dependent solver. The most important settings are in the **Time stepping**, in which the start time, time step, end time, relative tolerance, and absolute tolerance can be specified. Range (0, 2, 20000) denotes that the simulation starts from $t = 0$ to $t = 20,000$ with a time step of 2. The relative and absolute tolerances are crucial to obtain an accurate result. A smaller tolerance leads to a more accurate result; however, it takes longer to complete the simulation. For the present study, 1e-4 for relative tolerance and 1e-6 for absolute tolerance are good enough to obtain an accurate result. However, one has to run several tests to confirm that the results are mesh independent and not affected by the tolerance settings.

As mentioned, the simulation is forced to end when the mesh is significantly deformed. There is a setting in **Solver Parameters** to control stopping the simulation, named the **Stop condition**, as shown in Figure 3.28. The solver stops when the expression in the **Stop condition** edit field becomes negative. The default value for the stop condition is minqual1_ale-0.05. The ale module defines the variable minqual1_ale to estimate the minimum quality of the deformed mesh. The mesh quality defined in COMSOL is a value from 0 to 1, and a larger value represents better mesh quality. The minimum quality of the undeformed mesh at the beginning can be checked in **Mesh|Mesh Statistics**, as shown in Figure 3.29. In the present case, the minimum mesh quality is 0.7813. If the mesh dramatically deforms, it would affect the accuracy of the obtained results. Usually, the stop condition is set to be minqual1_ale-0.6 to avoid a significant mesh deformation.

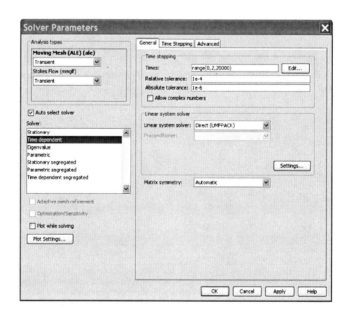

FIGURE 3.27   General tab in solver parameters. (Image courtesy of COMSOL, Inc.)

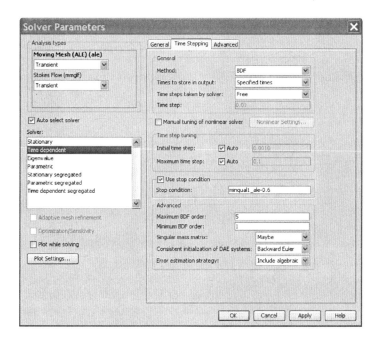

FIGURE 3.28   Time-stepping tab in solver parameters. (Image courtesy of COMSOL, Inc.)

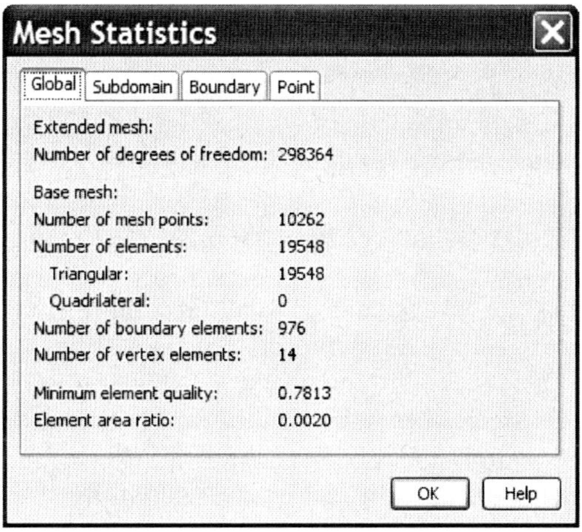

FIGURE 3.29   Mesh statistics. (Image courtesy of COMSOL, Inc.)

Finally, click the **Solve|Solve Problem** to run the simulation until the stop condition is satisfied. It is highly recommended to run the simulations in a 64-bit workstation with enough memory resource.

In **Postprocessing|Plot Parameters**, we are able to obtain several types of plots, as shown in Figure 3.30. For example, surface plot and streamline plot

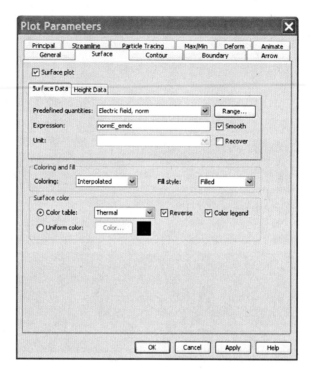

FIGURE 3.30   Surface plot in plot parameters. (Image courtesy of COMSOL, Inc.)

are two commonly used postprocessing techniques. Since it is a time-dependent problem, we need to pick up the time frame in the **General** tab first. Select electric field, norm and use reversed thermal color scheme in the **Surface** tab. Select the velocity field, use magnitude controlled as the streamline plot type, and set the density to 14 in the **Streamline** tab. Click **Apply** to get the two plots in a single figure, as shown in Figure 3.31.

One can also make an animation in the **Animate** tab. To examine the variables defined in **Physics|Global Equations**, use **Postprocessing|Global Variables Plot** to plot the variable of interest, for example, as shown in Figure 3.32. Click > to add the variables listed in **Predefined quantities** to **Quantities to Plot**. Select all the output times as the x axis and click **Apply** to obtain the plot, as shown in Figure 3.33. One can click the **Edit Plot** icon to edit this figure. Click the **Export Current Plot** to export the data into a separate file stored in the computer that can be postprocessed by other software.

As mentioned, a new geometry should be generated based on the deformed mesh to continue the computation. Open **Mesh|Create Geometry From Mesh** to obtain Figure 3.34. Select **Deformed mesh** in the ale frame and solution at the last time step $t = 262$ in the **Source** tab. Click **OK** to generate a new geometry based on the deformed mesh at $t = 262$. Figure 3.35 shows the undeformed and deformed mesh before and after **Create Geometry From Mesh**. Because the stop condition is set to

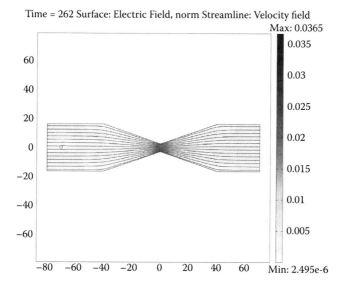

FIGURE 3.31   Surface plot of the electric field and streamlines of fluid field.

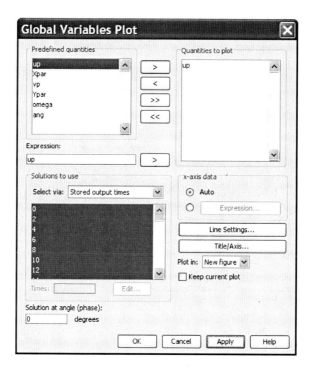

FIGURE 3.32   Global variables plot. (Image courtesy of COMSOL, Inc.)

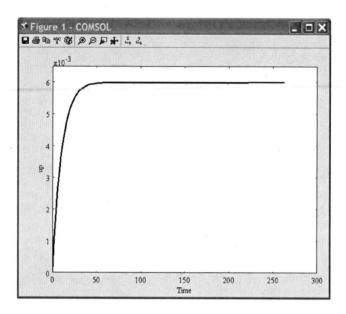

FIGURE 3.33   Time variation of the x component translational velocity.

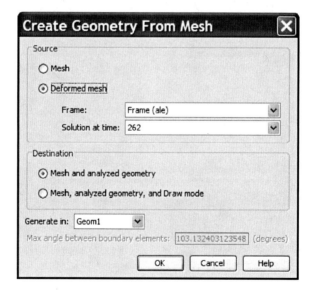

FIGURE 3.34   Create geometry from deformed mesh. (Image courtesy of COMSOL, Inc.)

be relatively high, the mesh on the left-hand side of the particle is slightly elongated without significant deformation, which helps to maintain the accuracy of the numerical results. Before restarting the simulation, the start time in the **Solve|Solver Parameters** can be revised to range (262, 2, 20000). Click **Solve|Restart**, which uses the previous solution at $t = 262$ as the initial value to continue solving. When

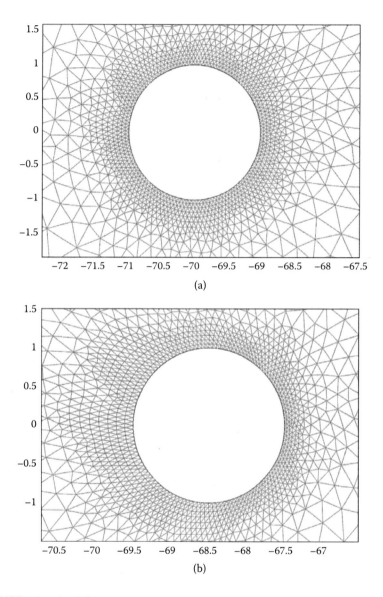

FIGURE 3.35   Undeformed (a) and deformed (b) meshes near the moving particle.

the solver stops again, regenerate the geometry and modify the start time to restart the solver. During one remeshing process, the particle approximately moves less than the particle diameter. If the particle needs to move a long distance, it might have hundreds of remeshing processes. Therefore, it is impossible to do the remeshing manually in the GUI of COMSOL Multiphysics 3.5a.

To carry out the automated remeshing, one has to generate a script M-file and run it in COMSOL with MATLAB®. To get the M-file, one can first set up all

the settings in the GUI of COMSOL Multiphysics 3.5a and run the simulation
to check if all the settings are correct. After that, do one remeshing in the GUI
of COMSOL Multiphysics 3.5a and run the simulation one more time. When the
solver stops again, click **File|Save As** to save all the settings and implementations
in the GUI into an M-file with an extension .m. Remove the useless implementations in the M-file and add some scripts to implement the automated remeshing.
The following full COMSOL MATLAB script M-file is based on a saved M-file
with some additive scripts. The authors' commentary beginning with "% NEW"
is given before the additive script.

```
%%%%%%%%%%%Converging-diverging channel.m%%%%%%%%%%%%%
flclear fem

% COMSOL version
clear vrsn
vrsn.name = 'COMSOL 3.4';
vrsn.ext = '';
vrsn.major = 0;
vrsn.build = 248;
vrsn.rcs = '$Name: $';
vrsn.date = '$Date: 2007/10/10 16:07:51 $';
fem.version = vrsn;

% NEW: Define electric field
electric=[15,22.5,30,37.5];

% NEW: FOR loop to go through all the electric fields
for i=1:length(electric)

% Constants
fem.const = {'eps_r','80', ...
             'eps0','8.854187817e-12', ...
             'rho','1e3', ...
             'Xp0','-70', ...
             'Yp0','0', ...
             'Masss','rho*pi*a^2', ...
             'Mass','Masss*Uc/eta/a', ...
             'Ir','1/2*Masss*a^2*Uc/eta/a^3', ...
             'a','10e-6', ...
             'zetar','2.5', ...
             'ep','32e-3', ...
             'eta','1.0e-3', ...
             'Uc','eps_r*eps0*ep*v0/(eta*a)', ...
             'v0',num2str(electric(i)), ...
             'Re','rho*Uc*a/eta', ...
             'depr','v0/ep'};

% Geometry
% NEW: Draw the converging-diverging channel
```

```
carr={curve2([0,40],[16.25,16.25]), ...
      curve2([40,80],[16.25,2.75]), ...
      curve2([80,120],[2.75,16.25]), ...
      curve2([120,150],[16.25,16.25]), ...
      curve2([150,150],[16.25,-16.25]), ...
      curve2([150,120],[-16.25,-16.25]), ...
      curve2([120,80],[-16.25,-2.75]), ...
      curve2([80,40],[-2.75,-16.25]), ...
      curve2([40,0],[-16.25,-16.25]), ...
      curve2([0,0],[-16.25,16.25])};
g2=geomcoerce('curve',carr);
g3=geomcoerce('solid',{g2});
g3=move(g3,[-80,0]);

% NEW: Draw the circular particle
g4=circ2('1','base','center','pos',{'-70','0'},'rot','0');
g5=geomcomp({g3,g4},'ns',{'g3','g4'},'sf','g3-g4','edge',
  'none');

% Analyzed geometry
clear s
s.objs={g5};
s.name={'CO2'};
s.tags={'g5'};

fem.draw=struct('s',s);
fem.geom=geomcsg(fem);

% NEW: Display geometry
geomplot(fem, 'Labelcolor','r','Edgelabels','on','submode',
  'off');

% NEW: Solve moving mesh
% Application mode 1
clear appl
appl.mode.class = 'MovingMesh';
appl.sdim = {'Xm','Ym','Zm'};
appl.shape = {'shlag(2,''lm1'')','shlag(2,''lm2'')','shlag(2,
  ''x'')','shlag(2,''y'')'};
appl.gporder = {30,4};
appl.cporder = 2;
appl.assignsuffix = '_ale';
clear prop
prop.smoothing='winslow';
prop.analysis='transient';
prop.allowremesh='on';
prop.origrefframe='ref';
appl.prop = prop;
clear bnd
bnd.defflag = {{1;1},{0;0}};
```

```
bnd.type = {'def','vel'};
bnd.veldefflag = {{0;0},{1;1}};
bnd.wcshape = [1;2];
bnd.veldeform = {{0;0},{'up+Ur';'vp+Vr'}};
bnd.name = {'Fixed','Particle'};
bnd.ind = [1,1,1,1,1,1,1,1,1,1,2,2,2,2];
appl.bnd = bnd;
clear equ
equ.gporder = 2;
equ.shape = [3;4];
equ.ind = [1];
appl.equ = equ;
fem.appl{1} = appl;

% NEW: Solve flow field using Stokes equations
% Application mode 2
clear appl
appl.mode.class = 'GeneralLaminarFlow';
appl.module = 'MEMS';
appl.shape = {'shlag(2,''lm3'')','shlag(2,''lm4'')','shlag(1,
  ''lm5'')','shlag(2,''u'')','shlag(2,''v'')','shlag(1,
  ''p'')'};
appl.gporder = {30,4,2};
appl.cporder = {2,1};
appl.assignsuffix = '_mmglf';
clear prop
prop.weakcompflow='Off';
prop.inerterm='Off';
clear weakconstr
weakconstr.value = 'on';
weakconstr.dim = {'lm3','lm4','lm5','lm6'};
prop.weakconstr = weakconstr;
prop.constrtype='non-ideal';
appl.prop = prop;
clear bnd
bnd.zeta = {-0.1,'-zetar/(eps_r*epsilon0_emdc)',-0.1,-0.1,
  -0.1,-0.1, ... -0.1};
bnd.intype = {'p','uv','uv','uv','uv','uv','uv'};
bnd.velType = {'U0in','U0in','U0in','u0','U0in','U0in',
  'U0in'};
bnd.eotype = {'mueo','zeta','mueo','mueo','mueo','mueo',
  'mueo'};
bnd.U0 = {0.0010,0,0,0,0,0,0};
bnd.wcgporder = 1;
bnd.E_x = {0,'Ex_emdc',0,0,0,0,0};
bnd.vw0 = {0,'Veof',0,'Veofp',0,0,0};
bnd.uwall = {0,0,0,'up+Ur+Ueofp',0,0,0};
bnd.E_y = {0,'Ey_emdc',0,0,0,0,0};
bnd.vwall = {0,0,0,'vp+Vr+Veofp',0,0,0};
bnd.uw0 = {0,'Ueof',0,'Ueofp',0,0,0};
```

```
bnd.walltype = {'noslip','semislip','noslip','lwall',
  'noslip','noslip','noslip'};
bnd.name = {'Inlet','Electroosmosis','Outlet','Particle',
  'Electrode','Ground', ... 'Symmetry'};
bnd.type = {'inlet','walltype','outlet','walltype',
  'walltype','walltype', ... 'sym'};
bnd.v0 = {0,0,0,'vp+Vr+Veofp',0,0,0};
bnd.wcshape = [1;2;3];
bnd.mueo = {7.0E-8,'mu_eo',7.0E-8,7.0E-8,7.0E-8,7.0E-8,7.
  0E-8};
bnd.isViscousSlip = {1,0,1,0,1,1,1};
bnd.u0 = {0,0,0,'up+Ur+Ueofp',0,0,0};
bnd.ind = [1,2,2,2,2,2,2,2,2,3,4,4,4,4];
appl.bnd = bnd;
clear equ
equ.shape = [4;5;6];
equ.cporder = {{1;1;2}};
equ.rho = 'Re';
equ.eta = 1;
equ.gporder = {{2;2;3}};
equ.thickness = 1;
equ.ind = [1];
appl.equ = equ;
fem.appl{2} = appl;

% NEW: Solve electrostatics using Laplace equation
% Application mode 3
clear appl
appl.mode.class = 'EmConductiveMediaDC';
appl.module = 'MEMS';
appl.assignsuffix = '_emdc';
clear prop
clear weakconstr
weakconstr.value = 'off';
weakconstr.dim = {'lm7'};
prop.weakconstr = weakconstr;
appl.prop = prop;
clear bnd
bnd.V0 = {1,0,0,0,0,0,0};
bnd.type = {'V','nJ0','V0','nJ0','nJ0','nJ0','nJ0'};
bnd.name = {'Electrode','Insulator','Ground',
  'Electroosmosis','Inlet','Outlet', ... 'Symmetry'};
bnd.ind = [3,2,2,2,2,2,2,2,2,1,2,2,2,2];
appl.bnd = bnd;
clear equ
equ.sigma = 1;
equ.name = 'default';
equ.ind = [1];
appl.equ = equ;
fem.appl{3} = appl;
```

```
fem.sdim = {{'Xm','Ym'},{'X','Y'},{'x','y'}};
fem.frame = {'mesh','ref','ale'};
fem.border = 1;

% Boundary settings
clear bnd
bnd.ind = [1,2,2,2,2,2,2,2,2,1,3,3,3,3];
bnd.dim = {'x','y','lm1','lm2','u','v','p','lmx_mmglf','lmy_
  mmglf', ... 'lm3','lm4','lm5','V'};

% Boundary expressions
bnd.expr = {'Ur',{'','','-omega*(y-Ypar)'}, ...
            'Vr',{'','','omega*(x-Xpar)'}, ...
            'Ueofp',{'','','-tEx_emdc'}, ...
            'Veofp',{'','','-tEy_emdc'}, ...
            'Ueof',{'','-zetar*tEx_emdc',''}, ...
            'Veof',{'','-zetar*tEy_emdc',''}};
fem.bnd = bnd;

% NEW: Boundary integration to calculate the hydrodynamic
  force
% NEW: and DEP force on the particle
% Coupling variable elements
clear elemcpl
% Integration coupling variables
clear elem
elem.elem = 'elcplscalar';
elem.g = {'1'};
src = cell(1,1);
clear bnd
bnd.expr = {{{},'-lm3'},{{},'-lm4'},{{},'(x-Xpar)*((tEx_
  emdc^2+tEy_emdc^2)*ny/2-lm4)-(y-Ypar)*((tEx_emdc^2+tEy_
  emdc^2)*nx/2-lm3)'},{{}, ...
  'depr*(tEx_emdc^2+tEy_emdc^2)*nx/2'},{{}, ...
  'depr*(tEx_emdc^2+tEy_emdc^2)*ny/2'}};
bnd.ipoints = {{{},'4'},{{},'4'},{{},'4'},{{},'4'},{{},
  '4'}};
bnd.frame = {{{},'mesh'},{{},'mesh'},{{},'mesh'},{{},
  'mesh'},{{},'mesh'}};
bnd.ind = {{'1','2','3','4','5','6','7','8','9','10'},{'11',
  '12','13', ... '14'}};
src{1} = {{},bnd,{}};
elem.src = src;
geomdim = cell(1,1);
geomdim{1} = {};
elem.geomdim = geomdim;
elem.var = {'F_x','F_y','Tr','Fd_x','Fd_y'};
elem.global = {'1','2','3','4','5'};
elem.maxvars = {};
elemcpl{1} = elem;
```

```
fem.elemcpl = elemcpl;
% ******************************

% NEW: Solve six ODEs to determine the particle's transla-
  tional velocity and rotational velocity
% ODE Settings
clear ode
ode.dim={'up','Xpar','vp','Ypar','omega','ang'};
ode.f={'Mass*upt-(F_x+Fd_x)','Xpart-up','Mass*vpt-(F_
  y+Fd_y)','Ypart-vp','Ir*omegat-Tr','angt-omega'};
ode.init={'0','Xp0','0','Yp0','0','0'};
ode.dinit={'0','0','0','0','0','0'};
fem.ode=ode;

% Multiphysics
fem=multiphysics(fem);

% NEW: Loop for continuous particle tracking
% NEW: Remeshing index
j=1;

% NEW: Define initial time
time_e=0;

% NEW: Define time step
time_step=2;

% NEW: Define data storage index
num=1;

% NEW: Define matrix for data storage
particle=zeros(10,11);

% NEW: Get the particle location
Xp=-70;
Yp=0;

% NEW: Define end condition of the computation (Loop control)
while Xp<63

% Initialize mesh
fem.mesh=meshinit(fem, ...
                  'hauto',5, ...

  'hmaxedg',[2,0.4,3,0.4,4,0.4,5,0.4,6,0.4,7,0.4,8,0.4,9,0.4,
   11,0.05,12,0.05,13,0.05,14,0.05],'hmaxsub',[1,1]);

% Extend mesh
fem.xmesh=meshextend(fem);
% ********************************************
```

```
if j==1
% Solve problem
fem.sol=femtime(fem, ...

'solcomp',{'lm5','ang','Ypar','p','lm1','lm4','V','v','lm3',
   'Xpar','up','vp','omega','u','y','lm2','lmy_mmglf',
   'lmx_mmglf','x'}, ...

'outcomp',{'lm5','ang','Ypar','p','lm1','lm4','V','v','lm3',
   'Xpar','up','vp','omega','u','Y','y','lm2','lmy_mmglf',
   'lmx_mmglf','x','X'}, ...
                    'tlist',[0:time_step:20000], ...
                    'tout','tlist', ...
                    'rtol',1e-4, ...
                    'atol',1e-6, ...
                    'stopcond','minqual1_ale-0.6');
else

% Mapping current solution to extended mesh
init = asseminit(fem,'init',fem0.sol,'xmesh',fem0.xmesh,
   'framesrc','ale','domwise','on');

% Solve problem
fem.sol=femtime(fem, ...
                    'init',init, ...

'solcomp',{'lm5','ang','Ypar','p','lm1','lm4','V','v','lm3',
   'Xpar','up','vp','omega','u','y','lm2','lmy_mmglf',
   'lmx_mmglf','x'}, ...

'outcomp',{'lm5','ang','Ypar','p','lm1','lm4','V','v','lm3',
   'Xpar','up','vp','omega','u','Y','y','lm2','lmy_mmglf',
   'lmx_mmglf','x','X'}, ...
                    'tlist',[time_e:time_step:20000], ...
                    'tout','tlist', ...
                    'rtol',1e-4, ...
                    'atol',1e-6, ...
                    'stopcond','minqual1_ale-0.6');
end

% Save current fem structure for restart purposes
fem0=fem;

% NEW: Get the last time from solution
time_e = fem.sol.tlist(end);

% NEW: Get the length of the time steps
Leng =length(fem.sol.tlist);

% Global variables plot
```

```
data=postglobalplot(fem,{'Xpar','Ypar','up','vp'}, ...
                'linlegend','on', ...
                'title','Particle', ...
                'Outtype','postdata', ...
                'axislabel',{'Time','Parameters'});

% NEW: Store the interested data
for n=2:Leng-1
    particle(num,1)=data.p(1,n);
    particle(num,2)=data.p(2,n);
    particle(num,3)=data.p(2,n+Leng);
    particle(num,4)=data.p(2,n+2*Leng);
    particle(num,5)=data.p(2,n+3*Leng);

% Change dimensionless velocity into dimensional velocity

 particle(num,6)=data.p(2,n+2*Leng)*1e3*80*8.854187817e
  -12*32e-3*electric(i)/(1e-3*1e-5);

 particle(num,7)=data.p(2,n+3*Leng)*1e3*80*8.854187817e
  -12*32e-3*electric(i)/(1e-3*1e-5);

% Integrate
I1=postint(fem,'-lm3', ...
            'unit','N/m', ...
            'dl',[11,12,13,14], ...
            'edim',1, ...
            'solnum',n);

% Integrate
I2=postint(fem,'-lm4', ...
            'unit','N/m', ...
            'dl',[11,12,13,14], ...
            'edim',1, ...
            'solnum',n);

% Integrate
I3=postint(fem,'depr*(Ex_emdc^2+Ey_emdc^2)*nx/2', ...
            'unit','N/m', ...
            'dl',[11,12,13,14], ...
            'edim',1, ...
            'solnum',n);

% Integrate
I4=postint(fem,'depr*(Ex_emdc^2+Ey_emdc^2)*ny/2', ...
            'unit','N/m', ...
            'dl',[11,12,13,14], ...
            'edim',1, ...
            'solnum',n);
```

```
            particle(num,8)=I1;
            particle(num,9)=I2;
            particle(num,10)=I3;
            particle(num,11)=I4;

 num=num+1;
end

% Plot solution
postplot(fem, ...
            'tridata',{'U_mmglf','cont','internal'}, ...
            'tridlim',[0 0.07], ...
            'trimap','jet(1024)', ...
            'solnum','end', ...
            'title','Surface: Velocity field', ...
            'geom','off', ...
            'axis',[-90,80,-40,40]);

% NEW: update the particle's location and orientation
Xp=data.p(2,Leng);
Yp=data.p(2,2*Leng);

% NEW: Time step selection in different regions
if(abs(Xp)<21)
  time_step=0.5;

elseif(abs(Xp)>20&&abs(Xp)<40)
  time_step=1;

else
  time_step=2;

end

% NEW: Output the COMSOL Multiphysics GUI .mph file if
  necessary
flsave(strcat('CD_h0_electric=',num2str(electric(i)/1.5),
  'KVm-1','_Xp=',num2str(Xp),'.mph'),fem);

% NEW: Write the data into a file
dlmwrite(strcat('CD_h0_electric=',num2str(electric(i)/1.5),
  'KVm-1.dat'),particle,',');

% Geometry
% Generate geom from mesh
fem = mesh2geom(fem, ...
                'frame','ale', ...
                'srcdata','deformed', ...
                'destfield',{'geom','mesh'}, ...
                'srcfem',1, ...
```

```
                    'destfem',1);

j=j+1;
end


end
```

3.4   RESULTS AND DISCUSSION

3.4.1   IN A STRAIGHT MICROCHANNEL

We first investigate electrophoresis of a spherical particle in a cylindrical tube. Due to the axisymmetric nature of this problem, we choose axial symmetry (2D) as the space dimension in the **Model Navigator**, as shown in Figure 3.36. The geometry of the axisymmetric model is shown in Figure 3.37, in which $R_m = 0$ is the axis. Table 3.3 lists all the constants used in the simulation. As the particle translocates along the axis of the tube, there is no lateral motion and rotational motion. As a result, we only solve the z- component translational velocity. Other settings are similar to those described in Section 3.3. However, the geometry of the axis also changes because of the particle translocation, as indicated in Figure 3.37. Therefore, the boundary condition of the axis for the ale module is no longer zero displacement. Figure 3.38 shows the boundary condition of the upper axis, whose r and z component mesh velocities are, respectively, 0 and vp*(1-s). Here, vp is the particle's z-component translational velocity, and s is the normalized boundary length with a range (0, 1). When one of the axis boundaries is selected, an arrow is

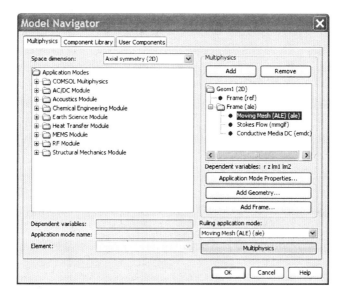

FIGURE 3.36   Settings in Multiphysics Model Navigator for axial symmetry model. (Image courtesy of COMSOL, Inc.)

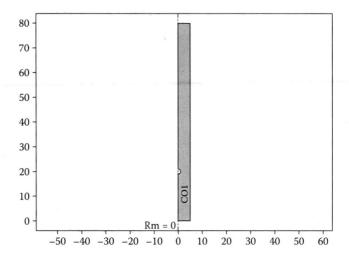

FIGURE 3.37 Axial symmetry model of a particle moving along the axis of a cylindrical tube.

TABLE 3.3
Constant Table

Variable	Value or Expression	Description
eps_r	80	Relative permittivity of fluid
eps0	8.854187817e-12 [F/m]	Permittivity of vacuum
rho	1e3 [kg/m^3]	Particle/fluid density
Rp0	0	Dimensionless initial r location of particle
Zp0	20	Dimensionless initial z location of particle
Masss	rho*4/3*pi*a^3	Dimensional particle mass
Mass	Masss*Uc/eta/a^2	Dimensionless particle mass
a	1E-06 [m]	Particle radius
zetar	0.375	Zeta potential ratio
ep	58e-3 [V]	Zeta potential of particle
eta	1e-3 [Pa*s]	Fluid viscosity
Uc	eps_r*eps0*ep* v0 /(eta*a)	Velocity scale
v0	10 [V]	Applied voltage
Re	rho*Uc*a/eta	Reynolds number

shown along the boundary. The s value of the start and end points of the arrow are, respectively, s = 0 and s = 1. Therefore, the r and z component mesh velocities of the lower axis boundary are, respectively, 0 and vp*s. More detailed implementations are listed in Table 3.4.

Figure 3.39 shows the electrophoretic velocity of a charged spherical particle of diameter d translating along the axis of a cylindrical tube of diameter a.

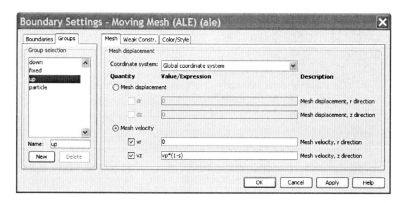

FIGURE 3.38   Boundary settings at the upper axis for the ale module. (Image courtesy of COMSOL, Inc.)

TABLE 3.4
Model Setup in the GUI of COMSOL Multiphysics 3.5a

Model Navigator	Select **Axial Symmetry (2D)** in space dimension and click **Multiphysics** button
	Select **COMSOL MultiphysicslDeformed MeshlMoving Mesh (ALE)lTransient analysis**. Click **Add** button
	Select **AC/DC ModulelStatics, ElectriclConductive Media DC**. Click **Add** button
	Select **MEMS ModulelMicrofluidicslStokes FlowlTransient analysis**. Click **Add** button
Option MenulConstants	Define constants in Table 3.3
Physics MenulSubdomain **Setting**	ale model
	Subdomain 1
	Free displacement
	emdc mode (external electric field)
	Subdomain 1
	$J_e = 0$; $Q_j = 0$; $\sigma = 1$
	mmglf mode
	Subdomain 1
	Tab Physics
	$\rho = Re$; $\eta = 1$
Physics MenulBoundary **Setting**	ale mode
	Particle surface: Mesh velocity vr = 0; vz = vp
	Upper axis: Mesh velocity vr = 0; vz = vp*(1-s)
	Lower axis: Mesh velocity vr = 0; vz = vp*s
	Other boundaries: Mesh displacement dx = 0 dy = 0

continued

TABLE 3.4 (CONTINUED)
Model Setup in the GUI of COMSOL Multiphysics 3.5a

	emdc mode Upper boundary: Ground Lower boundary: $V_0 = 1$ Other boundaries: Electric insulation
	mmglf mode Axis: Symmetry boundary|Axial symmetry Lower boundary: Inlet|Pressure, no viscous stress $P_0 = 0$ Upper boundary: Outlet|Pressure, no viscous stress $P_0 = 0$ Particle surface: Wall|Moving/leaking wall Uw = Ueofp Vw = vp+Veofp Channel wall: Wall|Moving/leaking wall Uw = Ueof Vw = Veof
Other Settings	Properties of ale model (**Physics** Menu|**Properties**) Smoothing method: Winslow Allow remeshing: On
	Properties of mmglf model (**Physics** Menu|**Properties**) Weak constraints: On Constraint type: Nonideal
	Deactivate the unit system (**Physics** Menu|**Model Setting**) Base unit system: None
Options Menu|**Integration** **Coupling**	Boundary selection: 6, 7 (Particle surface)

Variables|**Boundary**
  Variables
	Name: F_z Expression: -lm4; Integration order: 4; Frame:
	  Frame(mesh)

Options Menu|**Expressions**| **Boundary Expressions**	Boundary selection: 6, 7 (Particle surface) Name: Ueofp Expression: -tEr_emdc Name: Veofp Expression: -tEz_emdc
	Boundary selection: 5 (Channel wall) Name: Ueof Expression: -zetar*tEr_emdc Name: Veof Expression: -zetar*tEz_emdc
Physics Menu|**Global** **Equations**	Name: vp; Expression: Mass*vpt-F_z; Init (u): 0; Init (ut): 0 Name: zp; Expression: zpt-vp; Init (u): Zp0; Init (ut): 0
Mesh|**Free Mesh** **Parameters**	Tab subdomain Subdomain 1|Maximum element size: 0.4
	Tab boundary Channel wall: 2, 4, 5|**Maximum element size**: 0.2 Particle surface: 6, 7|**Maximum element size**: 0.04
Solve Menu|	**Solver Parameters** Tab General Times: Range (0, 3, 3000) Relative tolerance: 1e-4 Absolute tolerance: 1e-6

TABLE 3.4 (CONTINUED)
Model Setup in the GUI of COMSOL Multiphysics 3.5a

	Tab Timing Stepping
	Activate Use stop condition
	minqual1_ale-0.6
	Click = **Solve Problem**
Postprocessing Menul	Result check: zp, vp
	Global Variables PlotISelect all the predefined quantities, click >>
	Click Apply

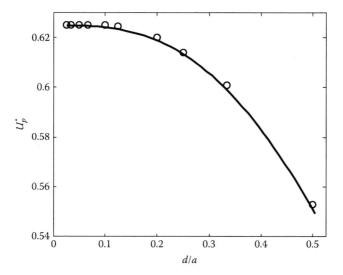

FIGURE 3.39 Dimensionless translational velocity of a sphere moving along the axis of a tube as a function of the ratio between the diameter of the sphere and that of the tube. The solid line and circles represent, respectively, the approximation solution from Keh and Anderson (1985) and our numerical results obtained by an axisymmetric model. (From Ai, Y., S. W. Joo, Y. Jiang, X. Xuan, and S. Qian. 2009. Transient electrophoretic motion of a charged particle through a converging-diverging microchannel: Effect of direct current-dielectrophoretic force. *Electrophoresis* 30:2499–2506. With permission from Wiley-VCH.)

The approximate solution of the electrophoretic velocity is available under thin EDL approximation (Keh and Anderson 1985):

$$U_p^* = \left[1 - 1.28987 \left(\frac{d}{a} \right)^3 + 1.89632 \left(\frac{d}{a} \right)^5 - 1.02780 \left(\frac{d}{a} \right)^6 + O \left(\frac{d}{a} \right)^8 \right] (1 - \gamma), \text{(3.24)}$$

where $\gamma = \zeta_w / \zeta_p$ denotes the ratio of the zeta potential of the channel wall to that of the particle.

The electrophoretic velocity shown in Figure 3.39 is normalized by $\varepsilon_f \varepsilon_0 \zeta_p E_z / \mu$, μ with E_z representing the electric field along the axis of the tube in the absence of the particle. The present numerical results are in good agreement with the approximate solution.

In this case, the DEP force is zero due to the axial symmetry of the electric field around the particle. Therefore, the DEP effect is not considered in this case. To validate the DEP force calculation in the present numerical model, we study the DEP force acting on a dielectric sphere of radius a near a planar wall, as shown in Figure 3.40. Obviously, the presence of the particle can significantly distort the electric field. As the electric field between the particle and the planar wall is enhanced, the induced negative DEP force acting on the particle is pointed away from the planar wall. Figure 3.41 shows the DEP force as a function of the dimensionless gap size, $\delta^* = d_p / a$, where d_p is the distance from the particle center to the planar wall. The DEP force is normalized by $\varepsilon_0 \varepsilon_f E_\infty^2 a^2 / 2$, in which E_∞ is the uniform electric field imposed parallel to the planar wall far away from the particle. Our numerical results are also in good agreement with the analytical results obtained by Young and Li (2005).

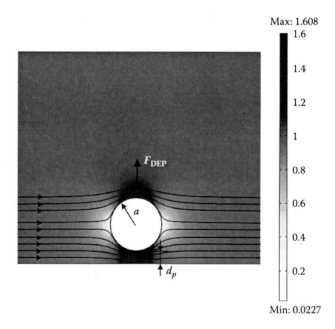

FIGURE 3.40 A spherical particle of radius a located d_p away from a planar wall. The gray levels denote the magnitude of the electric field; a dark color indicates a higher electric field. The induced negative DEP effect pushes the particle away from the planar wall. (From Ai, Y., S. W. Joo, Y. Jiang, X. Xuan, and S. Qian. 2009. Transient electrophoretic motion of a charged particle through a converging-diverging microchannel: Effect of direct current-dielectrophoretic force. *Electrophoresis* 30:2499–2506. With permission from Wiley-VCH.)

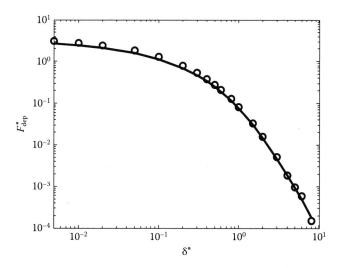

FIGURE 3.41 Dimensionless DEP force exerted on a sphere near a planar wall as a function of the dimensionless gap size. The solid line and circles represent, respectively, Young and Li's (2005) analytical solution and our numerical results obtained by a 3D model. (From Ai, Y., S. W. Joo, Y. Jiang, X. Xuan, and S. Qian. 2009. Transient electrophoretic motion of a charged particle through a converging-diverging microchannel: Effect of direct current-dielectrophoretic force. *Electrophoresis* 30:2499–2506. With permission from Wiley-VCH.)

Kang, Xuan et al. (2006) experimentally studied the electrokinetic motion of a charged spherical particle in a straight microchannel with a rectangular hurdle inside. Next, we use the 2D numerical model to reproduce this experimental work. The experimental study observed obvious lateral particle motion when the particle passed through the hurdle. Figure 3.42 compares the numerical predictions with the experimental data of the particle trajectories. For two 15.7-μm particles under a 5-kV/m electric field, the numerical results obtained by the 2D model considering the DEP effect are in good agreement with the experimental data. It is concluded that the 2D model is sufficient for capturing the DEP effect on the electrokinetic particle transport in microfluidics. If the DEP effect is ignored in the numerical model, the predicted particle trajectory is symmetric with respect to the hurdle, which dramatically deviates from the experimental observations. Apparently, the experimentally observed lateral particle motion is due to the DEP effect arising from the nonuniform electric field in the hurdle region. Therefore, the DEP effect must be taken into account in the numerical study of electrokinetic particle transport in microfluidics where nonuniform electric fields are present.

3.4.2 IN A CONVERGING-DIVERGING MICROCHANNEL

In this section, we investigate the electrokinetic motion of a circular particle in a converging-diverging microchannel. The dimensions of the converging-diverging section are taken from the fabricated device (Xuan, Xu, and Li 2005), with

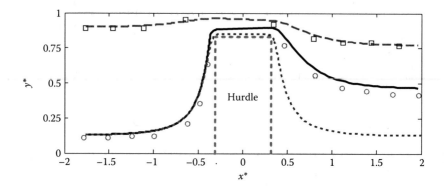

FIGURE 3.42 Particle trajectories through a microchannel with a rectangular hurdle in the middle. The solid and dashed lines represent the predicted particle trajectories considering the DEP force, the circles and squares represent the experimental data, and the dotted line represents the predicted particle trajectory of the lower particle without considering the DEP force. The x and y locations are both normalized by the channel width. (From Ai, Y., S. W. Joo, Y. Jiang, X. Xuan, and S. Qian. 2009. Transient electrophoretic motion of a charged particle through a converging-diverging microchannel: Effect of direct current-dielectrophoretic force. *Electrophoresis* 30:2499–2506. With permission from Wiley-VCH.)

$L_b = L_c = 400$ μm, $L_a = 400$ μm, and $L_d = 300$ μm. Therefore, the entire length of the microchannel is 1,500 μm. The widths of the uniform section and throat are, respectively, $a = 325$ μm and $b = 55$ μm. The applied electric field strength E is calculated by dividing the electric potential difference between the inlet and outlet over the total length of the microchannel. The dimensional initial lateral particle location is defined as $h^* = 2h/a$, where h is the distance between the center of the particle and the centerline of the microchannel.

3.4.2.1 Lateral Motion

Figure 3.43a compares the predicted particle trajectory through a converging-diverging microchannel with and without considering the DEP effect when $E = 10$ kV/m, $d = 20$ μm, $\zeta_p = -58$ mV, and $\gamma = 0.3$. The solid line (or circles) and dashed line (or squares) are, respectively, $h^* = -0.5$ and $h^* = 0.7$. If the DEP effect is ignored, the particle trajectory is symmetric with respect to the throat. However, the particle trajectory becomes asymmetric with respect to the throat when the DEP effect is considered. In addition, the particle is pushed toward the centerline of the microchannel after passing through the throat, which is explained later in the chapter. Figure 3.43b compares the numerical results obtained with considering DEP to the experimental results when $E = 15$ kV/m, $d = 10.35$ μm, $\zeta_p = -32$ mV, and $\gamma = 2.5$. Because the particle size is smaller than that used in Figure 3.43a and the DEP effect is proportional to the third power of the particle size, the particle in Figure 3.43b experiences smaller lateral motion. The size-based lateral particle motion has been experimentally demonstrated for particle separation (Barbulovic-Nad et al. 2006; Kang, Xuan, et al. 2006; Kang et al. 2008; Parikesit et al. 2008; Kang, Kang, et al. 2006; Lewpiriyawong, Yang, and Lam 2008).

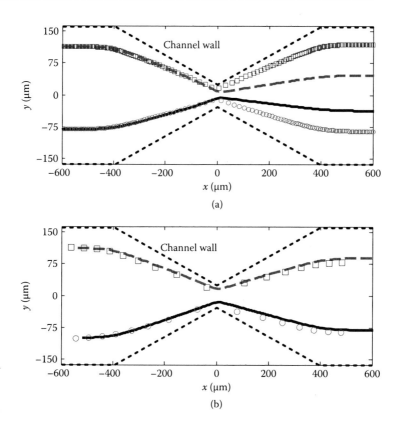

(a)

(b)

FIGURE 3.43 (a) Predicted particle trajectories with (solid and dashed lines) and without (circles and squares) considering the DEP force. $E = 10$ kV/m, $d = 20$ μm, $\zeta_p = -58$ mV, $a = 325$ μm, $b = 55$ μm, and $\gamma = 0.3$. (b) Predicted particle trajectories considering the DEP force (solid and dashed lines) compared with the experimental data (circles and squares). $E = 15$ kV/m, $d = 10.35$ μm, $\zeta_p = -32$ mV, $a = 325$ μm, $b = 55$ μm, and $\gamma = 2.5$. (From Ai, Y., S. W. Joo, Y. Jiang, X. Xuan, and S. Qian. 2009. Transient electrophoretic motion of a charged particle through a converging-diverging microchannel: Effect of direct current-dielectrophoretic force. *Electrophoresis* 30:2499–2506. With permission from Wiley-VCH.)

Figure 3.44 shows the distribution of the DEP force in the throat region evaluated by a point-dipole approximation (Pohl 1978) without considering the distortion of the electric field due to the presence of the particle. Because the particle experiences negative dielectrophoresis, the DEP force acting on the particle is pointed from a higher electric field to a lower electric field. Because the maximum electric field is in the throat region, the DEP force acting on the particle is always pointed away from it, as shown in Figure 3.44. Trajectory (a) in Figure 3.44 represents the numerical prediction without considering the DEP effect, which is identical to the streamlines of the flow field and the electric field originated from the initial particle location. The x-component DEP force is negative in the converging section and

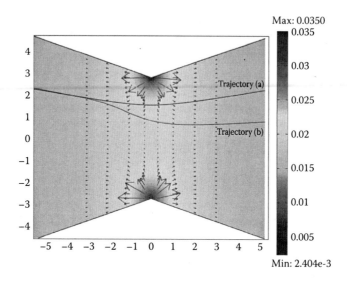

FIGURE 3.44 Distribution of the DEP force (arrows) around the throat of the converging-diverging microchannel. The gray levels represent the normalized electric field strength. The trajectories (a) and (b) represent, respectively, the predicted particle trajectories without and with considering the DEP force. (From Ai, Y., S. W. Joo, Y. Jiang, X. Xuan, and S. Qian. 2009. Transient electrophoretic motion of a charged particle through a converging-diverging microchannel: Effect of direct current-dielectrophoretic force. *Electrophoresis* 30:2499–2506. With permission from Wiley-VCH.)

becomes positive in the diverging section. However, the y component DEP force is always pointed toward the centerline of the microchannel, which in turn pushes the particle toward the centerline of the microchannel. When the particle is initially located at the centerline of the microchannel, the y component DEP force is zero due to the symmetric electric field along the y axis. As a result, the particle would not experience lateral motion when it moves along the centerline of the microchannel.

3.4.2.2 Effect of Electric Field

The DEP effect is proportional to the square of the electric field; electrophoresis and electroosmosis are both linearly proportional to the electric field. We thus investigate the effect of the applied electric field on particle transport along the centerline of the converging-diverging microchannel when $\zeta_p = -32$ mV and $\gamma = 2.5$. Figure 3.45 shows the ratio of the particle's translational velocity to that in the uniform upstream section, $\lambda_p = U_p / U_{up}$, under three different electric fields. The translational velocity ratio without considering the DEP effect is independent of the applied electric field and is symmetric with respect to the throat; thus it is included as a reference in Figure 3.45. When the DEP effect is taken into account, the velocity ratio becomes asymmetric and strongly dependent on the applied electric field. This is attributed to the dependence of the x component DEP force on the applied electric field. Figure 3.46 shows the x component electrophoretic

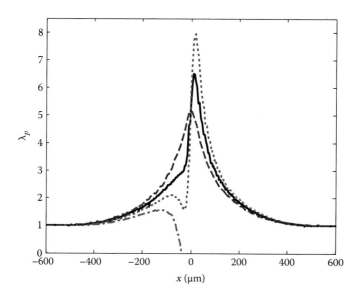

FIGURE 3.45   (a) Translational velocity ratio of a 20-μm particle along the centerline of the converging-diverging microchannel. $\zeta_p = -32$ mV, $a = 325$ μm, $b = 55$ μm, and $\gamma = 2.5$. The solid, dotted, and dash-dotted lines represent, respectively, the velocity ratio under an electric field of $E = 10$ kV/m, $E = 20$ kV/m, and $E = 35$ kV/m considering the DEP force. The symmetric dashed line represents the velocity ratio without considering the DEP force. (From Ai, Y., S. W. Joo, Y. Jiang, X. Xuan, and S. Qian. 2009. Transient electrophoretic motion of a charged particle through a converging-diverging microchannel: Effect of direct current-dielectrophoretic force. *Electrophoresis* 30:2499–2506. With permission from Wiley-VCH.)

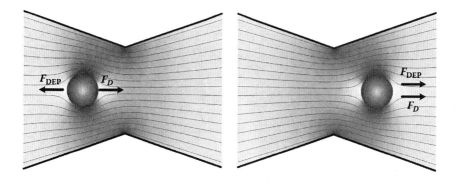

FIGURE 3.46   The x component electrophoretic force and DEP force acting on the particle in the converging (a) and diverging (b) sections.

force F_D and DEP force F_{DEP} acting on the particle when it is in the converging section and the diverging section, respectively. The x component electrophoretic force also drives the particle moving from left to right. However, the x component DEP force retards the particle motion in the converging section and accelerates the particle motion in the diverging section. Therefore, the particle velocity becomes asymmetric with respect to the throat. When the DEP effect is negligible, the particle velocity is nearly symmetric. When the DEP force is larger than the electrophoretic force under a high electric field, the DEP force in the converging section could prevent the particle from passing through the throat, as shown in Figure 3.45a. This particle-trapping phenomenon has also been observed experimentally by Kang, Xuan et al. (2006).

Figure 3.47 investigates the effect of the applied electric field on the lateral particle motion when it is initially located away from the centerline of the microchannel. The other conditions are $d = 20$ μm, $h^* = 0.5$, $\zeta_p = -58$ mV, and $\gamma = 0.3$. A higher electric field leads to a larger DEP effect, which in turn results in more significant lateral motion toward the centerline of the microchannel. Obviously, the converging-diverging microchannel can be used for particle focusing, which has been experimentally observed (Xuan, Raghibizadeh, and Li 2005) and successfully implemented in a straight channel with a pair of oil menisci (Thwar, Linderman, and Burns 2007).

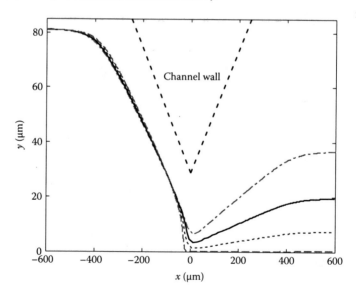

FIGURE 3.47 Particle trajectories of a 20-μm particle initially located at $h^* = 0.5$ under electric fields of $E = 10$ kV/m (dash-dotted line), $E = 15$ kV/m (solid line), $E = 20$ kV/m (dotted line), and $E = 25$ kV/m (dashed line). $\zeta_p = -58$ mV, $a = 325$ μm, $b = 55$ μm, and $\gamma = 0.3$. (From Ai, Y., S. W. Joo, Y. Jiang, X. Xuan, and S. Qian. 2009. Transient electrophoretic motion of a charged particle through a converging-diverging microchannel: Effect of direct current-dielectrophoretic force. *Electrophoresis* 30:2499–2506. With permission from Wiley-VCH.)

3.4.2.3   Effect of Particle Size

Because the DEP effect also depends on the particle size, Figure 3.48 studies the effect of particle size on its transport along the centerline of the channel when $\zeta_p = -58$ mV and $\gamma = 0.3$. The velocity ratio profile without considering DEP is included as a reference. The velocity ratio profile for a 10-μm particle is close to the reference profile, which implies a small DEP effect. As the particle size increases, the DEP effect becomes more pronounced. The velocity ratio profile for a 25-μm particle is clearly asymmetric with respect to the throat. If the particle size increases further, the DEP force in the converging section can prevent the particle from passing through the throat. Therefore, the converging-diverging microchannels could be used for particle trapping and sorting.

3.4.3   IN A Y-SHAPED MICROCHANNEL WITH A HURDLE

We consider two circular particles initially located upstream of a microchannel electrokinetically moving toward a Y junction with a hurdle in front, as shown in Figure 3.49a. When the particle is in the hurdle region, a significantly nonuniform electric field is induced around the particle, as shown in Figure 3.49b. Accordingly, a DEP force is exerted on the particle, which is directed away from

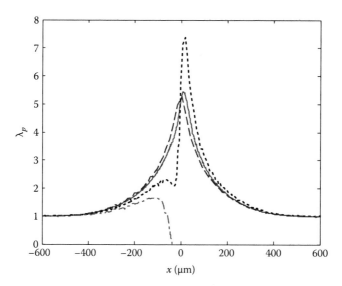

FIGURE 3.48   Translational velocity ratio of particles with diameter $d = 10$ μm (solid line), $d = 25$ μm (dotted line), and $d = 40$ μm (dash-dotted line) along the centerline of the converging-diverging microchannel under a 10 kV/m electric field. The symmetric dashed line represents the predicted velocity ratio without considering the DEP force. $\zeta_p = -58$ mV, $a = 325$ μm, $b = 55$ μm, and $\gamma = 0.3$. (From Ai, Y., S. W. Joo, Y. Jiang, X. Xuan, and S. Qian. 2009. Transient electrophoretic motion of a charged particle through a converging-diverging microchannel: Effect of direct current-dielectrophoretic force. *Electrophoresis* 30:2499–2506. With permission from Wiley-VCH.)

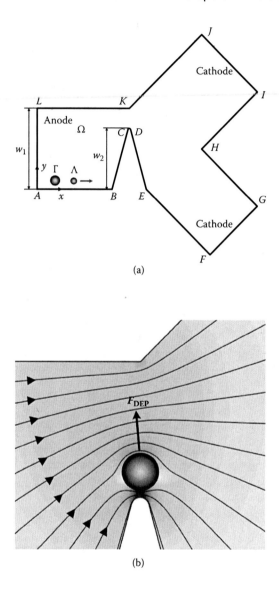

(a)

(b)

FIGURE 3.49 (a) A two-dimensional schematic view of two particles moving in a constricted microchannel. AL represents the inlet, while FG and IJ represent the outlets; (b) distribution and streamlines of the electric field around the constricted section in the presence of a particle. The gray levels indicate the electric field intensity, with the dark color representing high electric field. The arrow denotes the direction of the induced DEP force exerted on the particle.

the hurdle and in turn induces lateral particle motion. The geometry used in the numerical simulation is similar to an experimental microfluidic device (Kang et al. 2008) that has experimentally achieved the size-based particle separation.

Here, we demonstrate the numerical simulation of particle separation using the developed model. The numerical implementation is similar to that described in Section 3.3. Since two particles are considered in the present study, two sets of ODEs are required to solve the translation and rotation of the two particles. The zeta potential of the microchannel is assumed to be $\zeta_w = -80$ mV. The electric potentials at the two outlets are fixed to zero. The channel width w_1 is 300 μm.

3.4.3.1   Effect of Particle Size

Figure 3.50 shows the trajectories of two particles with diameters of 10 and 20 μm in the Y-shaped microchannel, indicating obvious particle separation. The electric potential on the anode (inlet AL) is 7 V, and the zeta potentials of the two particles are both −32 mV. The constriction ratio, defined as the ratio of the hurdle width to the channel width w_2/w_1, is 0.753. If the DEP effect is negligible, the particles should exactly follow the streamline of the electric field. Therefore, all the particles with the initial location as shown in Figure 3.50 should go to outlet FG. This is true for 10-μm particles because the DEP effect is not sufficient to induce a significant lateral motion. However, the 20-μm particle experiences a larger DEP force, which is strong enough to push it toward the upper outlet IJ. This kind of size-based particle separation has been successfully implemented in a real microfluidic device, and our numerical predictions are in qualitative agreement with the experimental observations (Kang et al. 2008).

3.4.3.2   Effect of Particle's Zeta Potential

Figure 3.51 shows the trajectories of two 20-μm particles with different zeta potentials in the Y-shaped microchannel. The electric potential and the constriction ratio are both the same as those in Section 3.4.3.1. However, the zeta potentials of the two particles are, respectively, 32 and −32 mV. In the present study, EOF dominates over particle electrophoresis. Therefore, EOF always transports the particles toward the Y junction. Therefore, the velocity of the particle with $\zeta_p = 32$ mV (particle A) is almost 2.3 times that with $\zeta_p = -32$ mV (particle B). As particle A moves faster than particle B, it experiences the DEP effect for less time than particle B. Although the two particles experience an identical DEP force, the lateral motion of particle A is smaller than that of particle B, which also leads to particle separation.

3.4.3.3   Effect of Electric Field

As mentioned, the applied electric field is of great importance in the DEP effect. Figure 3.52 shows the trajectories of two particles with diameters of 10 and 20 μm in the Y-shaped microchannel. The conditions are the same as those in Section 3.4.3.1 except the electric potential on the anode is reduced to 3 V. Because the electric field is reduced, the two particles experience smaller lateral motion than that when 7 V is applied on the anode. Although the 20-μm particle still experiences larger lateral

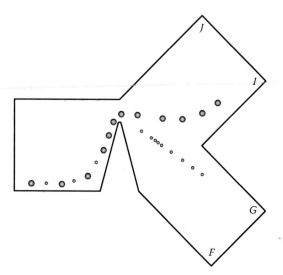

FIGURE 3.50 Trajectories of 10- and 20-μm particles through the constricted micro-channel. The electric potential applied on the inlet is 7 V. The zeta potentials of the two particles are both −32 mV. The constriction ratio w_2/w_1 is 0.753.

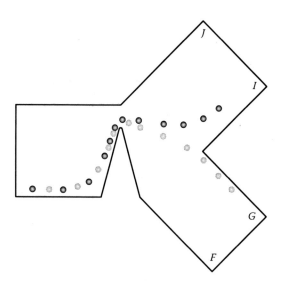

FIGURE 3.51 Trajectories of two 20-μm particles bearing different zeta potentials through the constricted microchannel. The electric potential applied on the inlet is 7 V. The constriction ratio w_2/w_1 is 0.753. The zeta potentials of the dark and the light particles are, respectively, 32 and −32 mV.

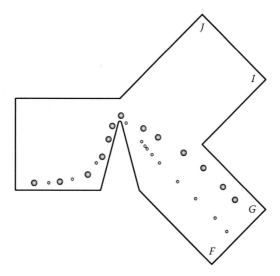

FIGURE 3.52 Trajectories of 10- and 20-µm particles through the constricted micro-channel. The electric potential applied on the inlet is 3 V. The zeta potentials of the two particles are both −32 mV. The constriction ratio w_2/w_1 is 0.753.

motion than the 10-µm particle, particle separation cannot be achieved under this electric field. Thus, one can adjust the applied electric potential on the anode to control the separation threshold of particle size. In addition, a higher electric potential leads to a smaller separation threshold of the particle size.

3.4.3.4 Effect of Channel Geometry

The constriction ratio w_2/w_1 can significantly affect the nonuniformity of the electric field, which is another important factor on particle separation. Figure 3.53 shows the trajectories of two particles with diameters of 10 and 20 µm in the Y-shaped microchannel. The conditions are the same as those in Section 3.4.3.1 except the constriction ratio is reduced to 0.513. As a result, the nonuniformity of the electric field is reduced, which leads to a smaller DEP effect. Therefore, the 20-µm particle cannot be separated from the 10-µm particle, as shown in Figure 3.53. To limit the joule heating problem in DC electrokinetics induced by high electric fields (Xuan 2008), reconfigurable oil droplets have been utilized to construct different constriction ratios to control particle separation (Barbulovic-Nad et al. 2006) and focusing (Thwar, Linderman, and Burns 2007) under a relatively low electric field.

3.4.4 IN AN L-SHAPED MICROCHANNEL

In this section, we numerically and experimentally investigate the electrokinetic particle transport in an L-shaped microchannel, which is commonly used to switch the transport direction of fluids and particles in microfluidic devices (Rhee and Burns 2008). Particle focusing in a serpentine microchannel composed of many L-shaped corners has been experimentally observed, which implies a significant

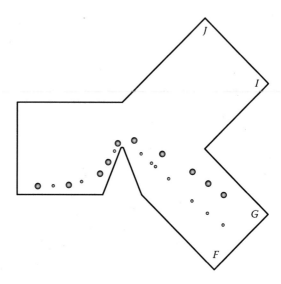

FIGURE 3.53 Trajectories of 10- and 20-μm particles through the constricted microchannel. The electric potential applied on the inlet is 7 V. The zeta potentials of the two particles are both −32 mV. The constriction ratio w_2/w_1 is 0.513.

DEP effect (Zhu et al. 2009). The purpose of this study is to investigate the DEP effect arising from the L-shaped corner on the electrokinetic particle transport.

3.4.4.1 Experimental Setup

Polystyrene particles with diameters of 4 and 10 μm (Molecular Probes, Inc., Eugene, OR) were used in the experimental study. The original particle suspension is extremely concentrated; appropriate dilution using 1 mM potassium chloride (KCl) solution is required to achieve the tracking of the transport of a single particle. A standard soft lithographic technique was employed to fabricate the L-shaped channel shown in Figure 3.54a using polydimethylsiloxane (PDMS) (Duffy et al. 1998). Briefly, SU-8 photoresist (Formulation 25, MicroChem Corp., Newton, MA) was first spin coated on a clean glass slide, followed by a two-step soft bake (65°C for 3 min and 95°C for 7 min), as shown in Figure 3.55a. Next, the photoresist film was exposed to ultraviolet light under a 3,500-dpi mask with a desired L-shaped geometry, followed by another two-step hard bake (65°C for 1 min and 95°C for 3 min). After the hard bake, a positive master was obtained by developing the photoresist for 4 min with commercial SU-8 developer solution, as shown in Figure 3.55b. Subsequently, two components of PDMS (prepolymer and curing agent; Sylgard184 Silicone Elastomer Kit, Dow Corning Corp., Freeland, MI) were mixed at a 10:1 ratio by weight with sufficient agitation. As a result of agitation, many air bubbles exist in the mixture. Therefore, the mixture was degassed in a vacuum chamber until no gas bubbles remained and then poured over the master and polymerized in a vacuum at 65°C for 4 h, as shown in Figure 3.55c. The cured PDMS with an L-shaped microchannel was then peeled from the master,

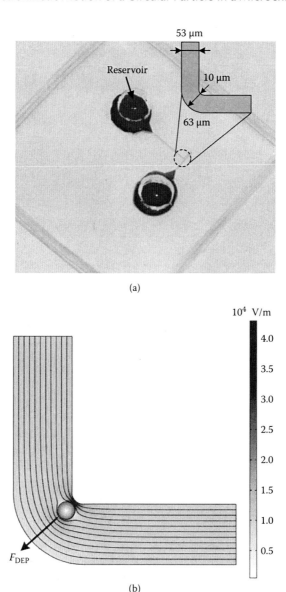

(a)

(b)

FIGURE 3.54 (a) Photograph of an L-shaped PDMS-based microchannel. The channel was filled with green food dye for clear demonstration. The inset is a schematic view of the channel with actual dimensions. The width of the channel is 53 µm, and the radii of the arc connections at the inner and outer corners are, respectively, 10 and 63 µm. (b) Distribution and streamlines of electric field (10 KV/m average) within the L-shaped channel in the presence of a particle. The arrow denotes the direction of the DC DEP force exerted on the particle. (From Ai, Y., S. Park, J. Zhu, X. Xuan, A. Beskok, and S. Qian. 2010. DC electrokinetic particle transport in an L-shaped microchannel. *Langmuir* 26 (4):2937–2944. With permission from ACS.)

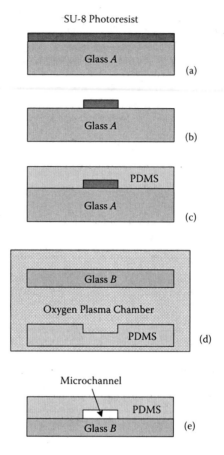

FIGURE 3.55 Microfabrication procedure of the PDMS/glass L-shaped microchannel.

and two holes were punched to serve as reservoirs. Later, the PDMS layer and another clean glass slide were treated by oxygen RF (radio frequency) plasma to develop hydroxyl groups (–OH) on the surfaces (Harrick Plasma Inc., Ithaca, NY), as shown in Figure 3.55d. Once the treated PDMS and glass were brought into contact, they formed permanent Si–O–Si bonding with loss of a water molecule, as shown in Figure 3.55e. Immediately after the bonding step, the diluted particle solution was added into one of the reservoirs and automatically driven into the microchannel by capillary force. As illustrated in Figure 3.54a, the microchannel was measured as 53 (±1) μm in width and 25 (±1) μm in depth. The length of the entire channel between the two reservoirs was 15 mm.

The fabricated PDMS/glass microchannel was loaded in an inverted optical microscope (Nikon Eclipse TE2000U, Nikon Instruments, Lewisville, TX), where the DC electrokinetic particle transport in the L-shaped microchannel was observed by a preequipped CCD (charge-coupled device) camera (Powerview™, TSI, Inc., Shoreview, MN). To get rid of the pressure-driven flow effect on the particle transport, the fluid heights in the two reservoirs were carefully balanced until particles inside the channel became stationary when the external electric field was off. Two 1-mm diameter electrodes were placed in the two reservoirs to generate an electric field along the microchannel using a DC power supply (Circuit Specialists Inc., Mesa, AZ). The particle location during the transport process was captured at a rate of 7.25 Hz with an exposure time of 100 μs and later extracted using the image-processing software ImageJ (National Institutes of Health, http://rsbweb.nih.gov/ij/). The reading error of the particle's center was about 2 pixels, corresponding to about 0.645 μm. Particle velocity was estimated by dividing the travel distance of particles over the time step in a series of successive images. The relative error of the particle velocity was less than about 4.8%. Finally, the electrokinetic mobility of particles can be calculated by dividing the particle velocity over the corresponding electric field applied.

3.4.4.2 Experimental Results

Figure 3.56 illustrates the superposed trajectory of a 10-μm particle moving through the L-shaped channel under electric fields of 6 kV/m (a) and 12 kV/m (b). The time interval between adjacent images is 0.14 s. The 10-μm particle experiences obvious lateral motion, especially under a higher electric field, when it passes the corner. Figure 3.54b shows the electric field around the particle when it is located in the corner. As a result, the induced negative DEP effect pushes the particle away from the inner corner, which is experimentally observed in Figure 3.56. As the DEP significantly decreases as the particle size shrinks, 4-μm particles almost followed the flow streamlines without an obvious lateral motion. In the experiments, all the 20-μm particles were moving within the 2/3 middle

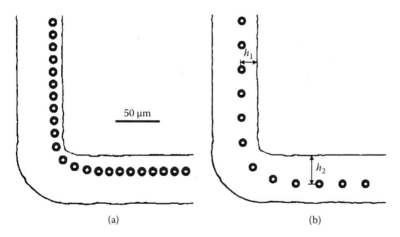

(a)   (b)

FIGURE 3.56 Trajectories of a 10-μm particle moving through the L-shaped channel under an electric field of 6 kV/m (a) and 12 kV/m (b). Time interval between adjacent particles is 0.14 s. (From Ai, Y., S. Park, J. Zhu, X. Xuan, A. Beskok, and S. Qian. 2010. DC electrokinetic particle transport in an L-shaped microchannel. *Langmuir* 26 (4):2937–2944. With permission from ACS.)

region of the microchannel upstream; the 4-μm particles could move closer to the channel wall. This is attributed to the DEP repulsive force as discussed in Section 3.4.1.

The distances from the center of the particle to the inner channel wall upstream and downstream are defined, respectively, as h_1 and h_2. Figure 3.57 depicts h_1 versus h_2, both of which are normalized by the channel width b, to show the degree of lateral motion under the two different electric fields (6 and 12 kV/m). Obviously, the 4-μm particles fall near the $h_1 = h_2$ line, implying no significant lateral motion due to a negligible DEP effect. The lateral motion of 10-μm particles is obvious and highly depends on the magnitude of the applied electric field. The lateral motion induced by the L-shaped corner can also be used to focus and separate particles in microfluidics.

3.4.4.3 Comparison between Experimental and Numerical Results

Next, we use the developed numerical model to reproduce the experimental study. The computational domain is shown in Figure 3.58; the circular particle is initially located upstream with a distance h_1 away from the inner channel wall. The average electrokinetic mobilities of 4 μm and 10 μm particles are, respectively, 4.0×10^{-8} m^2/(V•s) and 1.6×10^{-8} m^2/(V•s) by measuring the average velocities

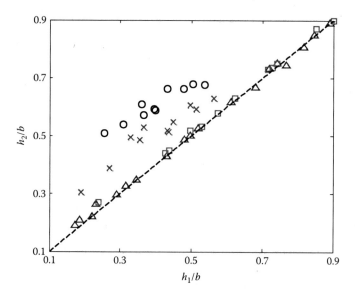

FIGURE 3.57 Trajectory shift for particles of different sizes under different electric field intensities. Circles and crosses represent, respectively, the trajectory shifts of 10-μm particles under electric fields of 12 and 6 kV/m. Squares and triangles represent, respectively, the trajectory shifts of 4-μm particles under 12 and 6 kV/m. The dashed line is a reference line corresponding to $h_1 = h_2$. (From Ai, Y., S. Park, J. Zhu, X. Xuan, A. Beskok, and S. Qian. 2010. DC electrokinetic particle transport in an L-shaped microchannel. *Langmuir* 26 (4):2937–2944. With permission from ACS.)

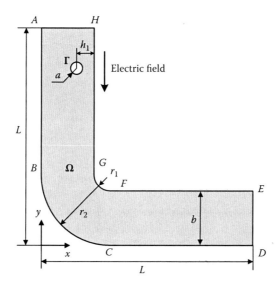

FIGURE 3.58 A two-dimensional schematic view of a circular particle of radius a migrating in an L-shaped microchannel. An external electric field **E** is applied between the inlet and outlet of the channel. (From Ai, Y., S. Park, J. Zhu, X. Xuan, A. Beskok, and S. Qian. 2010. DC electrokinetic particle transport in an L-shaped microchannel. *Langmuir* 26 (4):2937–2944. With permission from ACS.)

of particles in the straight section where the DEP effect is almost negligible. The particle electrokinetic mobility η, taking into account the boundary effect, is given as (Keh and Anderson 1985)

$$\eta = \left[1 - 0.267699 \left(\frac{a}{d} \right)^3 + 0.338324 \left(\frac{a}{d} \right)^5 - 0.04022 \left(\frac{a}{d} \right)^6 \right] \times \frac{\varepsilon_0 \varepsilon_f}{\mu} (\zeta_p - \zeta_w), \quad (3.25)$$

where a is the particle radius, and d is the perpendicular distance between the center of the particle and the channel wall. The nominal density of the polystyrene particles (1.05 g/ml) is slightly larger than the fluid density. Thus, the diluted particle solutions are sonicated prior to each experiment to eliminate particle sedimentation. Due to the DEP repulsive force arising from the particle and channel wall interaction, the particles are mostly moving in the middle region of the channel depth. The fluid viscosity and relative permittivity are, respectively, $\mu = 1.0 \times 10^{-3}$ kg/(m•s) and $\varepsilon_f = 80$. Based on the reported zeta potential of PDMS, $\zeta_w = -80$ mV (Kang, Xuan, et al. 2006; Venditti, Xuan, and Li 2006) and the measured particle mobility, the zeta potentials of the 4 μm and 10 μm particles were estimated as −56.8 and −22.0 mV, respectively, for use in the following numerical simulations.

Figure 3.59 shows the experimental particle trajectories and the corresponding numerical predictions obtained by the developed numerical model described

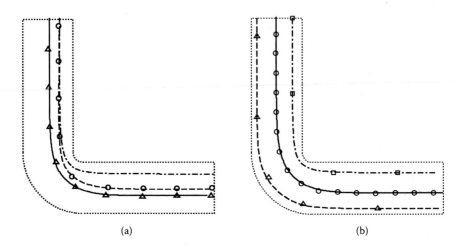

(a)   (b)

FIGURE 3.59 Comparisons between experimental (symbols) and predicted (lines) particle trajectories. (a) 10-μm particles located at $h_1/b = 0.27$ (circles, dashed line, and dash-dotted line) and $h_1/b = 0.47$ (triangles and solid line) under a 12-kV/m electric field. The dash-dotted line denotes the numerical prediction without DEP. (b) The 10-μm particle located at $h_1/b = 0.51$ (circles and solid line) under a 6-kV/m electric field, and 4-μm particles located at $h_1/b \approx 0.2$ (squares and dash-dotted line) and $h_1/b = 0.89$ (triangles and dashed line) under a 12-kV/m electric field. The DEP effect is considered in all the numerical predictions. (From Ai, Y., S. Park, J. Zhu, X. Xuan, A. Beskok, and S. Qian. 2010. DC electrokinetic particle transport in an L-shaped microchannel. *Langmuir* 26 (4):2937–2944. With permission from ACS.)

in Section 3.2. For the case of $h_1/b = 0.27$ under a 12 kV/m electric field, the numerical predictions agree with the experimental data when the DEP effect is taken into account. Once the DEP effect is ignored, the particle almost follows the streamline of the electric field originating from its initial location. Before the particle passes the corner, the numerical prediction without the DEP effect also agrees with the experimental results, which indicates a negligible DEP effect in the straight section. However, the numerical prediction without the DEP effect significantly deviates from the experimental result when the particle passes through the corner. It is concluded that the DEP effect must be considered when a significantly nonuniform electric field is generated around the particle. A good agreement is also achieved for $h_1/b = 0.47$ and $E = 12$ kV/m (Figure 3.59a).

Similarly, the numerical prediction of the 10-μm particle initially located at $h_1/b = 0.51$ under 6 kV/m shown in Figure 3.59b is also in good agreement with the corresponding experimental results. For 4-μm particles, good agreements between numerical predictions and experimental data are also obtained for $h_1/b \approx 0.2$ and $h_1/b = 0.89$ under a 12-kV/m electric field.

3.4.4.4 Particle Rotation

The DEP effect affects the particle's rotational motion in addition to its trajectory, as shown in Figure 3.60. Note that the angle is positive when the particle rotates

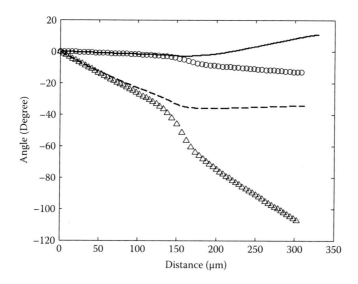

FIGURE 3.60   Rotation angles of two 10-µm particles initially located at $h_1/b = 0.27$ (dashed line and triangles) and $h_1/b = 0.47$ (solid line and circles) through the L-shaped channel under a 12-kV/m electric field. Symbols and lines represent, respectively, numerical predictions without and with DEP. (From Ai, Y., S. Park, J. Zhu, X. Xuan, A. Beskok, and S. Qian. 2010. DC electrokinetic particle transport in an L-shaped microchannel. *Langmuir* 26 (4):2937–2944. With permission from ACS.)

counterclockwise. Because the DEP effect is negligible in the straight section, the numerical predictions of the particle rotation with and without DEP are nearly identical. After the particle passes through the corner, the particle's angle considering the DEP effect significantly differs from that without considering it.

Figure 3.61 illustrates the transient rotation of two 10-µm particles initially located at $h_1/b = 0.12$ (zone A) and $h_1/b = 0.88$ (zone B) along their trajectories under a 12-kV/m electric field. Due to the presence of the dielectric particle, the electric field between the particle and the channel wall is enhanced. As a result, the Smoluckowski slip velocity on the particle surface next to the channel wall is higher than that on the other side, which induces a net torque on the particle. Thus, the rotational direction of the particle in zone A is clockwise, while the rotational direction of the particle in zone B is counterclockwise. In addition, the particle rotates faster when it is closer to the channel wall. When the particle is located at the centerline of the straight section, the particle cannot rotate at all. The particle initially located at $h_1/b = 0.12$ in zone A is pushed toward the centerline of the downstream microchannel, which accordingly reduces the rotational velocity. The particle initially located at $h_1/b = 0.88$ in zone B displays a slighter lateral motion after passing the corner, indicating less of a DEP effect at the outer corner. As a result, its rotational velocity downstream is higher than that for $h_1/b = 0.12$. It is suggested that the DEP effect could affect not only the particle's trajectory but also the particle's rotational dynamics.

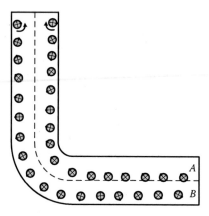

FIGURE 3.61 Rotation of two 10-μm particles initially located at $h_1/b = 0.12$ and 0.88 through the L-shaped channel under a 12-kV/m electric field. The crosses inside the particle and the dot on the particle surface are used for a clear demonstration of the particle's rotation. (From Ai, Y., S. Park, J. Zhu, X. Xuan, A. Beskok, and S. Qian. 2010. DC electrokinetic particle transport in an L-shaped microchannel. *Langmuir* 26 (4):2937–2944. With permission from ACS.)

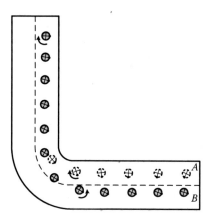

FIGURE 3.62 Rotation of a 10-μm particle initially located at $h_1/b = 0.26$ through the L-shaped channel under a 20-kV electric field. The solid and hollow particles represent, respectively, the numerical predictions with and without DEP. (From Ai, Y., S. Park, J. Zhu, X. Xuan, A. Beskok, and S. Qian. 2010. DC electrokinetic particle transport in an L-shaped microchannel. *Langmuir* 26 (4):2937–2944. With permission from ACS.)

Figure 3.62 shows the transient rotation of a 10-μm particle initially located at $h_1/b = 0.26$ in zone A through the L-shaped channel under a 20-kV/m electric field. When the DEP effect is considered in the numerical modeling, the particle is shifted from zone A to zone B after passing the corner. Therefore, the rotational direction is reversed once it crosses the centerline of the channel. When the DEP effect is ignored in the numerical modeling, the trajectory and rotation of the particle upstream are similar to those with DEP; thus they are not shown here. However, the particle remains in zone A after passing the corner, and the rotational direction is unchanged. When the electric field is strong enough, all the incoming particles from upstream could be shifted to zone B downstream, coming out with a consistent rotational direction.

3.4.4.5 Effect of Particle Size

Here, we investigate the effect of particle size on the particle trajectory through the L-shaped microchannel. Figure 3.63 shows the trajectories of three particles with different sizes (4 μm, 10 μm, and 15 μm in diameter) initially located at $h_1/b = 0.26$ in the upstream. The zeta potential of all three particles is −56.8 mV, corresponding to an electrokinetic mobility of 1.6×10^{-8} m²/(V•s). The 10-μm particle experiences more obvious lateral motion than the 4-μm particle, which has also been experimentally observed. Therefore, the L-shaped microchannel is also capable of the size-based particle separation. The 15-μm particle can also be separated from the 10-μm particle at the corner region. However, both particles become closer again in the downstream because of the repulsive DEP force that originates from the channel wall in the downstream. Therefore, one has to adjust the electric field strength and geometry to find the optimum configuration for a specific particle separation.

3.4.4.6 Effect of Electric Field

As mentioned, the lateral particle motion can be well controlled by the applied electric field. Figure 3.64 shows the focusing of two 4-μm particles (a) and two 10-μm particles (b) initially located at $h_1/b = 0.12$ and 0.88 in the upstream through

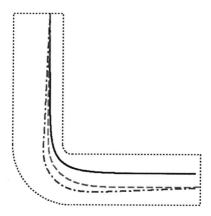

FIGURE 3.63 Trajectories of particles of 4 (solid line), 10 (dashed line) and 15 (dash-dotted line) μm in diameter through the L-shaped microchannel under a 20-kV/m electric field. The zeta potential of the particle is −56.8 mV, and the particle is initially located at $h_1/b = 0.26$ in the upstream. (From Ai, Y., S. Park, J. Zhu, X. Xuan, A. Beskok, and S. Qian. 2010. DC electrokinetic particle transport in an L-shaped microchannel. *Langmuir* 26 (4):2937–2944. With permission from ACS.)

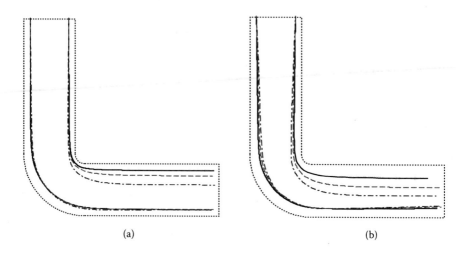

(a)   (b)

FIGURE 3.64   (a) Trajectories of two 4-µm particles initially located at $h_1/b = 0.12$ and 0.88 in the upstream under electric fields of 12 (solid line), 40 (dashed line), and 100 (dash-dotted line) kV/m; (b) trajectories of two 10-µm particles initially located at $h_1/b = 0.12$ and 0.88 in the upstream under electric fields of 6 (solid line), 12 (dashed line), and 20 (dash-dotted line) kV/m. (From Ai, Y., S. Park, J. Zhu, X. Xuan, A. Beskok, and S. Qian. 2010. DC electrokinetic particle transport in an L-shaped microchannel. *Langmuir* 26 (4):2937–2944. With permission from ACS.)

the L-shaped channel under different electric fields. The electrokinetic mobility of the 4-µm particle measured in the experiment was used in the numerical modeling, corresponding to a zeta potential of −22.0 mV. The 4-µm particles experienced negligible focusing under a 12-kV/m electric field and slight focusing under electric fields of 40 and 100 kV/m. The particle focusing can be characterized by the ratio of the particle flow width at the inlet to that at the outlet, w_1/w_2, which are, respectively, 1.17 and 1.59 under 40 and 100 kV/m. In contrast, a distinct focusing effect of the 10-µm particle with the experimentally derived electrokinetic mobility is observed in Figure 3.64b. The particle focusing ratios w_1/w_2 under 6, 12, and 20 KV/m are, respectively, 1.34, 2.06, and 3.89. Thus, the L-shaped microchannel can also be used for particle focusing.

3.5   CONCLUDING REMARKS

We numerically studied electrokinetic particle transport in a straight microchannel, a converging-diverging microchannel, a Y-shaped microchannel, and an L-shaped microchannel, with emphasis on the DEP effect using the predefined ALE method in COMSOL Multiphysics 3.5a. The DEP effect arising from the interaction between the nonuniform electric field and the dielectric particle must be taken into account in the numerical study of electrokinetic particle transport in microfluidic devices. The numerical simulations were in good agreement with the approximation solutions, other researchers' experimental results, and our own experimental results. The DEP effect strongly depends on the particle size, the applied electric field, and the geometry of the microchannel.

When the DEP effect is negligible, the velocity profile through a converging-diverging microchannel is symmetric with respect to the throat. Once the DEP effect becomes dominant, the velocity profile becomes asymmetric. If the DEP effect is strong enough, it could prevent the particle from passing through the converging-diverging microchannel, indicating a potential application of particle trapping. If the particle is not initially located at the centerline of the microchannel, the DEP effect will push the particle toward the middle region of the microchannel, where the DEP effect is minimized. As the DEP effect is proportional to the third power of particle size, it has been experimentally demonstrated for size-based particle separation, which has also been predicted by our developed numerical model. The DEP effect is also important in an L-shaped channel when the particle size and the applied electric field are large enough. Dielectrophoresis shows great potential for continuous trapping, focusing, and separation of particles in microfluidics.

REFERENCES

Ai, Y., S. W. Joo, Y. Jiang, X. Xuan, and S. Qian. 2009. Transient electrophoretic motion of a charged particle through a converging-diverging microchannel: Effect of direct current-dielectrophoretic force. *Electrophoresis* 30:2499–2506.

Ai, Y., S. Park, J. Zhu, X. Xuan, A. Beskok, and S. Qian. 2010. DC electrokinetic particle transport in an L-shaped microchannel. *Langmuir* 26 (4):2937–2944.

Barbulovic-Nad, I., X. Xuan, J. S. H. Lee, and D. Li. 2006. DC-dielectrophoretic separation of microparticles using an oil droplet obstacle. *Lab on a Chip* 6 (2):274–279.

Cummings, E. B., and A. K. Singh. 2003. Dielectrophoresis in microchips containing arrays of insulating posts: Theoretical and experimental results. *Analytical Chemistry* 75 (18):4724–4731.

Davison, S. M., and K. V. Sharp. 2008. Transient simulations of the electrophoretic motion of a cylindrical particle through a 90 degrees corner. *Microfluidics and Nanofluidics* 4 (5):409–418.

Dietrich, K., C. T. E. Jan, B. Albert van den, and B. M. S. Richard. 2008. Miniaturizing free-flow electrophoresis—A critical review. *Electrophoresis* 29 (5):977–993.

Duffy, D. C., J. C. McDonald, O. J. A. Schueller, and G. M. Whitesides. 1998. Rapid prototyping of microfluidic systems in poly(dimethylsiloxane). *Analytical Chemistry* 70 (23):4974–4984.

Gloria, O., S. Christina, B. K. Matthew, T. Anubhav, and C. Anuj. 2008. Electrophoretic migration of proteins in semidilute polymer solutions. *Electrophoresis* 29 (5):1152–1163.

Hawkins, B. G., A. E. Smith, Y. A. Syed, and B. J. Kirby. 2007. Continuous-flow particle separation by 3D insulative dielectrophoresis using coherently shaped, dc-biased, ac electric fields. *Analytical Chemistry* 79 (19):7291–7300.

Hu, H. H., D. D. Joseph, and M. J. Crochet. 1992. Direct simulation of fluid particle motions. *Theoretical and Computational Fluid Dynamics* 3 (5):285–306.

Hu, H. H., N. A. Patankar, and M. Y. Zhu. 2001. Direct numerical simulations of fluid-solid systems using the arbitrary Lagrangian-Eulerian technique. *Journal of Computational Physics* 169 (2):427–462.

Hughes, T. J. R., W. K. Liu, and T. K. Zimmermann. 1981. Lagrangian-Eulerian finite element formulation for incompressible viscous flows. *Computer Methods in Applied Mechanics and Engineering* 29 (3):329–349.

Kang, K. H., Y. J. Kang, X. C. Xuan, and D. Q. Li. 2006. Continuous separation of microparticles by size with direct current-dielectrophoresis. *Electrophoresis* 27 (3):694–702.

Kang, K. H., X. C. Xuan, Y. Kang, and D. Li. 2006. Effects of dc-dielectrophoretic force on particle trajectories in microchannels. *Journal of Applied Physics* 99 (6):064702.

Kang, Y., B. Cetin, Z. Wu, and D. Li. 2009. Continuous particle separation with localized AC-dielectrophoresis using embedded electrodes and an insulating hurdle. *Electrochimica Acta* 54 (6):1715–1720.

Kang, Y. J., D. Q. Li, S. A. Kalams, and J. E. Eid. 2008. DC-Dielectrophoretic separation of biological cells by size. *Biomedical Microdevices* 10 (2):243–249.

Keh, H. J., and J. L. Anderson. 1985. Boundary effects on electrophoretic motion of colloidal spheres. *Journal of Fluid Mechanics* 153:417–439.

Lapizco-Encinas, B. H., and M. Rito-Palomares. 2007. Dielectrophoresis for the manipulation of nanobioparticles. *Electrophoresis* 28 (24):4521–4538.

Lapizco-Encinas, B. H., B. A. Simmons, E. B. Cummings, and Y. Fintschenko. 2004a. Dielectrophoretic concentration and separation of live and dead bacteria in an array of insulators. *Analytical Chemistry* 76 (6):1571–1579.

Lapizco-Encinas, B. H., B. A. Simmons, E. B. Cummings, and Y. Fintschenko. 2004b. Insulator-based dielectrophoresis for the selective concentration and separation of live bacteria in water. *Electrophoresis* 25 (10–11):1695–1704.

Leopold, K., B. Dieter, and K. Ernst. 2004. Capillary electrophoresis of biological particles: Viruses, bacteria, and eukaryotic cells. *Electrophoresis* 25 (14):2282–2291.

Lewpiriyawong, N., C. Yang, and Y. C. Lam. 2008. Dielectrophoretic manipulation of particles in a modified microfluidic H filter with multi-insulating blocks. *Biomicrofluidics* 2 (3):034105.

Liang, L. T., Y. Ai, J. J. Zhu, S. Qian, and X. C. Xuan. 2010. Wall-induced lateral migration in particle electrophoresis through a rectangular microchannel. *Journal of Colloid and Interface Science* 347 (1):142–146.

Parikesit, G. O. F., A. P. Markesteijn, O. M. Piciu, A. Bossche, J. Westerweel, I. T. Young, and Y. Garini. 2008. Size-dependent trajectories of DNA macromolecules due to insulative dielectrophoresis in submicrometer-deep fluidic channels. *Biomicrofluidics* 2 (2):024103.

Pohl, H. A. 1978. *Dielectrophresis*. Cambridge, UK: Cambridge University Press.

Qian, S., and S. W. Joo. 2008. Analysis of self-electrophoretic motion of a spherical particle in a nanotube: Effect of nonuniform surface charge density. *Langmuir* 24 (9):4778–4784.

Qian, S., S. W. Joo, W. Hou, and X. Zhao. 2008. Electrophoretic motion of a spherical particle with a symmetric nonuniform surface charge distribution in a nanotube. *Langmuir* 24 (10):5332–5340.

Qian, S., A. H. Wang, and J. K. Afonien. 2006. Electrophoretic motion of a spherical particle in a converging-diverging nanotube. *Journal of Colloid and Interface Science* 303 (2):579–592.

Rhee, M., and M. A. Burns. 2008. Microfluidic assembly blocks. *Lab on a Chip* 8 (8):1365–1373.

Rodriguez, M. A., and D. W. Armstrong. 2004. Separation and analysis of colloidal/nano-particles including microorganisms by capillary electrophoresis: A fundamental review. *Journal of Chromatography B* 800 (1–2):7–25.

Thwar, P. K., J. J. Linderman, and M. A. Burns. 2007. Electrodeless direct current dielectrophoresis using reconfigurable field-shaping oil barriers. *Electrophoresis* 28 (24):4572–4581.

Unni, H. N., H. J. Keh, and C. Yang. 2007. Analysis of electrokinetic transport of a spherical particle in a microchannel. *Electrophoresis* 28 (4):658–664.

Venditti, R., X. C. Xuan, and D. Q. Li. 2006. Experimental characterization of the temperature dependence of zeta potential and its effect on electroosmotic flow velocity in microchannels. *Microfluidics and Nanofluidics* 2 (6):493–499.

Xuan, X. C. 2008. Joule heating in electrokinetic flow. *Electrophoresis* 29 (1):33–43.

Xuan, X. C., R. Raghibizadeh, and D. Li. 2005. Wall effects on electrophoretic motion of spherical polystyrene particles in a rectangular poly(dimethylsiloxane) microchannel. *Journal of Colloid and Interface Science* 296:743–748.

Xuan, X. C., B. Xu, and D. Q. Li. 2005. Accelerated particle electrophoretic motion and separation in converging-diverging microchannels. *Analytical Chemistry* 77 (14):4323–4328.

Xuan, X. C., C. Z. Ye, and D. Q. Li. 2005. Near-wall electrophoretic motion of spherical particles in cylindrical capillaries. *Journal of Colloid and Interface Science* 289 (1):286–290.

Ye, C. Z., and D. Q. Li. 2004a. Electrophoretic motion of two spherical particles in a rectangular microchannel. *Microfluidics and Nanofluidics* 1 (1):52–61.

Ye, C. Z., and D. Q. Li. 2004b. 3-D transient electrophoretic motion of a spherical particle in a T-shaped rectangular microchannel. *Journal of Colloid and Interface Science* 272 (2):480–488.

Ye, C. Z., X. C. Xuan, and D. Q. Li. 2005. Eccentric electrophoretic motion of a sphere in circular cylindrical microchannels. *Microfluidics and Nanofluidics* 1 (3):234–241.

Ying, L. M., S. S. White, A. Bruckbauer, L. Meadows, Y. E. Korchev, and D. Klenerman. 2004. Frequency and voltage dependence of the dielectrophoretic trapping of short lengths of DNA and dCTP in a nanopipette. *Biophysical Journal* 86 (2):1018–1027.

Young, E. W. K., and D. Q. Li. 2005. Dielectrophoretic force on a sphere near a planar boundary. *Langmuir* 21 (25):12037–12046.

Zhu, J., T.-R. Tzeng, G. Hu, and X. Xuan. 2009. DC dielectrophoretic focusing of particles in a serpentine microchannel. *Microfluidics and Nanofluidics* 7 (6):751–756.

Zhu, J., and X. Xuan. 2009. Dielectrophoretic focusing of particles in a microchannel constriction using DC-biased AC electric fields. *Electrophoresis* 30 (15):2668–2675.

4 Electrokinetic Transport of Cylindrical-Shaped Cells in a Straight Microchannel

Electrokinetic particle manipulation using only electric fields without moving parts has been extensively used in lab-on-a-chip devices for particle characterization, trapping, focusing, separation, sorting, and assembly (Li 2004; Keh and Anderson 1985; Ye and Li 2004; Ye, Xuan, and Li 2005; Unni, Keh, and Yang 2007). The success of these electrically controlled microfluidic devices for particle manipulation relies on a comprehensive understanding of both fluid and particle behavior in these devices. Most existing theoretical (Li 2004; Keh and Anderson 1985; Ye and Li 2004; Ye, Xuan, and Li 2005) and experimental (Xuan, Raghibizadeh, and Li 2005; Xuan, Xu, and Li 2005; Zhu and Xuan 2009; Kang et al. 2006, 2008; Zhu et al. 2009; Xuan, Ye, and Li 2005) studies have been performed exclusively on spherical particles.

However, many particles used in microfluidic applications, such as the elongated biological entities, are nonspherical. In comparison, there are relatively few systematic studies of electrokinetic transport of nonspherical particles, especially in a confined geometry. In this chapter, electrokinetic transport of cylindrical-shaped cells under direct current (DC) electric fields in a straight microfluidic channel is theoretically and experimentally investigated with emphasis on the dielectrophoretic (DEP) effect on the particle's orientation with respect to the external electric field. A two-dimensional (2D) multiphysics model, composed of the Stokes and continuity equations for the fluid flow, the Laplace equation for the electric field externally imposed, Newton's equation for particle translation, and the Euler equation for the particle's rotational motion, is employed to explore the dynamics of the electrokinetic motion of cylindrical-shaped particles in a microfluidic channel.

Since the electrical double layer (EDL) thickness is much smaller than the particle size and the characteristic length of the microchannel, the fluid motion within the EDL is described by the Helmholtz–Smoluchowski velocity at the edge of the EDL. The model takes into account the full fluid-particle-electric field interactions under the thin EDL approximation and accounts for the DEP force acting on the particle stemming from the spatially nonuniform electric field surrounding the particle.

The fully coupled system is numerically solved using the arbitrary Lagrangian–Eulerian (ALE) finite element method. The numerical predictions are in quantitative agreement with the obtained experimental results. It is found that the DEP effect should be taken into account to study the electrokinetic transport of cylindrical-shaped particles even in a straight microchannel with a uniform cross-sectional area. The results show that cylindrical-shaped particles experience oscillatory motions under relatively low imposed electric fields and are aligned with their longest axes parallel to the imposed electric field under relatively high electric fields due to the DEP effect.

4.1   INTRODUCTION

An object, such as a particle, a glass plate, and a polydimethylsiloxane (PDMS) slide, brought into contact with an aqueous medium, typically acquires a surface charge and forms an EDL on its surface. On application of an external electric field tangential to this surface, the free ions within the EDL move toward the electrode with an opposite charge. These moving ions drag the liquid along with them, yielding a fluid motion called electroosmosis relative to a fixed object. If the object is a dispersed particle, the electrostatic force on a charged particle generates a net motion called electrophoresis relative to the liquid medium. A neutral particle can still move in the presence of a spatially nonuniform electric field due to the difference in the electrical properties of the liquid and particle, which is called dielectrophoresis. Fluid electroosmosis, particle electrophoresis, and particle dielectrophoresis typically coexist in various electrokinetic particle manipulations, including particle separation, assembly, sorting, focusing, and characterization, in microfluidic devices (Kang and Li 2009). Due to the coexistence of the various electrokinetic phenomena in the electrical control of particle transport, the success of these electrically controlled microfluidic devices depends on a comprehensive understanding of both fluid and particle motions in microfluidics.

Numerical modeling of electrokinetic particle motion in microfluidic devices provides a convenient and fast method for verifying new ideas that otherwise would require enormous effort. Compared to the extensive numerical studies of electrophoresis of spherical latex particles, there are relatively few studies of electrokinetic transport of cylindrical-shaped particles. Ye et al. (2002), Hsu et al. (2008; Hsu and Kuo 2006), and Liu et al. (Liu, Bau, and Hu 2004; Liu, Qian, and Bau 2007) studied the translation of a finite cylinder concentrically and eccentrically positioned along the axis of a tube using a quasi-static method without considering the particle rotation. Davison and Sharp used a transient model to study the electrokinetic motion of a cylindrical particle in a tube (2006, 2007) and in an L-shaped microchannel (2008). The model takes into account the fluid-particle-electric field interactions; however, it did not consider the DEP effect on particle transport. Ignoring the induced DEP motion can cause significant errors in the velocity, trajectory, and orientation of the particle, which has been demonstrated in Chapter 3. A DEP-induced alignment of nanowires and carbon nanotubes (Evoy et al. 2004; Makaram et al. 2007; Chang and

Hong 2009; Raychaudhuri et al. 2009; Lao et al. 2006; Monica et al. 2008) to external electric fields was experimentally observed, indicating a significant DEP effect on the motion of cylindrical particles subjected to external electric fields. Considering the DEP effect, Winter and Welland (2009) predicted that nonspherical particles are always aligned with their longest axis parallel to the electric field using a transient model, which does not consider the distortions of the electric and flow fields by the presence of the particle. This approximation can also lead to deviations from the experimentally observed particle behaviors.

In this chapter, transient electrokinetic motion of cylindrical cells in a straight microchannel under DC electric fields is experimentally and numerically investigated. The experimental results are used to validate the mathematical model, which is composed of the Stokes and continuity equations for the flow field, the Laplace equation for the imposed electric field, and the Newton and Euler equations for the particle's translational and rotational motion, respectively. The model considers the fluid-particle-electric field interactions as well as the induced DEP force exerted on the particle. The Maxwell stress tensor (MST) method is used to calculate the DEP force. The ALE finite element method is used to solve the fully coupled system numerically. Section 4.2 briefly describes the experimental setup; Section 4.3 introduces the mathematical model. Detailed numerical implementation of the model in COMSOL® is given in Section 4.4. Some typical experimental and numerical results are discussed in Section 4.5; the final section provides concluding remarks.

4.2 EXPERIMENTAL SETUP

Desmodesmus cf. *quadricauda*, a green alga of the Chlorophyceae, was grown in culture solution under fluorescent light (cool white plus, 6,500 lux, continuous illumination) and aerated with high-efficiency particulate air (HEPA) filtered air. Algae were then fixed in 4% formaldehyde in 0.1 M phosphate buffer (pH 7.4) for 12 h at 4°C and rinsed three times in 0.1 M phosphate buffer prior to use in the experiments. Figure 4.1 depicts the scanning electron microscopic (SEM) image of three *Desmodesmus* cf. *quadricauda* cells whose shape is similar to a cylinder capped by two hemispheres. The cells are suspended in 0.1 M KCl solution. Sodium dodecyl sulfate (SDS; Fisher Scientific, Fair Lawn, NJ) was added at a volume ratio of 0.5% to the solution for reducing the particle adhesions to channel walls.

The PDMS/glass microfluidic channel with a rectangular cross section was fabricated using the standard soft lithographic technique (Duffy et al. 1998). Detailed microfabrication procedures are presented in Chapter 3. The length, width, and depth of the fabricated microchannel are, respectively, 10 mm, 50 (±1) μm and 25 (±1) μm. The diameter of the reservoirs is 10 mm. A DC electric field was generated by applying a DC voltage drop between two platinum electrodes of 1.25 mm in diameter that were placed into the two fluid reservoirs filled with the particle solution. The imposed electric field was supplied by a DC power supply (Circuit Specialists, Inc., Mesa, AZ). Pressure-driven flow was carefully eliminated before each experiment by balancing the liquid heights in the two

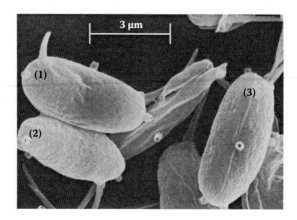

FIGURE 4.1   SEM micrograph of three *Desmodesmus* cf. *quadricauda* unicells. (From Ai, Y., A. Beskok, D. T. Gauthier, S. W. Joo, and S. Qian. 2009. DC electrokinetic transport of cylindrical cells in straight microchannels. *Biomicrofluidics* 3:044110 with permission from the American Institute of Physics.)

reservoirs. The particle's motion was visualized and recorded at a rate of 7.25 Hz via an inverted microscope imaging system (Nikon Eclipse TE2000U equipped with a Powerview™ CCD [charge-coupled device] camera, Lewisville, TX). The captured images were further processed using ImageJ (National Institutes of Health, http://rsbweb.nih.gov/ij/) to extract the location and orientation of the cells at each time step. The reading error of a given cell's location and angle were, respectively, about 0.645 μm (±2 pixels) and ±2°. The translational velocity was calculated by dividing the travel distance between adjacent cells over the time step in a series of successive images.

4.3   MATHEMATICAL MODEL

As discussed in Chapter 3, a 2D model is sufficient to capture the essential physics of the electrokinetic particle transport in microfluidics. In addition, we experimentally observed that most of the cells were well focused, and the cell length did not vary much during the entire transport process. It is suggested that the cell's translation and rotation mainly happen on the channel length-width plane (xy plane shown in Figure 4.2a). As a result, a 2D model is used in the present study to investigate the electrokinetic transport of nonspherical algal cells in a straight microchannel.

Desmodesmus cf. *quadricauda*, shown in Figure 4.1, is similar to a cylinder capped with two hemispheres. Therefore, we study the electrokinetic motion of a cylindrical particle with a uniform zeta potential ζ_p in a straight microchannel filled with an incompressible and Newtonian aqueous solution of density ρ, dynamic viscosity μ, and relative permittivity ε_f, as shown in Figure 4.2. The computational domain is bounded by the inlet AB, outlet CD, channel walls AD and BC, and the particle surface Γ. A Cartesian spatial frame (x, y) with its origin at the center of the inlet is used in the present study. The radius and the length of

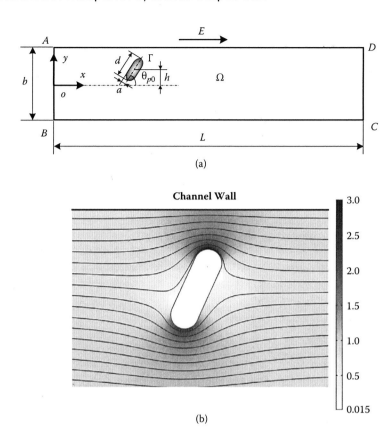

(a)

(b)

FIGURE 4.2 (a) A two-dimensional schematic view of a cylindrical particle in a straight microchannel. An external electric field was applied between the inlet AB and the outlet CD; (b) distribution and streamlines of the electric field within the microchannel in the presence of a cylindrical particle. The gray levels indicate the electric field intensity normalized by the electric field intensity in the absence of the particle, with the darker level representing a higher electric field. (From Ai, Y., A. Beskok, D. T. Gauthier, S. W. Joo, and S. Qian. 2009. DC electrokinetic transport of cylindrical cells in straight microchannels. *Biomicrofluidics* 3:044110 with permission from the American Institute of Physics.)

the particle are, respectively, a and d. Initially, it is located at (x_{p0}, y_{p0}) and presenting an angle θ_{p0} with respect to the x axis. An electric field is applied parallel to the x axis to drive the electrokinetic motion. The channel of length L and width b also bears a uniform zeta potential ζ_w and thus generates an electroosmotic flow (EOF) inside the microchannel.

As discussed in Chapter 3, the thin EDL approximation is valid for microscale electrokinetics. As a result, the net charge within the computational domain is zero, and the electric potential is thus governed by the Laplace equation

$$\nabla^2 \phi = 0 \qquad \text{in } \Omega, \qquad (4.1)$$

The electric potentials on segments AB and CD are, respectively, $\phi = \phi_0$ and $\phi = 0$. All the other boundaries are assumed to be insulating, $\mathbf{n} \bullet \nabla \phi = 0$, where $\mathbf{n}$ is the unit normal vector pointing from the boundary surface into the fluid. The initial value of the electric potential is zero.

The inertial effect is neglected in the present study because of a very small Reynolds number. Thus, the flow field is governed by the Stokes equations:

$$\rho \frac{\partial \mathbf{u}}{\partial t} - \mu \nabla^2 \mathbf{u} + \nabla p = 0 \qquad \text{in } \Omega, \tag{4.2}$$

$$\nabla \bullet \mathbf{u} = 0 \qquad \text{in } \Omega. \tag{4.3}$$

The initial values of the flow velocity and the pressure are all zero. The pressures at the two openings of the channel are both zero. The EOF arising from the charged channel wall is described by the Smoluchowski slip velocity:

$$\mathbf{u} = \frac{\varepsilon_0 \varepsilon_f \zeta_w}{\mu} (\mathbf{I} - \mathbf{nn}) \bullet \nabla \phi \qquad \text{on BC and AD}. \tag{4.4}$$

In this equation, ε_0 is the permittivity of the vacuum, and the quantity $(\mathbf{I} - \mathbf{nn}) \bullet \nabla \phi$ defines the electric field tangential to the charged surface, with $\mathbf{I}$ denoting the second-order unit tensor. The fluid velocity on the particle surface is described as

$$\mathbf{u} = \frac{\varepsilon_0 \varepsilon_f \zeta_p}{\mu} (\mathbf{I} - \mathbf{nn}) \bullet \nabla \phi + \mathbf{U}_p + \omega_p \times (\mathbf{x}_s - \mathbf{x}_p) \qquad \text{on } \Gamma. \tag{4.5}$$

For Equation (4.5), $\mathbf{x}_s$ and $\mathbf{x}_p$ are, respectively, the position vector of the surface and center of the particle. $\mathbf{U}_p$ and ω_p are, respectively, the particle's translational velocity and rotational velocity, which are, respectively, determined by

$$m_p \frac{d\mathbf{U}_p}{dt} = \mathbf{F}_H + \mathbf{F}_{DEP}, \tag{4.6}$$

and

$$I_p \frac{d\omega_p}{dt} = \int (\mathbf{x}_s - \mathbf{x}_p) \times (\mathbf{T}^H \bullet \mathbf{n} + \mathbf{T}^E \bullet \mathbf{n}) d\Gamma. \tag{4.7}$$

Here, m_p and I_p are, respectively, the mass and moment of inertia of the particle. The detailed calculation of the moment of inertia is shown in the Appendix. $\mathbf{F}_H$ and $\mathbf{F}_{DEP}$ defined in Equation (4.6) are, respectively, the hydrodynamic force and the DEP force acting on the particle,

$$\mathbf{F}_H = \int \mathbf{T}^H \bullet \mathbf{n} d\Gamma = \int [-p\mathbf{I} + \mu(\nabla \mathbf{u} + \nabla \mathbf{u}^T)] \bullet \mathbf{n} d\Gamma, \tag{4.8}$$

$$\mathbf{F}_{DEP} = \int \mathbf{T}^E \bullet \mathbf{n}d\Gamma = \int \varepsilon_0\varepsilon_f\left[\mathbf{EE} - \frac{1}{2}(\mathbf{E}\bullet\mathbf{E})\mathbf{I}\right]\bullet\mathbf{n}d\Gamma. \tag{4.9}$$

In this equation, $\mathbf{T}^H$ and $\mathbf{T}^E$ are, respectively, the hydrodynamic and Maxwell stress tensors.

The particle radius a, as the length scale; the zeta potential of the channel wall ζ_w, as the potential scale; $U_0 = \varepsilon_0\varepsilon_f\zeta_w^2/(\mu a)$ as the velocity scale; and $\mu U_0/a$ as the pressure scale are selected to normalize all the previous governing equations:

$$\nabla^{*2}\phi^* = 0 \qquad \text{in } \Omega, \tag{4.10}$$

$$\text{Re}\frac{\partial\mathbf{u}^*}{\partial t^*} - \nabla^{*2}\mathbf{u}^* + \nabla^*p^* = 0 \qquad \text{in } \Omega, \tag{4.11}$$

$$\nabla^*\bullet\mathbf{u}^* = 0 \qquad \text{in } \Omega, \tag{4.12}$$

and the corresponding boundary conditions,

$$\mathbf{n}\bullet\nabla^*\phi^* = 0 \qquad \text{on BC, AD and } \Gamma, \tag{4.13}$$

$$\phi^* = \frac{\phi_0}{\zeta_w} \qquad \text{on AB,} \tag{4.14}$$

$$\mathbf{u}^* = (\mathbf{I} - \mathbf{nn})\bullet\nabla^*\phi^* \qquad \text{on BC and AD,} \tag{4.15}$$

$$\mathbf{u}^* = \mathbf{U}_p^* + \omega_p^*\times(\mathbf{x}_s^* - \mathbf{x}_p^*) + \gamma(\mathbf{I} - \mathbf{nn})\bullet\nabla^*\phi^* \qquad \text{on BC and AD.} \tag{4.16}$$

In these equations, $\text{Re} = \rho U_0 a/\mu$, and $\gamma = \zeta_p/\zeta_w$ is the ratio of the zeta potential of the particle to that of the channel wall.

The force and torque are, respectively, normalized by μU_0 and $a\mu U_0$, yielding the dimensionless equations of particle motion:

$$m_p^*\frac{d\mathbf{U}_p^*}{dt^*} = \int(\mathbf{T}^{H*}\bullet\mathbf{n} + \mathbf{T}^{E*}\bullet\mathbf{n})d\Gamma^*, \tag{4.17}$$

$$I_p^*\frac{d\omega_p^*}{dt^*} = \int\left(\mathbf{x}_s^* - \mathbf{x}_p^*\right)\times\left(\mathbf{T}^{H*}\bullet\mathbf{n} + \mathbf{T}^{E*}\bullet\mathbf{n}\right)d\Gamma^*, \tag{4.18}$$

where the mass and the moment of inertia are, respectively, normalized by $a\mu/U_0$ and $a^3\mu/U_0$,

$$\mathbf{T}^{H*} = -p^*\mathbf{I} + \left(\nabla^*\mathbf{u}^* + \nabla^*\mathbf{u}^{*T}\right),$$

and

$$\mathbf{T}^{E*} = \mathbf{E}^*\mathbf{E}^* - \frac{1}{2}\left(\mathbf{E}^* \bullet \mathbf{E}^*\right)\mathbf{I}.$$

4.4   NUMERICAL IMPLEMENTATION IN COMSOL

To reduce the computational effort, the width and length of the channel in the computation are, respectively, 50 and 300 μm. The densities of the fluid and the particle are assumed identical, $\rho = 1.0 \times 10^3$ kg/m^3, to neglect the effect of gravity. The other parameters are listed in Table 4.1. The implementation of the present mathematical model in COMSOL Multiphysics® 3.5a is similar to that described in Chapter 3 with a different geometry. Table 4.2 shows detailed instructions

TABLE 4.1
Constant Table

Variable	Value or Expression	Description
eps_r	80	Relative permittivity of fluid
eps0	8.854187817e-12 [F/m]	Permittivity of vacuum
rho	1e3 [kg/m^3]	Particle/fluid density
Xp0	10	Dimensionless initial x location of particle
Yp0	0	Dimensionless initial y location of particle
Masss1	rho*pi*a^2	Dimensional particle mass I
Masss2	rho*2*a*(l-2*a)	Dimensional particle mass II
Mass	(Masss1+Masss2)*Uc/eta/a	Dimensionless particle mass
a	2.25E-06 [m]	Particle radius
l	1.35E-05 [m]	Particle length
Ir	(Masss1*(0.5*a^2+(l-2*a)^2/4+(l-2*a)*4*a/ (3*pi))+Masss2*1/12*((l-2*a)^2+4*a^2))*Uc/ eta/a^3	Dimensionless moment of inertia
zetar	0.8	Zeta potential ratio
ew	80e-3 [V]	Zeta potential of channel wall
eta	1e-3 [Pa*s]	Fluid viscosity
Uc	eps_r*eps0*ew* ew/(eta*a)	Velocity scale
v0	5.0625 [V]	Applied voltage
Re	rho*Uc*a/eta	Reynolds number

TABLE 4.2
Model Setup in the GUI of COMSOL Multiphysics 3.5a

Model Navigator	Select **2D** in space dimension and click **Multiphysics** button.
	Select **COMSOL Multiphysics|Deformed Mesh|Moving Mesh (ALE)|Transient analysis**. Click **Add** button.
	Select **COMSOL Multiphysics|Electromagnetics|Conductive Media DC**. Click **Add** button.
	Select **MEMS Module|Microfluidics|Stokes Flow|Transient analysis**. Click **Add** button.
Option Menu|**Constants**	Define variables in Table 4.1.
Physics Menu|**Subdomain Setting**	ale model
	Subdomain 1
	Free displacement
	emdc mode (external electric field)
	Subdomain 1
	$J_e = 0$; $Q_j = 0$; $d = 1$; $\sigma = 1$
	mmglf mode
	Subdomain 1
	Tab Physics
	$\rho = \mathrm{Re}$; $\eta = 1$; Thickness = 1
Physics Menu|**Boundary Setting**	ale mode
	Particle surface: Mesh velocity vx = up+Ur
	vy = vp+Vr
	Other boundaries: Mesh displacement dx = 0 dy = 0
	emdc mode
	Left boundary: V_0 = v0/ew
	Right boundary: Ground
	Other boundaries: Electric insulation
	mmglf mode
	Left boundary: Inlet|Pressure, no viscous stress $P_0 = 0$
	Right boundary: Outlet|Pressure, no viscous stress $P_0 = 0$
	Particle surface: Wall|Moving/leaking wall Uw = up+Ur+Ueofp
	Vw = vp+Vr+Veofp
	Channel wall: Wall|Moving/leaking wall Uw = Ueof Vw = Veof
Other Settings	Properties of ale model (**Physics** Menu|**Properties**)
	Smoothing method: Winslow
	Allow remeshing: On
	Properties of mmglf model (**Physics** Menu|**Properties**)
	Weak constraints: On
	Constraint type: Nonideal
	Deactivate the unit system (**Physics** Menu|**Model Setting**)
	Base unit system: None

continued

TABLE 4.2 (CONTINUED)
Model Setup in the GUI of COMSOL Multiphysics 3.5a

Options Menu|**Integration Coupling Variables|Boundary Variables**	Boundary selection: 4, 5, 7, 8, 9, 10 (Particle surface)
	Name: F_x Expression: -lm3; Integration order: 4; Frame: Frame(mesh)
	Name: F_y Expression: -lm4; Integration order: 4; Frame: Frame(mesh)
	Name: Fd_x Expression: depr*(tEx_emdc^2+tEy_emdc^2)*nx/2; Integration order: 4; Frame: Frame(mesh)
	Name: Fd_y Expression: depr*(tEx_emdc^2+tEy_emdc^2)*ny/2; Integration order: 4; Frame: Frame(mesh)
	Name:Tr Expression:(x-Xpar)*(depr*(tEx_emdc^2+tEy_emdc^2)*/2-lm4)-(y-Ypar)*(depr*(tEx_emdc^2+tEy_emdc^2)*nx/2-lm3); Integration order: 4; Frame: Frame(mesh)
Options Menu| **Expressions|Boundary Expressions**	Boundary selection: 4, 5, 7, 8, 9, 10 (Particle surface)
	Name: Ueofp Expression: zetar*tEx_emdc
	Name: Veofp Expression: zetar*tEy_emdc
	Name: Ur Expression: -omega*(y-Ypar)
	Name: Vr Expression: omega*(x-Xpar)
	Boundary selection: 2, 3 (Channel wall)
	Name: Ueof Expression: tEx_emdc
	Name: Veof Expression: tEy_emdc
Physics Menu| **Global Equations**	Name: up; Expression: Mass*upt-(F_x+Fd_x); Init (u): 0; Init (ut): 0
	Name: Xpar; Expression: Xpart-up; Init (u): Xp0; Init (ut): 0
	Name: vp; Expression: Mass*vpt-(F_y+Fd_y); Init (u): 0; Init (ut): 0
	Name: Ypar; Expression: Ypart-vp; Init (u): Yp0; Init (ut): 0
	Name: omega; Expression: Ir*omegat-Tr; Init (u): 0; Init (ut): 0
	Name: ang; Expression: angt-omega; Init (u): 0; Init (ut): 0
Mesh|Free Mesh Parameters	Tab subdomain
	Subdomain 1|Maximum element size: 1
	Tab boundary
	Particle surface: 4, 5, 7, 8, 9, 10|**Maximum element size**: 0.1
	Channel wall: 2, 3|**Maximum element size**: 0.4
Solve Menu|	**Solver Parameters**
	Tab General
	Times: Range(0, 0.5, 50000)
	Relative tolerance: 1e-4
	Absolute tolerance: 1e-6
	Tab Timing Stepping
	Activate Use stop condition
	minqual1_ale-0.6
	Click = **Solve Problem**
Postprocessing Menu|	Result check: up, Xpar, vp, Ypar, omega, ang
	Global Variables Plot|Select all the predefined quantities, click >>
	Click Apply

for setting up the dimensionless model in the graphical user interface (GUI) of COMSOL Multiphysics 3.5a. The full MATLAB script M-file capable of automated remeshing is as follows:

```
%%%%%%%%%%Algal cell in a straight channel.m%%%%%%%%%%%%%

flclear fem

% NEW: Define particle radius
r=2.25e-6;

% NEW: Define particle length
len=6*r;

% NEW: Define aspect ratio
ratio=(len-2*r)/r;

% NEW: Define the dimensionless channel length (300 um)
L=225;

% NEW: Define the dimensionless channel width (50 um)
b=22.22;

% NEW: Define the electric field intensity (V/m)
E=10000;

% NEW: Define the electric potential difference (V)
phi=E*r*L;

% NEW: Define the particle angle
angle=[0,15,30,45,60,75,90];


% NEW: FOR loop to go through all the particle angles
for i=1:length(angle)

% NEW: Define the initial particle location
Xin=10;
Yin=0;

% Constants
fem.const = {'eps_r','80', ...
             'eps0','8.854187817e-12', ...
             'rho','1e3', ...
             'Xp0', num2str(Xin), ...
             'Yp0', num2str(Yin), ...
             'Masss1','rho*pi*a^2', ...
             'Masss2','rho*2*a*(1-2*a)', ...
             'Mass','(Masss1+Masss2)*Uc/eta/a', ...
```

```
                      'a',num2str(r), ...
                      'l',num2str(len), ...

'Ir',' (Masss1*(0.5*a^2+(l-2*a)^2/4+(l-2*a)*4*a/
  (3*pi))+Masss2*1/12*((l-2*a)^2+4*a^2))*Uc/eta/a^3', ...
                      'zetar','0.8', ...
                      'ew','80e-3', ...
                      'eta','1.0e-3', ...
                      'Uc','eps_r*eps0*ew*ew/(eta*a)', ...
                      'v0',num2str(phi), ...
                      'Re','rho*Uc*a/eta'};

% Geometry
% NEW: Draw the straight channel
g1=rect2(num2str(L),num2str(b),'base','center','pos',{num2st
  r(L/2),'0'},'rot','0');

% NEW: Draw the cylindrical particle
g2=rect2(num2str(ratio),'2','base','center','pos',
  {'10','0'},'rot','0');
g3=circ2('1','base','center','pos',{'8','0'},'rot','0');
g4=circ2('1','base','center','pos',{'12','0'},'rot','0');
g5=geomcomp({g2,g3,g4},'ns',{'g2','g3','g4'},'sf',
  'g2+g3+g4','edge','none');
g6=geomcomp({g5},'ns',{'g5'},'sf','g5','edge','none');
g7=geomdel(g6);
g7=rotate(g7,angle(i)*pi/180,[10,0]);
g8=geomcomp({g1},'ns',{'g1'},'sf','g1','edge','none');
g9=geomcomp({g8,g7},'ns',{'g8','g7'},'sf','g8-g7',
  'edge','none');

% Analyzed geometry
clear s
s.objs={g9};
s.name={'CO3'};
s.tags={'g9'};

fem.draw=struct('s',s);
fem.geom=geomcsg(fem);

% NEW: Display geometry
geomplot(fem, 'Labelcolor','r','Edgelabels','on','submode',
  'off');
% (Default values are not included)
% NEW: Solve moving mesh
% Application mode 1
clear appl
appl.mode.class = 'MovingMesh';
appl.sdim = {'Xm','Ym','Zm'};
```

```
appl.shape = {'shlag(2,''lm1'')','shlag(2,''lm2'')','shlag(2,
  ''x'')','shlag(2,''y'')'};
appl.gporder = {30,4};
appl.cporder = 2;
appl.assignsuffix = '_ale';
clear prop
prop.smoothing='winslow';
prop.analysis='transient';
prop.allowremesh='on';
prop.origrefframe='ref';
appl.prop = prop;
clear bnd
bnd.defflag = {{1;1},{0;0}};
bnd.veldeform = {{0;0},{'up+Ur';'vp+Vr'}};
bnd.wcshape = [1;2];
bnd.name = {'Fixed','Particle'};
bnd.type = {'def','vel'};
bnd.veldefflag = {{0;0},{1;1}};
bnd.ind = [1,1,1,2,2,1,2,2,2,2];
appl.bnd = bnd;
clear equ
equ.gporder = 2;
equ.shape = [3;4];
equ.ind = [1];
appl.equ = equ;
fem.appl{1} = appl;

% NEW: Solve flow field using Navier-Stokes equations
% Application mode 2
clear appl
appl.mode.class = 'GeneralLaminarFlow';
appl.module = 'MEMS';
appl.shape = {'shlag(2,''lm3'')','shlag(2,''lm4'')',
  'shlag(1,''lm5'')','shlag(2,''u'')','shlag(2,''v'')',
  'shlag(1,''p'')'};
appl.gporder = {30,4,2};
appl.cporder = {2,1};
appl.assignsuffix = '_mmglf';
clear prop
prop.weakcompflow='Off';
prop.inerterm='Off';
clear weakconstr
weakconstr.value = 'on';
weakconstr.dim = {'lm3','lm4','lm5','lm6'};
prop.weakconstr = weakconstr;
prop.constrtype='non-ideal';
appl.prop = prop;
clear bnd
bnd.uw0 = {0,'Ueof',0,0,0,0,0};
```

```
bnd.type = {'inlet','walltype','walltype','outlet',
  'walltype','walltype', ... 'sym'};
bnd.vwall = {0,0,'vp+Vr+Veofp',0,0,0,0};
bnd.vw0 = {0,'Veof',0,0,0,0,0};
bnd.E_y = {0,'Ey_emdc',0,0,0,0,0};
bnd.E_x = {0,'Ex_emdc',0,0,0,0,0};
bnd.wcshape = [1;2;3];
bnd.wcgporder = 1;
bnd.intype = {'p','uv','uv','uv','uv','uv','uv'};
bnd.U0 = {0.001,0,0,0,0,0,0};
bnd.walltype = {'noslip','semislip','lwall','noslip',
  'noslip','noslip','noslip'};
bnd.uwall = {0,0,'up+Ur+Ueofp',0,0,0,0};
bnd.eotype = {'mueo','zeta','mueo','mueo','mueo','mueo',
  'mueo'};
bnd.isViscousSlip = {1,0,1,1,1,1,1};
bnd.name = {'Inlet','Electroosmosis','Particle','Outlet',
  'Electrode','Ground', ... 'Symmetry'};
bnd.mueo = {7e-8,'mu_eo',7e-8,7e-8,7e-8,7e-8,7e-8};
bnd.zeta = {-0.1,'-zetar/(eps_r*epsilon0_emdc)',-0.1,-0.1,
  -0.1,-0.1, ... -0.1};
bnd.ind = [1,2,2,3,3,4,3,3,3,3];
appl.bnd = bnd;
clear equ
equ.eta = 1;
equ.gporder = {{2;2;3}};
equ.epsilonr = 'eps_r';
equ.thickness = 1;
equ.rho = 'Re';
equ.cporder = {{1;1;2}};
equ.shape = [4;5;6];
equ.ind = [1];
appl.equ = equ;
fem.appl{2} = appl;

% NEW: Solve electrostatics using Laplace equation
% Application mode 3
clear appl
appl.mode.class = 'EmConductiveMediaDC';
appl.module = 'MEMS';
appl.assignsuffix = '_emdc';
clear prop
clear weakconstr
weakconstr.value = 'off';
weakconstr.dim = {'lm7'};
prop.weakconstr = weakconstr;
appl.prop = prop;
clear bnd
bnd.V0 = {'v0/ew',0,0,0,0,0,0};
bnd.type = {'V','nJ0','V0','nJ0','nJ0','nJ0','nJ0'};
```

```
bnd.name = {'Electrode','Insulator','Ground',
  'Electroosmosis','Inlet','Outlet', ... 'Symmetry'};
bnd.ind = [1,2,2,2,2,3,2,2,2,2];
appl.bnd = bnd;
clear equ
equ.sigma = 1;
equ.name = 'default';
equ.d = 1;
equ.ind = [1];
appl.equ = equ;
fem.appl{3} = appl;
fem.sdim = {{'Xm','Ym'},{'X','Y'},{'x','y'}};
fem.frame = {'mesh','ref','ale'};
fem.border = 1;

% Boundary settings
clear bnd
bnd.ind = [1,2,2,3,3,1,3,3,3,3];
bnd.dim = {'x','y','lm1','lm2','u','v','p','lmx_mmglf','lmy_
  mmglf', ... 'lm3','lm4','lm5','V'};

% Boundary expressions
bnd.expr = {'Ur',{'','','-omega*(y-Ypar)'}, ...
            'Vr',{'','','omega*(x-Xpar)'}, ...
            'Ueof',{'','tEx_emdc',''}, ...
            'Veof',{'','tEy_emdc',''}, ...
            'Ueofp',{'','','zetar*tEx_emdc'}, ...
            'Veofp',{'','','zetar*tEy_emdc'}};
fem.bnd = bnd;

% Coupling variable elements
clear elemcpl
% Integration coupling variables
clear elem
elem.elem = 'elcplscalar';
elem.g = {'1'};
src = cell(1,1);
clear bnd
bnd.expr = {{{},'-lm3'},{{},'-lm4'},{{}, ...


'(x-Xpar)*((tEx_emdc^2+tEy_emdc^2)*ny/2-lm4)-(y-Ypar)*((tEx_
  emdc^2+tEy_emdc^2)*nx/2-lm3)'}, ...
  {{},'(tEx_emdc^2+tEy_emdc^2)*nx/2'},{{}, ...
  '(tEx_emdc^2+tEy_emdc^2)*ny/2'}};
bnd.ipoints = {{{},'4'},{{},'4'},{{},'4'},{{},'4'},{{},'4'}};
bnd.frame = {{{},'mesh'},{{},'mesh'},{{},'mesh'},{{},
  'mesh'},{{},'mesh'}};
bnd.ind = {{'1','2','3','6'},{'4','5','7','8','9','10'}};
src{1} = {{},bnd,{}};
```

```
elem.src = src;
geomdim = cell(1,1);
geomdim{1} = {};
elem.geomdim = geomdim;
elem.var = {'F_x','F_y','Tr','Fd_x','Fd_y'};
elem.global = {'1','2','3','4','5'};
elem.maxvars = {};
elemcpl{1} = elem;
fem.elemcpl = elemcpl;

% ODE Settings
clear ode
ode.dim={'up','Xpar','vp','Ypar','omega','ang'};
ode.f={'Mass*upt-(F_x+Fd_x)','Xpart-up','Mass*vpt-(F_
  y+Fd_y)','Ypart-vp','Ir*omegat-Tr','angt-omega'};
ode.init={'0','Xp0','0','Yp0','0',num2str(angle(i)*pi/180)};
ode.dinit={'0','0','0','0','0','0'};
fem.ode=ode;

% Multiphysics
fem=multiphysics(fem);

% NEW: Loop for continuous particle tracking
% NEW: Remeshing index
j=1;

% NEW: Define initial time
time_e=0;

% NEW: Define data storage index
num=1;

% NEW: Define matrix for data storage
particle=zeros(10,7);

% NEW: Get the particle location
Xp=Xin;
Yp=Yin;

% NEW: Define end condition of the computation (Loop control)
while Xp<215

% Initialize mesh
fem.mesh=meshinit(fem, ...
                  'hauto',5, ...
                  'hmaxedg',[4,0.1,5,0.1,7,0.1,8,0.1,9,0.1,
                  10,0.1,3,0.4,2,0.4], ...
                  'hmaxsub',[1,1]);
% Extend mesh
```

```
fem.xmesh=meshextend(fem);
% ***********************************************

if j==1
% Solve problem
fem.sol=femtime(fem, ...

'solcomp',{'lm4','lm3','lm2','lm1','lmx_mmglf','lm5','Xpar',
  'ang','up','vp','omega','v','u','V','Ypar','p','lmy_
  mmglf','y','x'}, ...

'outcomp',{'lm4','lm3','lm2','lm1','lmx_mmglf','lm5','Xpar',
  'ang','up','vp','omega','v','u','V','Ypar','p','Y','X','
  lmy_mmglf','y','x'}, ...
            'tlist',[0:0.5:50000], ...
            'tout','tlist', ...
            'rtol',1e-4, ...
            'atol',1e-6, ...
            'stopcond','minqual1_ale-0.6');
else
% Mapping current solution to extended mesh
init = asseminit(fem,'init',fem0.sol,'xmesh',fem0.xmesh,
  'framesrc','ale','domwise','on');

% Solve problem
fem.sol=femtime(fem, ...
                'init',init,...
'solcomp',{'lm4','lm3','lm2','lm1','lmx_mmglf','lm5','Xpar',
  'ang','up','vp','omega','v','u','V','Ypar','p','lmy_
  mmglf','y','x'}, ...
'outcomp',{'lm4','lm3','lm2','lm1','lmx_mmglf','lm5','Xpar',
  'ang','up','vp','omega','v','u','V','Ypar','p','Y','X',
  'lmy_mmglf','y','x'}, ...
            'tlist',[time_e:0.5:50000], ...
            'tout','tlist', ...
            'rtol',1e-4, ...
            'atol',1e-6, ...
            'stopcond','minqual1_ale-0.6');
end

% Save current fem structure for restart purposes
fem0=fem;

% NEW: Save current fem for postprocessing purposes
solution{j}=fem;

% NEW: Get the last time from solution
time_e = fem.sol.tlist(end);

% NEW: Get the length of the time steps
```

```
Leng =length(fem.sol.tlist);

% Global variables plot
data=postglobalplot(fem,{'ang','omega','Xpar','Ypar','up',
                        'vp'}, ...
                        'title','Particle', ...
                        'Outtype','postdata', ...
                        'axislabel',{'Time','Parameters'});

% Store angle data
for n=1:Leng
        particle(num,1)=data.p(1,n);
        particle(num,2)=data.p(2,n);
        particle(num,3)=data.p(2,n+Leng);
        particle(num,4)=data.p(2,n+2*Leng);
        particle(num,5)=data.p(2,n+3*Leng);
        particle(num,6)=data.p(2,n+4*Leng);
        particle(num,7)=data.p(2,n+5*Leng);

    num=num+1;
end

% NEW: Get the particle location
Xp=data.p(2,3*Leng);
Yp=data.p(2,4*Leng);

% Plot solution
postplot(fem, ...
        'tridata',{'normE_emdc','cont','internal'}, ...
        'trimap','jet(1024)', ...
        'tridlim',[0 2.5*phi/80e-3/L], ...
        'solnum','end', ...
        'title','Surface: Electric field', ...
        'geom','off', ...
        'axis',[-5,140,-10,10]);

% NEW: Output the COMSOL Multiphysics GUI .mph file if
  necessary
flsave(strcat('Alga_E10K_New_angle=',num2str(angle(i)),'_',
  num2str(j),'.mph'),fem);

% NEW: Write the data into a file
dlmwrite(strcat('Alga_E10K_New_angle=',num2str(angle(i)),
  '.dat'),particle,',');

% Geometry
% Generate geom from mesh
fem = mesh2geom(fem, ...
                'frame','ale', ...
                'srcdata','deformed', ...
```

```
          'destfield',{'geom','mesh'}, ...
          'srcfem',1, ...
          'destfem',1);

j=j+1;
end

% NEW: Make animation
postmovie(solution, ...
          'tridata',{'normE_emdc','cont','internal'}, ...
          'tridlim',[0 2.5*phi/80e-3/L], ...
          'trimap','jet(1024)', ...
          'title','Surface: Dimensionless electric field',
          ...
          'geom','off', ...
          'axis',[-5,140,-10,10], ...
          'Filename',strcat('E10K_
           angle=',num2str(angle(i)),'_.avi'), ...
          'Width',800, ...
          'Height',600, ...
          'fps',20);

end
```

4.5 RESULTS AND DISCUSSION

An approximation solution of the electrokinetic velocity of a cylindrical particle along the axis of a tube under the thin EDL assumption is given as (Liu, Bau, and Hu 2004)

$$U_p = \frac{\varepsilon_0 \varepsilon_f E_z}{\mu(1+\lambda^2)}(\zeta_p - \zeta_w),$$
(4.19)

where E_z is the axial electric field in the absence of particles, and λ is the ratio of the particle radius to the tube radius. We first run the electrokinetic motion of the algal cells in a capillary and get the steady particle velocity U_p when it is parallel to the electric field along the axis of the capillary. Based on the zeta potential of PDMS, $\zeta_w = -80$ mV (Kang et al. 2006; Venditti, Xuan, and Li 2006), and the viscosity and permittivity of fluid at room temperature (i.e., $\rho = 1000$ kg/m^3, and $\varepsilon_f = 80$), the averaged zeta potential of *Desmodesmus* cf. *quadricauda* was estimated to be $\zeta_p = -42$ mV. The electric field inside the microchannel is not uniform due to the presence of the particle, as shown in Figure 4.2b. In the following discussion, the electric field intensities are obtained by dividing the electric potential difference over the entire length of the channel.

4.5.1 EXPERIMENTAL RESULTS

Figures 4.3a, 4.3c, and 4.3e show, respectively, the superposed trajectory of *Desmodesmus* cf. *quadricauda* cells in the straight microchannel under three

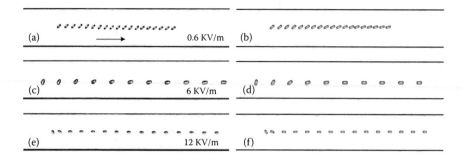

FIGURE 4.3 Trajectories of cylindrical particles moving from left to right in a straight microchannel. The particle trajectories were obtained by superposing sequential images of the same particle into a single figure. The left images (a), (c), and (e) are experimental observations under different electric fields, while the right images (b), (d), and (f) are the corresponding numerical predictions. Time interval between adjacent particles in Figures 4.3a and 4.3b was 0.7 s; the time interval in the other figures was 0.14 s. (From Ai, Y., A. Beskok, D. T. Gauthier, S. W. Joo, and S. Qian. 2009. DC electrokinetic transport of cylindrical cells in straight microchannels. *Biomicrofluidics* 3:044110 with permission from the American Institute of Physics.)

different electric fields, $E = 0.6$, 6, and 12 kV/m. Under a relatively low electric field, $E = 0.6$ kV/m; the particle rotates very slightly as it translates. When the applied electric field increases to $E = 6$ kV/m, the particle is aligned with its longest axis parallel to the external electric field. The alignment becomes even faster when the electric field is increased to $E = 12$ kV/m. The properties of the cells are listed in Table 4.3. Figures 4.3b, 4.3d, and 4.3f are the corresponding numerical predictions, which capture the particle motion in the experiments mentioned.

We further quantitatively compare the predicted translational velocity to the experimental data, indicating good agreement in Figure 4.4a. The predicted translational velocity without considering the DEP effect is nearly the same as that obtained with considering the DEP effect and thus is not shown here. Figure 4.4b shows the quantitative comparison between the predicted particle angle and the experimental data. As mentioned, the variation of the particle angle observed in the experiment is very small under $E = 0.6$ kV/m, which is close to the numerical prediction without DEP. The numerical prediction with DEP seems to overpredict the angle variation, which may be due to the slight shape mismatch between the actual cell and the cylindrical shape used in the numerical simulation. Under $E = 6$ kV/m, the numerical prediction with DEP can well recover the decrease in the particle angle owing to the alignment observed in the experiments. If the DEP effect is ignored in the numerical simulation, the particle angle does not decrease. It is obvious that the particle alignment phenomenon is induced by the DEP effect. When the electric field is increased to $E = 12$ kV/m, the cell is aligned faster in the numerical simulation, which is in good agreement with the experimental results. It is concluded that the DEP effect is of great importance in the electrokinetic transport of nonspherical particles even in a uniform microchannel.

TABLE 4.3
Properties of Cells in Figure 4.3

Property	Figure 4.3a	Figure 4.3c	Figure 4.3e
Cell radius (μm)	1.88	2.25	1.6
Cell length (μm)	8.4	8.46	6.4
Initial location (μm, μm)	(52.27, 1.29)	(28.69, −3.69)	(43.85, 0.14)
Initial angle (°)	40	75	122
Zeta potential (mV)	−49.6	−38.0	−65.0

Source: From Ai, Y., A. Beskok, D. T. Gauthier, S. W. Joo, and S. Qian. 2009. DC electrokinetic transport of cylindrical cells in straight microchannels. *Biomicrofluidics* 3:044110 with permission from the American Institute of Physics.

As the DEP effect is proportional to the square of the electric field, it is too small to align the particle under a low electric field. Once the DEP effect becomes dominant, the particle alignment phenomenon comes into play, which has been widely used to manipulate and assemble synthetic nanowires and carbon nanotubes onto electrodes (Evoy et al. 2004; Makaram et al. 2007; Chang and Hong 2009; Raychaudhuri et al. 2009; Lao et al. 2006; Monica et al. 2008).

4.5.2 EFFECT OF CHANNEL WALL

The following studies are performed by the verified numerical model. As the equations are dimensionless, all the following results are presented in a dimensionless manner as well. The dimensionless initial location of the particle is (10, 0). The initial angle of the particle is 60° except for the case studying the effect of the particle's initial angle. The zeta potential of the channel is $\zeta_w = -80$ mV.

In general, electrokinetic transport in microfluidics is usually confined in microchannels. Therefore, the boundary effect might play an important role. Figure 4.5 shows the superposed particle trajectory in a straight channel with four different channel widths with the consideration of DEP. The electric field intensity, aspect ratio of the particle, and the zeta potential ratio are, respectively, $E^* = 0.0169$ ($E = 0.6$ kV/m), $d/a = 6$, and $\gamma = 0.525$. When the channel width is $b^* = 10$, Figure 4.5a shows that the particle experiences an obvious oscillatory motion. As the electric field is very low, the DEP effect is limited, and the particle motion is governed by electrophoresis and electroosmosis. Since the particle's aspect ratio is close to the channel width, the presence of the particle strongly distorts the electric field, as shown in Figure 4.2b. The fluctuation of the electric field between the particle and the channel wall is responsible for the oscillatory motion. This kind of motion has also been predicted in a previous numerical study without considering the DEP effect (Davison and Sharp 2007). As the channel width increases, the amplitude of the oscillatory motion accordingly decreases, as shown in Figures 4.5b, 4.5c,

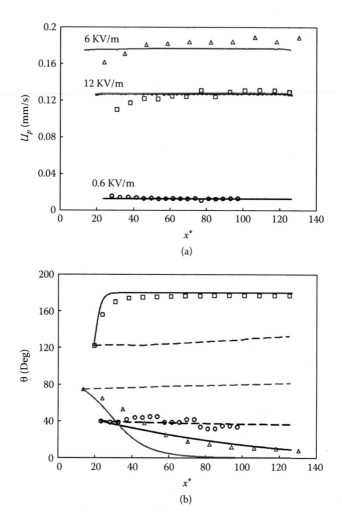

FIGURE 4.4   Comparison of translational velocity (a) and angle (b) between experimental results and numerical predictions. Circles, triangles, and squares are the experiments of Figures 4.3a, 4.3c, and 4.3e, respectively. Solid and dashed lines are the corresponding numerical predictions with and without considering the DEP effect, respectively. (From Ai, Y., A. Beskok, D. T. Gauthier, S. W. Joo, and S. Qian. 2009. DC electrokinetic transport of cylindrical cells in straight microchannels. *Biomicrofluidics* 3:044110 with permission from the American Institute of Physics.)

and 4.5d. Furthermore, a longer traveling distance is required to complete one cycle of the oscillatory motion in a wider channel.

Figure 4.6 shows the superposed particle trajectory under the same conditions as in Figure 4.5 except the electric field is increased 10 times to $E^* = 0.169$. Because the DEP effect becomes dominant, the particle is aligned parallel to the external electric field quickly and stays in this orientation. The boundary effect

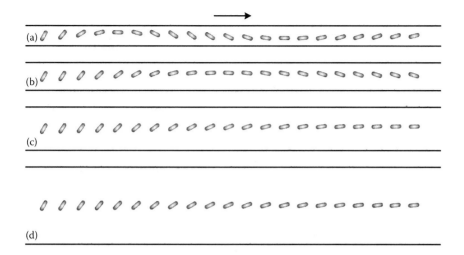

FIGURE 4.5 Sequential images of rotation and translation of a cylindrical particle in a straight channel with different channel widths. The arrow denotes the translational direction of the particle. The simulation conditions were $E^* = 0.0169$, $d/a = 6$, $\gamma = 0.525$. (a) $b^* = 10$; (b) $b^* = 14$; (c) $b^* = 22.22$; (d) $b^* = 40$. (From Ai, Y., A. Beskok, D. T. Gauthier, S. W. Joo, and S. Qian. 2009. DC electrokinetic transport of cylindrical cells in straight microchannels. *Biomicrofluidics* 3:044110 with permission from the American Institute of Physics.)

on the particle orientation becomes negligible. Figure 4.7 shows the variation of the particle's orientation along the microchannel under two different electric fields. The other conditions are $d/a = 6$, $b^* = 10$, and $\gamma = 0.525$. As explained, the particle experiences an oscillatory motion when the applied electric field is relatively low. However, the amplitude of the oscillatory motion gradually attenuates because of the DEP effect. Under a relatively high electric field, the particle becomes aligned to the electric field when the DEP effect is taken into account in the modeling. However, if the DEP effect is neglected in the modeling, the numerical prediction shows that the particle also experiences an oscillatory motion. In addition, the oscillatory amplitude does not attenuate. Thus, the DEP effect must be taken into account in the electrokinetic motion of nonspherical particles in microfluidics.

4.5.3 Effect of Electric Field

The DEP effect mainly depends on the applied electric field. Apparently, the electric field can significantly affect particle motion, especially particle orientation. Figure 4.8 shows the variation of the particle orientation along the microchannel under different electric fields. The simulation conditions are $d/a = 4$, $b^* = 22.22$, and $\gamma = 0.525$. When the applied electric field is very low, $E^* = 0.0028$, the rotational motion is mainly determined by electrophoresis and electroosmosis, which lead to a slight decrease in particle orientation. As the electric field

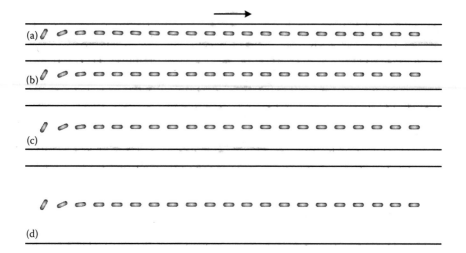

FIGURE 4.6 Sequential images of rotation and translation of a cylindrical particle in a straight channel with different channel widths. The arrow denotes the translational direction of the particle. The simulation conditions were $E^* = 0.169$, $d/a = 6$, $\gamma = 0.525$. (a) $b^* = 10$; (b) $b^* = 14$; (c) $b^* = 22.22$; (d) $b^* = 40$. (From Ai, Y., A. Beskok, D. T. Gauthier, S. W. Joo, and S. Qian. 2009. DC electrokinetic transport of cylindrical cells in straight microchannels. *Biomicrofluidics* 3:044110 with permission from the American Institute of Physics.)

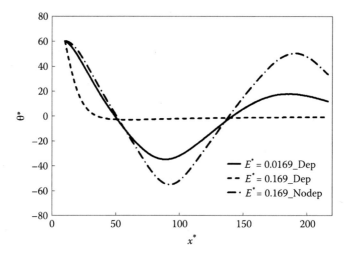

FIGURE 4.7 Orientation variations of a cylindrical particle along the centerline of the microchannel. The simulation conditions were $d/a = 6$, $b^* = 10$, and $\gamma = 0.525$. (From Ai, Y., A. Beskok, D. T. Gauthier, S. W. Joo, and S. Qian. 2009. DC electrokinetic transport of cylindrical cells in straight microchannels. *Biomicrofluidics* 3:044110 with permission from the American Institute of Physics.)

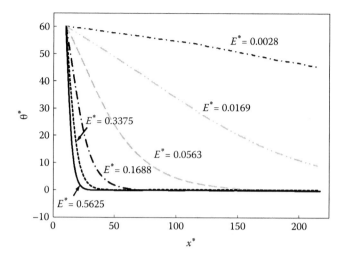

FIGURE 4.8 Orientation variations of a cylindrical particle along the centerline of the microchannel under different electric fields. The simulation conditions were $d/a = 4$, $b^* = 22.22$, and $\gamma = 0.525$. (From Ai, Y., A. Beskok, D. T. Gauthier, S. W. Joo, and S. Qian. 2009. DC electrokinetic transport of cylindrical cells in straight microchannels. *Biomicrofluidics* 3:044110 with permission from the American Institute of Physics.)

increases, the DEP effect increases faster than electrophoresis and electroosmosis. As a result, the decrease in the particle orientation becomes more rapid. If the applied electric field increases further, the DEP effect becomes dominant on the particle's rotational motion. Therefore, the particle is quickly aligned to the external electric field. Usually, a high electric field is applied to achieve fast alignment of synthetic nanowires and carbon nanotubes in the experiments.

4.5.4 Effect of Zeta Potential Ratio

Figures 4.9a and 4.9b show, respectively, the effect of the zeta potential ratio on the particle orientation along the microchannel under a low, $E^* = 0.0028$ ($E = 0.1$ kV/m), and a high, $E^* = 0.28$ ($E = 10$ kV/m), electric field. The other simulation conditions are $d/a = 6$ and $b^* = 10$. Under $E^* = 0.0028$, the oscillatory motion depends on the zeta potential ratio, as shown in Figure 4.9a. In all the studies, the particle is actually driven by EOF as the zeta potential of the channel wall is higher than that of the particle. Hence, electrophoresis always retards the particle transport in the present study. A low zeta potential ratio implies a higher EOF effect. As a result, the amplitude of the oscillatory motion increases and the period of the oscillatory motion decreases under a smaller zeta potential ratio. Under $E^* = 0.28$, all the particles will be aligned to the external electric field. However, a smaller zeta potential ratio leads to higher particle mobility, which requires a longer traveling distance to complete the alignment, as shown in Figure 4.9b.

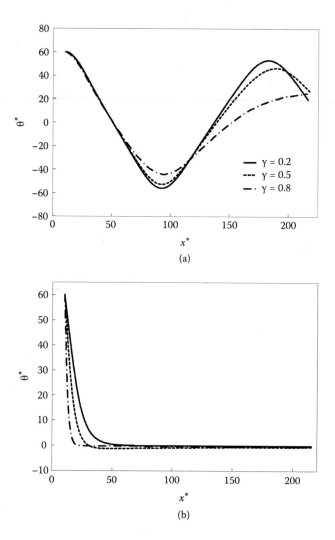

FIGURE 4.9 Orientation variations of a cylindrical particle with different zeta potential ratios along the centerline of the microchannel. The simulation conditions were $d/a = 6$, $b^* = 10$, and $\gamma = 0.525$. The lines in Figure 4.9b are in the same legend as Figure 4.9a. (a) $E^* = 0.0028$; (b) $E^* = 0.28$. (From Ai, Y., A. Beskok, D. T. Gauthier, S. W. Joo, and S. Qian. 2009. DC electrokinetic transport of cylindrical cells in straight microchannels. *Biomicrofluidics* 3:044110 with permission from the American Institute of Physics.)

4.5.5 Effect of Particle's Aspect Ratio

Figures 4.10a and 4.10b show, respectively, the effect of the particle's aspect ratio on the particle orientation along the microchannel under a low, $E^* = 0.0028$, and a high, $E^* = 0.28$, electric field when $b^* = 22.22$ and $\gamma = 0.2$. Under $E^* = 0.0028$, the particle's rotational motion is governed by electrophoresis and EOF. A particle with a larger aspect ratio has a longer arm of force, which leads to larger torque

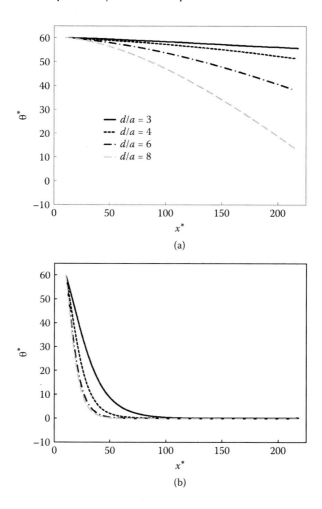

FIGURE 4.10 Orientation variations of a cylindrical particle with different aspect ratios along the centerline of the microchannel. The simulation conditions were $b^* = 22.22$ and $\gamma = 0.2$. The lines in Figure 4.10b are in the same legend as Figure 4.10a. (a) $E^* = 0.0028$; (b) $E^* = 0.28$. (From Ai, Y., A. Beskok, D. T. Gauthier, S. W. Joo, and S. Qian. 2009. DC electrokinetic transport of cylindrical cells in straight microchannels. *Biomicrofluidics* 3:044110 with permission from the American Institute of Physics.)

acting on the particle. Thus, a particle with a larger aspect ratio experiences faster rotation. It is expected that a particle with a larger aspect ratio would experience a more significant oscillatory motion. When the cylindrical particle degrades to a circular particle ($d/a = 2$), the particle does not rotate because it moves along the centerline of the microchannel. Under $E^* = 0.28$, a particle with a larger aspect ratio also aligns to the electric field faster than that with lower aspect ratio, as shown in Figure 4.10b. In general, synthetic nanowires and carbon nanotubes have very large aspect ratios, which could be aligned to the electric field quickly.

4.5.6 EFFECT OF PARTICLE'S INITIAL ANGLE

Figure 4.11 shows the effect of the particle's initial orientation on its transport under a low, $E^* = 0.0028$, and a high, $E^* = 0.28$, electric field when $d/a = 6$, $b^* = 22.22$, and $\gamma = 0.2$. Under $E^* = 0.0028$, the angle of the particle gradually decreases except for the zero initial angle case. Because the channel width is much larger than the

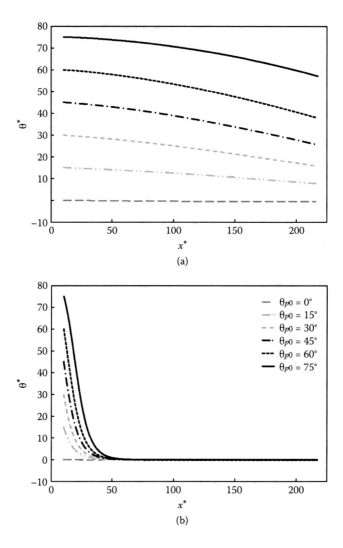

FIGURE 4.11 Orientation variations of a cylindrical particle with different initial angles along the centerline of the microchannel. The simulation conditions were $d/a = 6$, $b^* = 22.22$, and $\gamma = 0.2$. The lines in Figure 4.11a are in the same legend as Figure 4.11b. (a) $E^* = 0.0028$; (b) $E^* = 0.28$. (From Ai, Y., A. Beskok, D. T. Gauthier, S. W. Joo, and S. Qian. 2009. DC electrokinetic transport of cylindrical cells in straight microchannels. *Biomicrofluidics* 3:044110 with permission from the American Institute of Physics.)

particle's aspect ratio, the decrease in the particle orientation should be the beginning of the oscillatory motion. Under $E^* = 0.28$, all the particles are quickly aligned to the external electric field independent of its initial orientation.

4.6 CONCLUDING REMARKS

Electrokinetic transport of cylindrical algal cells in a straight microchannel has been experimentally and numerically studied. A good agreement between the experiments and the numerical simulations was achieved, which demonstrated the capability of the numerical model in predicting the electrokinetic particle transport in microfluidics. Experimental observations revealed that cylindrical algal cells are always aligned with their longest axes parallel to the external electric field when the electric field strength is relatively high, which has been widely used to align and assemble nanowires (Evoy et al. 2004; Makaram et al. 2007; Chang and Hong 2009; Raychaudhuri et al. 2009; Lao et al. 2006; Monica et al. 2008) and biological tissues (Pethig et al. 2008). The numerical modeling proves that the particle alignment phenomenon is attributed to the DEP effect, which must be considered in the modeling of electrokinetic transport of nonspherical particles even in a uniform channel. A higher electric field leads to a faster particle alignment. When the particle's aspect ratio is close to the channel width and the DEP effect is negligible under a low electric field, the cylindrical particle experiences an oscillatory motion.

APPENDIX

Consider a cylindrical-shaped particle shown in Figure 4.2 as a cuboid and two semicylinders with a unit length in the z direction. The moment of inertia of the cuboid shown in Figure A4.1 with respect to the z axis is given as

$$I_z = \frac{1}{12} m_1 \left[(2a)^2 + (d - 2a)^2 \right], \tag{A4.1}$$

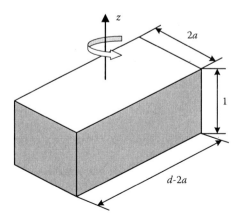

FIGURE A4.1 Moment of inertia of a cuboid.

where $m_1 = \rho(d - 2a)(2a)$ is the mass of the cuboid, with ρ denoting the particle density.

To calculate the moment of inertia of the semicylinder with respect to the z axis, the center of mass of the semicircle shown in Figure A4.2 must be known. An incremental area of mass dm (gray in Figure A4.2) of the semicircle can be expressed as

$$dm = \rho r d\theta dr, \tag{A4.2}$$

in which θ ranges from 0 to π, and r ranges from 0 to R. The mass of the semicylinder with a unit length in the z direction is given as

$$m_2 = \int_0^R \int_0^\pi \rho r d\theta dr = \rho \frac{\pi R^2}{2}. \tag{A4.3}$$

Because the semicircle is symmetric with respect to the y axis (Figure A4.2), the x coordinate of the center of mass is zero. Assume the y coordinate of the center of mass is y_{cm}; the following equation should be satisfied:

$$m_2 \bullet y_{cm} = \int y \bullet dm. \tag{A4.4}$$

Substituting Equation (A4.2) and $y = r \sin \theta$ into Equation (A4.4), Equation (A4.4) can be rewritten as

$$\rho \frac{\pi R^2}{2} \bullet y_{cm} = \int_0^R \int_0^\pi \rho r^2 \sin \theta d\theta dr, \tag{A4.5}$$

from which we can get $y_{cm} = \dfrac{4R}{3\pi}$

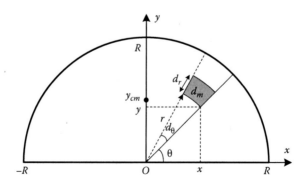

FIGURE A4.2   Schematic view of a semicircle.

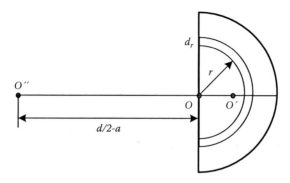

FIGURE A4.3 Parallel axis theorem of a semicircle.

The moment of inertia of the semicylinder with respect to the z axis through point O is

$$I_O = \int_0^a r^2 \, dm = \rho \int_0^a r^2 \pi r \, dr = \frac{1}{2} m_2 a^2, \tag{A4.6}$$

where $m_2 = \frac{1}{2} \rho \pi a^2$ is the mass of the semicylinder. According to the parallel axis theorem, which can determine the moment of inertia of a rigid body with respect to any axis, we can obtain the following equation:

$$I_O = I_{O'} + m_2 (OO')^2, \tag{A4.7}$$

where $I_{O'}$ is the moment of inertia with respect to the axis through the center of mass. From Equation (A4.7), we can obtain

$$I_{O'} = \frac{1}{2} m_2 a^2 - m_2 (\frac{4a}{3\pi})^2. \tag{A4.8}$$

Therefore, the moment of inertia with respect to the axis through point O'' is expressed as

$$I_{O''} = I_{O'} + m_2 (\frac{4a}{3\pi} + \frac{d}{2} - a)^2 = m_2 \left[\frac{a^2}{2} + (\frac{d}{2} - a)^2 + \frac{4a}{3\pi} (d - 2a) \right]. \tag{A4.9}$$

The total moment of inertia of the cylindrical-shaped particle with respect to the z axis through its center of mass is written as

$$I_{total} = \frac{1}{12} m_1 \left[(2a)^2 + (d - 2a)^2 \right] + 2m_2 \left[\frac{a^2}{2} + (\frac{d}{2} - a)^2 + \frac{4a}{3\pi} (d - 2a) \right]. \tag{A4.10}$$

REFERENCES

Ai, Y., A. Beskok, D. T. Gauthier, S. W. Joo, and S. Qian. 2009. DC electrokinetic transport of cylindrical cells in straight microchannels. *Biomicrofluidics* 3:044110.

Chang, Y. K., and F. C. N. Hong. 2009. The fabrication of ZnO nanowire field-effect transistors combining dielectrophoresis and hot-pressing. *Nanotechnology* 20 (23):235202.

Davison, S. M., and K. V. Sharp. 2006. Boundary effects on the electrophoretic motion of cylindrical particles: Concentrically and eccentrically-positioned particles in a capillary. *Journal of Colloid and Interface Science* 303 (1):288–297.

Davison, S. M., and K. V. Sharp. 2007. Transient electrophoretic motion of cylindrical particles in capillaries. *Nanoscale and Microscale Thermophysical Engineering* 11 (1–2):71–83.

Davison, S. M., and K. V. Sharp. 2008. Transient simulations of the electrophoretic motion of a cylindrical particle through a 90 degrees corner. *Microfluidics and Nanofluidics* 4 (5):409–418.

Duffy, D. C., J. C. McDonald, O. J. A. Schueller, and G. M. Whitesides. 1998. Rapid prototyping of microfluidic systems in poly(dimethylsiloxane). *Analytical Chemistry* 70 (23):4974–4984.

Evoy, S., N. DiLello, V. Deshpande, A. Narayanan, H. Liu, M. Riegelman, B. R. Martin, B. Hailer, J. C. Bradley, W. Weiss, T. S. Mayer, Y. Gogotsi, H. H. Bau, T. E. Mallouk, and S. Raman. 2004. Dielectrophoretic assembly and integration of nanowire devices with functional CMOS operating circuitry. *Microelectronic Engineering* 75 (1):31–42.

Hsu, J. P., Z. S. Chen, D. J. Lee, S. Tseng, and A. Su. 2008. Effects of double-layer polarization and electroosmotic flow on the electrophoresis of a finite cylinder along the axis of a cylindrical pore. *Chemical Engineering Science* 63 (18):4561–4569.

Hsu, J. P., and C. C. Kuo. 2006. Electrophoresis of a finite cylinder positioned eccentrically along the axis of a long cylindrical pore. *Journal of Physical Chemistry B* 110 (35):17607–17615.

Kang, K. H., X. C. Xuan, Y. Kang, and D. Li. 2006. Effects of dc-dielectrophoretic force on particle trajectories in microchannels. *Journal of Applied Physics* 99 (6):064702.

Kang, Y. J., and D. Q. Li. 2009. Electrokinetic motion of particles and cells in microchannels. *Microfluidics and Nanofluidics* 6 (4):431–460.

Kang, Y. J., D. Q. Li, S. A. Kalams, and J. E. Eid. 2008. DC-dielectrophoretic separation of biological cells by size. *Biomedical Microdevices* 10 (2):243–249.

Keh, H. J., and J. L. Anderson. 1985. Boundary effects on electrophoretic motion of colloidal spheres. *Journal of Fluid Mechanics* 153:417–439.

Lao, C. S., J. Liu, P. X. Gao, L. Y. Zhang, D. Davidovic, R. Tummala, and Z. L. Wang. 2006. ZnO nanobelt/nanowire Schottky diodes formed by dielectrophoresis alignment across Au electrodes. *Nano Letters* 6 (2):263–266.

Li, D. 2004. *Electrokinetics in Microfluidics*. New York: Elsevier Academic Press.

Liu, H., H. H. Bau, and H. H. Hu. 2004. Electrophoresis of concentrically and eccentrically positioned cylindrical particles in a long tube. *Langmuir* 20 (7):2628–2639.

Liu, H., S. Z. Qian, and H. H. Bau. 2007. The effect of translocating cylindrical particles on the ionic current through a nanopore. *Biophysical Journal* 92 (4):1164–1177.

Makaram, P., S. Selvarasah, X. G. Xiong, C. L. Chen, A. Busnaina, N. Khanduja, and M. R. Dokmeci. 2007. Three-dimensional assembly of single-walled carbon nanotube interconnects using dielectrophoresis. *Nanotechnology* 18 (39):395204.

Monica, A. H., S. J. Papadakis, R. Osiander, and M. Paranjape. 2008. Wafer-level assembly of carbon nanotube networks using dielectrophoresis. *Nanotechnology* 19 (8):085303.

Pethig, R., A. Menachery, E. Heart, R. H. Sanger, and P. J. S. Smith. 2008. Dielectrophoretic assembly of insulinoma cells and fluorescent nanosensors into three-dimensional pseudo-islet constructs. *IET Nanobiotechnology* 2 (2):31–38.

Raychaudhuri, S., S. A. Dayeh, D. L. Wang, and E. T. Yu. 2009. Precise semiconductor nanowire placement through dielectrophoresis. *Nano Letters* 9 (6):2260–2266.

Unni, H. N., H. J. Keh, and C. Yang. 2007. Analysis of electrokinetic transport of a spherical particle in a microchannel. *Electrophoresis* 28 (4):658–664.

Venditti, R., X. C. Xuan, and D. Q. Li. 2006. Experimental characterization of the temperature dependence of zeta potential and its effect on electroosmotic flow velocity in microchannels. *Microfluidics and Nanofluidics* 2 (6):493–499.

Winter, W. T., and M. E. Welland. 2009. Dielectrophoresis of non-spherical particles. *Journal of Physics D–Applied Physics* 42 (4):045501.

Xuan, X. C., R. Raghibizadeh, and D. Li. 2005. Wall effects on electrophoretic motion of spherical polystyrene particles in a rectangular poly(dimethylsiloxane) microchannel. *Journal of Colloid and Interface Science* 296:743–748.

Xuan, X. C., B. Xu, and D. Q. Li. 2005. Accelerated particle electrophoretic motion and separation in converging-diverging microchannels. *Analytical Chemistry* 77 (14):4323–4328.

Xuan, X. C., C. Z. Ye, and D. Q. Li. 2005. Near-wall electrophoretic motion of spherical particles in cylindrical capillaries. *Journal of Colloid and Interface Science* 289 (1):286–290.

Ye, C. Z., and D. Q. Li. 2004. 3-D transient electrophoretic motion of a spherical particle in a T-shaped rectangular microchannel. *Journal of Colloid and Interface Science* 272 (2):480–488.

Ye, C. Z., D. Sinton, D. Erickson, and D. Q. Li. 2002. Electrophoretic motion of a circular cylindrical particle in a circular cylindrical microchannel. *Langmuir* 18 (23):9095–9101.

Ye, C. Z., X. C. Xuan, and D. Q. Li. 2005. Eccentric electrophoretic motion of a sphere in circular cylindrical microchannels. *Microfluidics and Nanofluidics* 1 (3):234–241.

Zhu, J., T.-R. Tzeng, G. Hu, and X. Xuan. 2009. DC dielectrophoretic focusing of particles in a serpentine microchannel. *Microfluidics and Nanofluidics* 7 (6):751–756.

Zhu, J., and X. Xuan. 2009. Dielectrophoretic focusing of particles in a microchannel constriction using DC-biased AC electric fields. *Electrophoresis* 30 (15):2668–2675.

5 Shear- and Electrokinetics-Induced Particle Deformation in a Slit Channel

Numerous biological particles are flexible and dramatically deform in microfluidics, which offers a potential approach to interrogate the structural properties of biological particles for disease diagnosis. In this chapter, the shear- and electrokinetics-induced deformations of hyperelastic particles confined in a slit microchannel are numerically investigated using COMSOL Multiphysics® 3.5a. In a shear-driven flow, the circular particle initially located at the centerline of the channel is deformed as a perfect ellipse when the inertial effect is negligible. Under a direct current (DC) electric field directed from right to left, a negatively charged rod-like particle deforms as a C shape as it electrophoretically moves from left to right when it is initially perpendicular to the applied electric field. The shear force due to the nonuniform Smoluchowski slip velocity on the particle surface is responsible for particle deformation. In addition, the dielectrophoretic (DEP) effect could enhance the deformation as the electric field around the particle becomes asymmetric with respect to the center of the particle. When the particle is not perpendicular to the electric field imposed, a net torque stemming from the DEP effect rotates and aligns the particle with its longest axis parallel to the applied electric field, which is in qualitative agreement with the experimental observation presented in Chapter 4. As the nonuniformity of the electric field around the particle becomes weaker when the particle is aligned, the deformation is accordingly released. The numerical predictions conclude that the DEP effect is of great importance and must be considered in the modeling of electrokinetic motion of a deformable particle in microfluidics.

5.1 INTRODUCTION

Numerical modeling of electrokinetic transport of rigid particles in microfluidics has been demonstrated in Chapters 3 and 4. However, many biological particles such as red blood cells (RBCs) and vesicles are flexible and can significantly deform under a stress. Study of particle deformation in microfluidics is helpful in understanding the structural properties of complicated biological entities. It has been found that the health status of RBCs can be interrogated based on their deformability, which implies a potential disease diagnosis technique (Dondorp et al. 1999). The variation of deformability between healthy and unhealthy RBCs in a microchannel has been experimentally observed (Abkarian, Faivre, and Stone 2006; Abkarian et al. 2008). Comprehensive studies of the deformation of RBCs in a microchannel subjected to pressure-driven flows have been conducted (Korin, Bransky, and Dinnar 2007; Tomaiuolo et al. 2009; Tomaiuolo et al. 2011). A lab-on-a-chip device with a capillary network has been developed and fabricated for RBC deformability diagnosis in clinical applications (Chen et al. 2010).

Besides the increasing experimental studies of particle deformation in microfluidics, numerical modeling is necessary to gain insight into the deformation of soft biological particles. This topic is a typical fluid-structure interaction (FSI) problem, in which the interactive effect on both the fluid flow and solid deformation should be taken into account simultaneously for accurate prediction. The arbitrary Lagrangian–Eulerian (ALE) algorithm is one of the most accurate approaches to study the FSI problem (Hu, Joseph, and Crochet 1992; Hu, Patankar, and Zhu 2001). Most existing numerical studies focused on the particle deformation subjected to pressure-driven or shear-driven flows (Eggleton and Popel 1998; Secomb, Styp-Rekowska, and Pries 2007; Doddi and Bagchi 2009; Gao and Hu 2009; MacMeccan et al. 2009; Sugiyama et al. 2011). However, numerical studies of particle deformation due to electrokinetic effects are rare.

A recent experimental study revealed that electrokinetics-driven semiflexible rod-like microtubules deform into a U shape when they are perpendicular to the applied electric field (van den Heuvel et al. 2008). The electrokinetics-induced deformation of a long elastic particle suspended in an unbounded medium was numerically studied (Swaminathan, Gao, and Hu 2010). However, the boundary effect and the DEP effect were neglected in the study. As concluded in Chapters 3 and 4, the DEP effect plays an important role in the electrokinetic particle motion in microfluidics and must be taken into account in the numerical modeling.

In this chapter, the shear-induced and electrokinetics-induced particle deformations are both numerically studied using COMSOL Multiphysics 3.5a. The first numerical model solves the flow field and structural deformation using the ALE algorithm; while the latter simultaneously solves the electric field together with the two physical fields mentioned. The shear-induced particle deformation has been studied by Gao and Hu (2009) using their own ALE code. Therefore, we mainly focus on the implementation of this model in COMSOL Multiphysics 3.5a without detailed discussion. In the study of the electrokinetics-induced particle deformation, we mainly focus on the boundary and DEP effects on the particle deformation, which were neglected in the previous study by Swaminathan, Gao, and Hu (2010).

5.2 SHEAR-INDUCED PARTICLE DEFORMATION

5.2.1 MATHEMATICAL MODEL

We consider the deformation of a hyperelastic particle Ω_p subjected to a shear-driven flow between two parallel plates. The fluid domain Ω_f is schematically

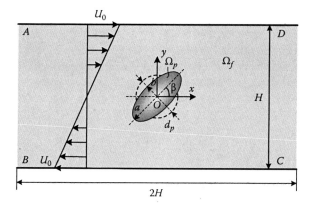

FIGURE 5.1 Schematics of a deformable particle subjected to a shear-driven flow between two parallel plates.

shown in Figure 5.1. The upper and lower plates are moving in opposite directions with an identical velocity U_0. The circular particle of diameter d_p is initially located at the centerline of the channel filled with an incompressible Newtonian viscous fluid. The height and length of the channel are, respectively, H and $2H$. The flow field inside the computation domain is governed by the Navier–Stokes equations:

$$\rho_f \frac{\partial \mathbf{u}}{\partial t} + \rho_f (\mathbf{u} \bullet \nabla)\mathbf{u} = -\nabla p + \mu \nabla^2 \mathbf{u} \qquad \text{in } \Omega_f, \tag{5.1}$$

$$\nabla \bullet \mathbf{u} = 0 \qquad \text{in } \Omega_f. \tag{5.2}$$

In these equations, ρ_f is the fluid density; $\mathbf{u}$ is the flow velocity; p is the pressure; and μ is the dynamic viscosity of the fluid. The initial flow velocity and the initial pressure are both zero. Normal flow with pressure $p = 0$ is imposed at the inlet (line segment AB) and outlet (line segment CD). The fluid velocity on the particle surface is $\mathbf{u} = \partial w / \partial t$, in which w is the displacement of the deformable particle governed by

$$\rho_p \frac{\partial^2 \mathbf{w}}{\partial t^2} - \nabla \bullet (\sigma(\mathbf{w})) = 0 \qquad \text{in } \Omega_p. \tag{5.3}$$

Here, ρ_p is the density of the deformable particle, and $\sigma(\mathbf{w})$ is the Cauchy stress as a function of the displacement. Consider the hyperelastic particle as an incompressible neo-Hookean material that could be mathematically described using the strain energy density function (Sugiyama et al. 2011):

$$W = \frac{G_0}{2}(I_C - 3). \tag{5.4}$$

In this equation, G_0 is the shear modulus, and $I_C = tr(\mathbf{C})$ is the first invariant of the right Cauchy–Green tensor, $\mathbf{C} = \mathbf{F}^T\mathbf{F}$, where $\mathbf{F} = \nabla_{\mathbf{X}}\mathbf{w} + \mathbf{I}$ is the deformation gradient tensor with $\mathbf{X}$ denoting the reference position. The corresponding Cauchy stress of the neo-Hookean material is expressed as

$$\sigma(\mathbf{w}) = J^{-1}\mathbf{PF}^T, \tag{5.5}$$

where J is the determinant of the deformation gradient tensor $\mathbf{F}$, and $\mathbf{P} = \dfrac{\partial W}{\partial \nabla_{\mathbf{X}}\mathbf{w}}$ is the first Piola–Kirchhoff stress.

The traction force on the particle-fluid interface is continuous, written as

$$\sigma(\mathbf{w}) \bullet \mathbf{n}_p = \sigma_f \bullet \mathbf{n}_f, \tag{5.6}$$

where $\mathbf{n}_p$ and $\mathbf{n}_f$ are, respectively, the unit normal vector directed from the particle surface into the fluid in the reference frame and the spatial frame; $\sigma_f = -p\mathbf{I} + \mu(\nabla\mathbf{u} + \nabla\mathbf{u}^T)$ is the hydrodynamic stress tensor. Note that the surface tension at the interface is neglected in the current study. The initial displacement of the solid phase is zero.

5.2.2 NUMERICAL IMPLEMENTATION

In this section, we demonstrate how to set up the mathematical model in COMSOL Multiphysics 3.5a and solve it using the ALE method. First, open COMSOL Multiphysics 3.5a to obtain the Model Navigator, as shown in Figure 5.2. To solve this problem, the structural mechanics module must be installed in COMSOL Multiphysics 3.5a. The present two-dimensional (2D) model assumes the z direction is much larger than the x and y directions, which refers to a plane strain problem. The plane strain for hyperelastic material is located at **Structural Mechanics Module|Plane Strain|Viscoelastic transient initialization**, which is defined in the reference frame and first added in the multiphysics model. When adding **COMSOL Multiphysics|Deformed Mesh|Moving Mesh (ALE)|Transient analysis**, the software will automatically generate a spatial frame, called ale. Finally, **COMSOL Multiphysics|Incompressible Navier–Stokes|Transient analysis** is added to the spatial frame, as shown in Figure 5.2. Draw a rectangle with a dimension 160×80 μm and locate its center at $(0, 0)$. Later, draw a circle at the center of the rectangle. The radius of the circle is 10 μm. The computational domain and the corresponding mesh are shown in Figure 5.3. Define the constants listed in Table 5.1 in COMSOL Multiphysics 3.5a.

The subdomain settings for the plane strain model, called smpn, are shown in Figure 5.4. Subdomain 1 is the fluid domain, which is thus named fluid and deactivated in the smpn model. Subdomain 2 is the particle domain and is named solid in the subdomain setting. Shear modulus and bulk modulus characterize, respectively, the stiffness and the compressibilty of materials. Considering the particle as an incompressible neo-Hookean material, the bulk modulus in

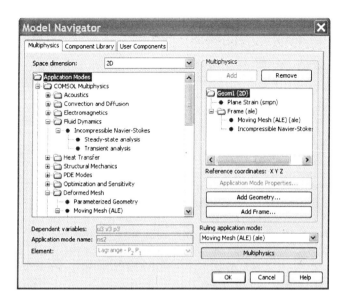

FIGURE 5.2   Setup in model navigator. (Image courtesy of COMSOL, Inc.)

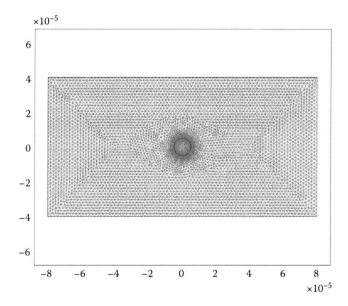

FIGURE 5.3   Computational domain and mesh.

TABLE 5.1

Constant Table

Variable	Value or Expression	Description
U0	4*eta*Re/rho_f/dp	Wall velocity
shear	2*U0/H	Shear rate
H	8*dp	Channel height
rho_s	1e3 [kg/m^3]	Particle density
eta	1e-3 [Pa*s]	Fluid viscosity
rho_f	1e3 [kg/m^3]	Fluid density
s_modulus	2*U0*eta/H/Ca	Shear modulus
b_modulus	10000 [Pa]	Bulk modulus
dp	10e-6 [m]	Particle diameter
Re	0.125	Reynolds number
Ca	0.3	Capillary number

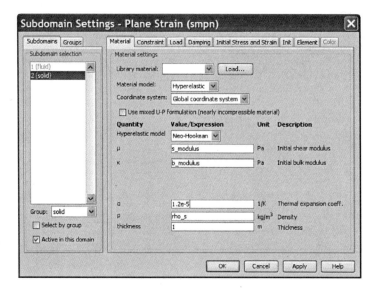

FIGURE 5.4   Subdomain settings for the smpn module. (Image courtesy of COMSOL, Inc.)

the present study is much higher than the shear modulus to minimize the volume change. Since temperature variation is not taken into account, the thermal expansion coefficiency is not important and remains at the default.

Figure 5.5 shows the detailed boundary settings for the smpn module. Segments ABCD, confining the fluids in the computational domain, are deactivated in the boundary settings for the smpn module. The constraint of the particle surface is set to be free. The type of load on the particle surface is distributed load. The x and y component loads on the particle surface are, respectively, −lm5*dvol_ale/

dvol and −lm6*dvol_ale/dvol, where lm5 and lm6 are, respectively, the x and y component hydrodynamic stress predefined in COMSOL for a highly accurate calculation. In FSI problems, the hydrodynamic force arising from the fluid flow is defined in the spatial frame; the structural mechanical force is defined in the reference frame. The expression dvol_ale/dvol basically transfers the stress defined in the spatial frame to the reference frame.

Figure 5.6 shows the subdomain settings for the ale module. The fluid subdomain is free displacement; the solid subdomain is set to be physics-induced displacement, whose x and y component displacements are, respectively, u and v.

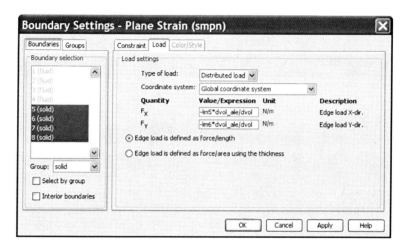

FIGURE 5.5 Boundary settings on the particle surface for the smpn module. (Image courtesy of COMSOL, Inc.)

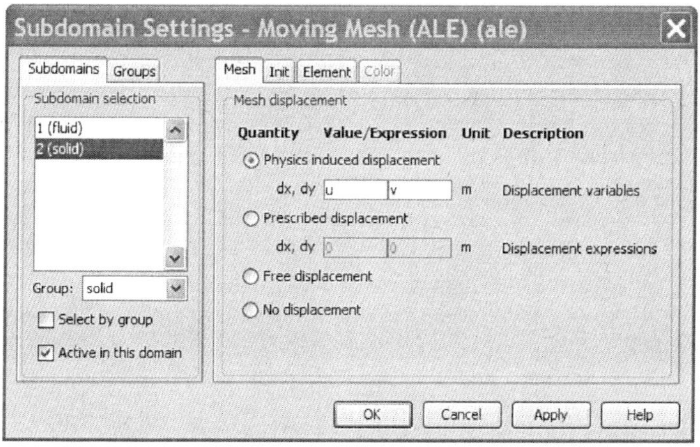

FIGURE 5.6 Subdomain settings for the ale module. (Image courtesy of COMSOL, Inc.)

The x and y component displacements on the particle surface are, respectively, u and v as well, which is shown in Figure 5.7. The x and y component displacements on the other boundaries are zero, as shown in Figure 5.8. The remeshing option for long-term particle tracking is turned on.

Figure 5.9 shows the subdomain settings for the ns module, in which the solid subdomain is deactivated. The fluid velocity at the particle surface equals the velocity due to the particle deformation, as shown in Figure 5.10. The fluid velocities at the upper and the lower boundaries are, respectively, U0 and −U0. The pressures at the inlet and outlet are both zero. The weak constraints lm5 and lm6 are turned on to obtain a more accurate calculation of the hydrodynamic force. All these settings are summarized in Table 5.2.

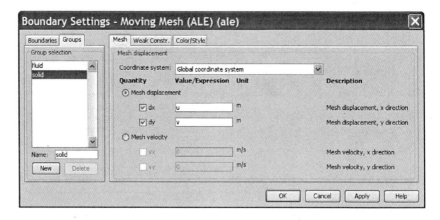

FIGURE 5.7   Boundary settings on the particle surface for the ale module. (Image courtesy of COMSOL, Inc.)

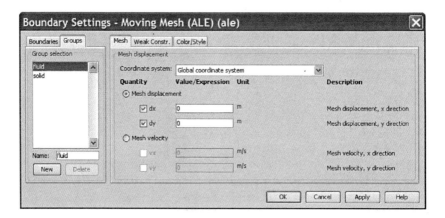

FIGURE 5.8   Boundary settings at outer boundaries for the ale module. (Image courtesy of COMSOL, Inc.)

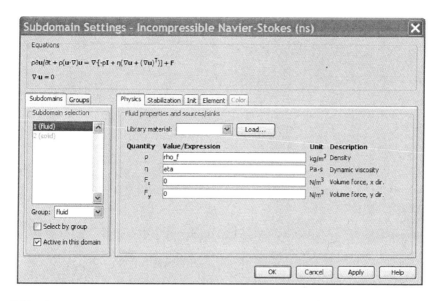

FIGURE 5.9   Subdomain settings for the ns module. (Image courtesy of COMSOL, Inc.)

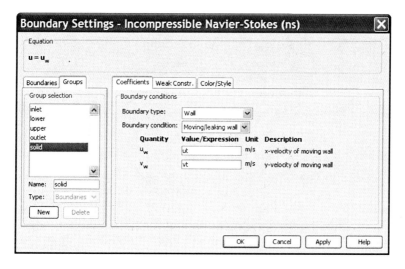

FIGURE 5.10   Boundary settings on the particle surface for the ns module. (Image courtesy of COMSOL, Inc.)

TABLE 5.2
Model Setup in the GUI of COMSOL Multiphysics 3.5a

Model Navigator	Select **2D** in space dimension and click **Multiphysics** button.
	Select **Structural Mechanics Module|Plane Strain|Viscoelastic transient initialization**. Click **Add** button.
	Select **COMSOL Multiphysics|Deformed Mesh|Moving Mesh (ALE)|Transient analysis**. Click **Add** button.
	Select **COMSOL Multiphysics|Incompressible Navier–Stokes| Transient analysis**. Remove the predefined variables in the **Dependent variables** and enter u2 v2 p2. Click **Add** button.
Option Menu| **Constants**	Define variables in Table 5.1
Physics Menu| **Subdomain Setting**	smpn mode
	Deactivate Subdomain 1 (Fluid)
	Subdomain 2 (Solid)
	Material model: Hyperelastic Hyperelastic model: Neo-hookean
	μ = s_modulus; κ = b_modulus; α = 1.2e-5; ρ = rho_s; thickness = 1
	ale model
	Subdomain 1
	Free displacement
	Subdomain 2
	Physics induced displacement
	dx = u; dy = v
	ns mode
	Subdomain 1
	ρ = rho_f; η = eta; Fx = 0; Fy = 0
	Deactivate Subdomain 2
Physics Menu| **Boundary Setting**	smpn mode
	Particle surface
	Tab Constraint
	Constrain condition: Free
	Tab Load
	Type of load: Distributed load
	Fx = −lm5*dvol_ale/dvol; Fy = −lm6*dvol_ale/dvol
	Other boundaries are inactive
	ale mode
	Particle surface
	dx = u; dy = v
	Other boundaries
	dx = 0; dy = 0
	ns mode
	Left boundary: Inlet|Pressure, no viscous stress P_0 = 0
	Right boundary: Outlet| Pressure, no viscous stress P_0 = 0

TABLE 5.2 (CONTINUED)
Model Setup in the GUI of COMSOL Multiphysics 3.5a

	Surface of particle I: Wall| Moving/leaking wall Uw = ut; Vw = vt
	Upper boundary: Wall| Moving/leaking wall Uw = U0; Vw = 0
	Lower boundary: Wall| Moving/leaking wall Uw = −U0; Vw = 0
Other Settings	Properties of ale model (**Physics Menu|Properties**)
	Smoothing method: Winslow
	Allow remeshing: On
	Properties of ns model (**Physics Menu|Properties**)
	Weak constraints: On
	Constraint type: Non-ideal
Mesh|Free Mesh	Tab subdomain
Parameters	Subdomain 1|Maximum element size: 2e-6
	Subdomain 2|Maximum element size: 1e-6
	Tab boundary
	Particle surface: 5, 6, 7, 8|**Maximum element size**: 0.2e-6
Solve Menu|	**Solver Parameters**
	Tab General
	Times: Range(0, 5e-5, 1e-2)
	Relative tolerance: 1e-3
	Absolute tolerance: 1e-4
	Tab Timing Stepping
	Activate Use stop condition
	minqual1_ale-0.6
	Click = **Solve Problem**
Postprocessing Menu|	Result check: Particle shape
	Domain Plot Parameters|Line/Extrusion
	Boundary select: 5, 6, 7, 8
	y-axis data: y; x-axis data: x
	Click Apply

Once the graphical user interface (GUI) file (.mph file) is generated, it could be saved as an M-file for further modifications. The revised M-file capable of automated remeshing is as follows:

```
%%%%%%%%%%%%%%%Shear-induced Particle
  Deformation.m%%%%%%%%%%%%%%%%%%%%

% COMSOL Multiphysics Model M-file
% Generated by COMSOL 3.5a (COMSOL 3.5.0.603, $Date:
  2008/12/03 17:02:19 $)

flclear fem

% COMSOL version
```

```
clear vrsn
vrsn.name = 'COMSOL 3.5';
vrsn.ext = 'a';
vrsn.major = 0;
vrsn.build = 603;
vrsn.rcs = '$Name: $';
vrsn.date = '$Date: 2008/12/03 17:02:19 $';
fem.version = vrsn;

% NEW: Define Ca number
Ca_s=[0.02,0.08,0.2,0.5,0.3,0.4,0.65];

% NEW: FOR loop to go through all the Ca number
for i=1:length(Ca_s)

% NEW: num2str can transfer the defined Ca number to the
  constant in COMSOL
% Constants
fem.const = {'U0','4*eta*Re/rho_f/dp', ...
             'shear','2*U0/H', ...
             'H','8*dp', ...
             'rho_s','1e3[kg/m^3]', ...
             'eta','1e-3[Pa*s]', ...
             'rho_f','1e3[kg/m^3]', ...
             's_modulus','2*U0*eta/H/Ca', ...
             'b_modulus','10000[Pa]', ...
             'dp','10[um]', ...
             'Re','0.125', ...
             'Ca',num2str(Ca_s(i))};

% Geometry
g1=rect2('160e-6','80e-6','base','center','pos',{'0','0'},
  'rot','0');
g3=circ2('5E-6','base','center','pos',{'0','0'},'rot','0');

% Geometry objects
clear s
s.objs={g1,g3};
s.name={'R1','C1'};
s.tags={'g1','g3'};

fem.draw=struct('s',s);
fem.geom=geomcsg(fem);

% (Default values are not included)

% NEW: Solve plane strain of neo-hookean material
% Application mode 1
clear appl
appl.mode.class = 'SmePlaneStrain';
```

```
appl.sdim = {'X','Y','Z'};
appl.module = 'SME';
appl.gporder = 4;
appl.cporder = 2;
appl.assignsuffix = '_smpn';
clear prop
prop.analysis='time';
prop.deformframe='ref';
prop.frame='ref';
appl.prop = prop;
clear bnd
bnd.loadtype = {'area','length'};
bnd.name = {'fluid','solid'};
bnd.Fx = {0,'-lm5*dvol_ale/dvol'};
bnd.Fy = {0,'-lm6*dvol_ale/dvol'};
bnd.ind = [1,1,1,1,2,2,2,2];
appl.bnd = bnd;
clear equ
equ.kappa = {1e10,'b_modulus'};
equ.rho = {7850,'rho_s'};
equ.mu = {8e5,'s_modulus'};
equ.materialmodel = {'iso','hyper'};
equ.name = {'fluid','solid'};
equ.usage = {0,1};
equ.ind = [1,2];
appl.equ = equ;
fem.appl{1} = appl;

% NEW: Solve moving mesh
% Application mode 2
clear appl
appl.mode.class = 'MovingMesh';
appl.sdim = {'Xm','Ym','Zm'};
appl.shape = {'shlag(2,''lm3'')','shlag(2,''lm4'')','shlag
   (2,''x'')','shlag(2,''y'')'};
appl.gporder = {30,4};
appl.cporder = 2;
appl.assignsuffix = '_ale';
clear prop
prop.smoothing='winslow';
prop.analysis='transient';
prop.allowremesh='on';
prop.origrefframe='ref';
clear weakconstr
weakconstr.value = 'on';
weakconstr.dim = {'lm3','lm4'};
prop.weakconstr = weakconstr;
appl.prop = prop;
clear bnd
bnd.defflag = {{1;1}};
```

```
bnd.wcshape = [1;2];
bnd.name = {'fluid','solid'};
bnd.deform = {{0;0},{'u';'v'}};
bnd.ind = [1,1,1,1,2,2,2,2];
appl.bnd = bnd;
clear equ
equ.gporder = 2;
equ.physexpr = {{0;0},{'u';'v'}};
equ.name = {'fluid','solid'};
equ.shape = [3;4];
equ.type = {'free','phys'};
equ.ind = [1,2];
appl.equ = equ;
fem.appl{2} = appl;


% NEW: Solve flow field using Navier-Stokes equations
% Application mode 3
clear appl
appl.mode.class = 'FlNavierStokes';
appl.dim = {'u2','v2','p2','nxw','nyw'};
appl.shape = {'shlag(2,''lm5'')','shlag(2,''lm6'')',
  'shlag(1,''lm7'')','shlag(2,''u2'')','shlag(2,''v2'')',
  'shlag(1,''p2'')'};
appl.gporder = {30,4,2};
appl.cporder = {2,1};
appl.assignsuffix = '_ns';
clear prop
clear weakconstr
weakconstr.value = 'on';
weakconstr.dim = {'lm5','lm6','lm7'};
prop.weakconstr = weakconstr;
prop.constrtype='non-ideal';
appl.prop = prop;
clear bnd
bnd.type = {'inlet','walltype','walltype','inlet',
  'walltype'};
bnd.walltype = {'noslip','lwall','lwall','noslip','lwall'};
bnd.uwall = {0,'-U0','U0',0,'ut'};
bnd.vwall = {0,0,0,0,'vt'};
bnd.name = {'inlet','lower','upper','outlet','solid'};
bnd.outtype = {'p','p','p','uv','p'};
bnd.u0 = {0,-0.2,0.2,0,0};
bnd.wcshape = [1;2;3];
bnd.velType = {'U0in','u0','u0','U0in','U0in'};
bnd.U0in = {0,1,1,0,1};
bnd.intype = {'p','uv','uv','p','uv'};
bnd.ind = [1,2,3,4,5,5,5,5];
appl.bnd = bnd;
clear equ
equ.eta = {'eta',1};
```

```
equ.gporder = {{2;2;3}};
equ.name = {'fluid','solid'};
equ.rho = {'rho_f',1};
equ.cporder = {{1;1;2}};
equ.shape = [4;5;6];
equ.cdon = {0,1};
equ.sdon = {0,1};
equ.usage = {1,0};
equ.ind = [1,2];
appl.equ = equ;
fem.appl{3} = appl;
fem.sdim = {{'Xm','Ym'},{'X','Y'},{'x','y'}};
fem.frame = {'mesh','ref','ale'};
fem.border = 1;
clear units;
units.basesystem = 'SI';
fem.units = units;

% Multiphysics
fem=multiphysics(fem);

% NEW: Loop for continuous particle tracking
% NEW: Remeshing index
j=1;

% NEW: Define initial time
time_e=0;

% NEW: Define time_step
time_step=0.5e-4;

% NEW: Define end condition of the computation (Loop
  control)
% NEW: Usually two calculations are enough to get the
  equilibrium particle shape
while j<3

% Initialize mesh
fem.mesh=meshinit(fem, ...
                  'hauto',5, ...
                  'hmaxedg',[5,0.2e-6,6,0.2e-6,7,0.2e-6,8,
                  0.2e-6], ...
                  'hmaxsub',[1,2e-6,2,1e-6]);

% Extend mesh
fem.xmesh=meshextend(fem);

if j==1

% Solve problem
fem.sol=femtime(fem, ...
```

```
'solcomp',{'lm4','v','lm3','u','u2','lm7','p2','lm6','lm5',
  'v2','y','x'}, ...
'outcomp',{'lm4','lm3','u2','p2','lm7','lm6','v2','lm5','v',
  'u','Y','X','y','x'}, ...
            'blocksize','auto', ...
            'tlist',[colon(time_e,time_step,0.01) ], ...
            'rtol',1e-3, ...
            'tout','tlist', ...
            'atol',1e-4, ...
            'maxorder',2, ...
            'stopcond','minqual1_ale-0.5', ...
            'uscale','none');

else

% NEW: Map solution in the deformed mesh into the new geom-
  etry with undeformed mesh
init = asseminit(fem,'init',fem0.sol,'xmesh',fem0.xmesh,'blo
  cksize','auto','framesrc','ale','domwise','on');

% Solve problem
fem.sol=femtime(fem, ...
                'init',init, ...
'solcomp',{'lm4','v','lm3','u','u2','lm7','p2','lm6','lm5',
  'v2','y','x'}, ...
'outcomp',{'lm4','lm3','u2','p2','lm7','lm6','v2','lm5','v',
  'u','Y','X','y','x'}, ...
            'blocksize','auto', ...
            'tlist',[colon(time_e,time_step,0.1) ], ...
            'rtol',1e-3, ...
            'tout','tlist', ...
            'atol',1e-4, ...
            'maxorder',2, ...
            'stopcond','minqual1_ale-0.5', ...
            'uscale','none');
end

% Save current fem structure for restart purposes
fem0=fem;

% NEW: Get the last time from solution
time_e = fem.sol.tlist(end);

% NEW: Output the COMSOL Multiphysics GUI .mph file if
  necessary
flsave(strcat('Shear_driven_hu_Ca=',num2str(Ca_s(i)),'_j=',
  num2str(j),'.mph'),fem);

j=j+1;

% Generate geom from mesh
```

```
fem = mesh2geom(fem, ...
                'frame','ale', ...
                'srcdata','deformed', ...
                'destfield',{'geom','mesh'}, ...
                'srcfem',1, ...
                'destfem',1);
end

end
```

5.2.3 RESULTS AND DISCUSSION

The particle deformation is determined by two dimensionless parameters: the Reynolds number $Re = \rho_f \gamma d_p^2 / \mu$ and the capillary number $Ca = \gamma \mu / G_0$, defined as the ratio of the viscous force to the elastic force. Here, $\gamma = 2U_0/H$ is the shear rate. When the Reynolds number is very low, the inertial effect on the particle deformation is negligible, and the deformed particle can be fitted into a perfect ellipse (Gao and Hu 2009):

$$\frac{(x\cos\beta + y\sin\beta)^2}{a^2} + \frac{(-x\sin\beta + y\cos\beta)^2}{b^2} = 1, \tag{5.7}$$

where β is the angle between the x axis and the major axis of the particle; a and b are, respectively, the semimajor and semiminor axis. Figure 5.11 shows that the deformed particle can be fitted to an ellipse described by Equation (5.7). A larger Ca number implies a smaller shear modulus, which in turn leads to more significant deformation. A new parameter $D = (a - b)/(a + b)$ is defined to quantify the stretch ratio of the particle. Figure 5.12 shows that our numerical results agree with the results obtained by Gao and Hu (2009) using their own ALE code.

We further simulate the deformable particle–particle interaction in a shear flow, as shown in Figure 5.13, which has also been studied by Gao and Hu (2009). The center-to-center distance is $2g_0 = H$. The channel height is four times the particle radius. Figure 5.14 shows the locations and shapes of the two particles under different time steps corresponding to Re = 0.1 and Ca = 0.25. Initially, the shapes of the two particles are exactly circular. Due to the shear-driven flow, the two particles are moving toward each other, as shown in Figures 5.14a–5.14d. When the two particles encounter each other, they experience "rollover" behavior, as shown in Figures 5.14e and 5.14f. Finally, the two particles continue moving in the opposite directions, as shown in Figures 5.14g and 5.14h. The numerical predictions are similar to Gao and Hu's results (2009), which are thus not discussed in detail.

5.3 ELECTROKINETIC-INDUCED PARTICLE DEFORMATION

5.3.1 MATHEMATICAL MODEL

Figure 5.15 schematically shows a 2D incompressible hyperelastic particle Ω_p confined between two parallel plates filled with an incompressible Newtonian

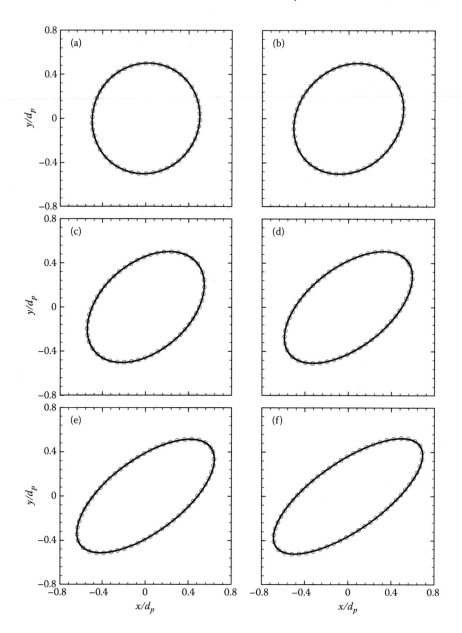

FIGURE 5.11 Particle shape at steady state when Re = 0.125: (a) $Ca = 0.02$; (b) $Ca = 0.08$; (c) $Ca = 0.2$; (d) $Ca = 0.3$; (e) $Ca = 0.4$; (f) $Ca = 0.5$. Solid line and circles represent, respectively, our numerical results and the fitted ellipses given by Gao and Hu (2009).

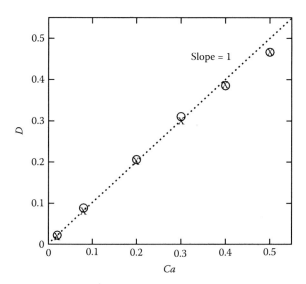

FIGURE 5.12 Stretch ratio $D = (a - b)/(a + b)$ as a function of Ca. Circles and crosses represent, respectively, our numerical results and the results obtained by Gao and Hu (2009). (From Ai, Y., M. Benjamin, A. Sharma, and S. Qian S. 2010. Electrokinetic motion of a deformable particle: Dielectrophoretic effect. *Electrophoresis* 32:2282–2291 with permission from Wiley-VCH.)

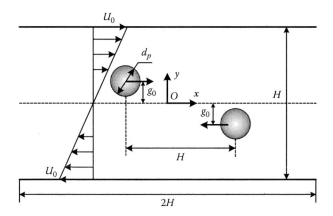

FIGURE 5.13 Schematics of deformable particle–particle interaction in a shear-driven flow.

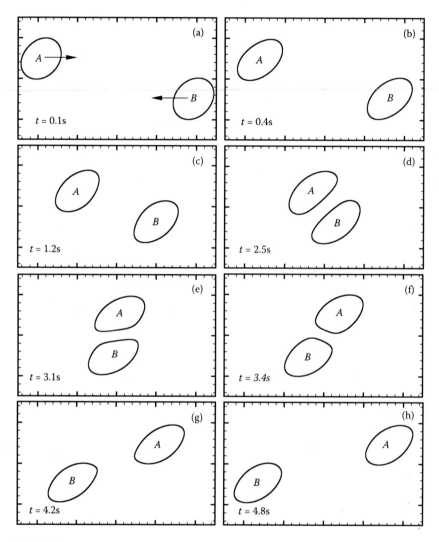

FIGURE 5.14 Particle–particle interactions in a shear-driven flow under different time steps.

fluid Ω_f. The initial shape of the particle is a rod of length L_p capped with two semicircles of radius a at both ends. The particle's center of mass is initially located at (x_{p0}, y_{p0}), and its longest axis initially presents an angle θ_{p0} counterclockwise with respect to the centerline of the microchannel. An electric field $\mathbf{E}$ is applied across the microchannel to generate the electrokinetic particle motion.

The thin EDL approximation is used in the present model for microscale electrokinetics. As a result, the net charge density in the fluid domain Ω_f is

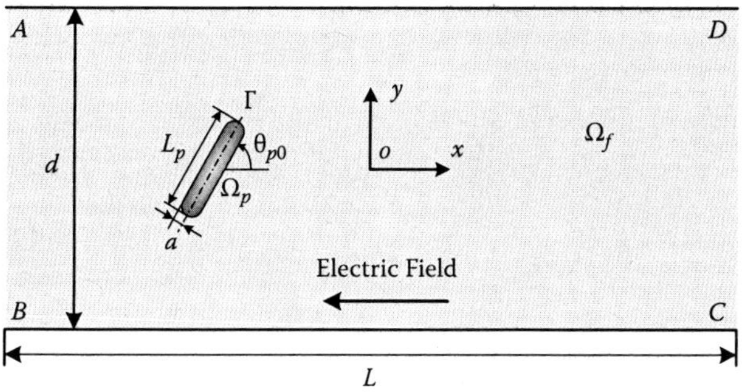

FIGURE 5.15 Schematics of the electrokinetic motion of a deformable cylindrical particle in a straight microchannel. (From Ai, Y., M. Benjamin, A. Sharma, and S. Qian S. 2010. Electrokinetic motion of a deformable particle: Dielectrophoretic effect. *Electrophoresis* 32:2282–2291 with permission from Wiley-VCH.)

zero, and the distribution of the electric potential is governed by the Laplace equation:

$$\nabla^2\phi = 0 \quad \text{in } \Omega_f. \tag{5.8}$$

The electric potentials on segments AB and CD are, respectively, $\phi = 0$ and $\phi = \phi_0$. All the other boundaries are assumed to be insulating, $\mathbf{n}_f \bullet \nabla\phi = 0$. The initial electric potential is zero.

As the Reynolds number in the present study is very small (i.e., typically less than 0.001), the inertial terms in the Navier–Stokes equations are negligible. The fluid motion is thus modeled by the continuity equation and the Stokes equations, given as

$$\nabla \bullet \mathbf{u} = 0 \quad \text{in } \Omega_f, \tag{5.9}$$

and

$$\rho_f \frac{\partial \mathbf{u}}{\partial t} = -\nabla p + \mu\nabla^2\mathbf{u} \quad \text{in } \Omega_f. \tag{5.10}$$

The initial flow velocity and the initial pressure are all zero. The pressures at the two openings of the fluid domain Ω_f are both zero. If the two parallel walls are charged, an EOF is generated next to the charged boundary, which can be described by the Smoluchowski slip velocity:

$$\mathbf{u} = \frac{\varepsilon_0\varepsilon_f\zeta_w}{\mu}(\mathbf{I} - \mathbf{n}_f\mathbf{n}_f)\bullet\nabla\phi \quad \text{on BC and AD,} \tag{5.11}$$

where ζ_w is the zeta potential of the rigid wall, and $\mathbf{I}$ is the second-order unit tensor. The quantity $(\mathbf{I} - \mathbf{n}_f\mathbf{n}_f)\bullet\nabla\phi$ defines the electric field tangential to the charged surface. As the particle is also charged, the fluid velocity on the particle surface consists of the Smoluchowski slip velocity and the velocity related to the particle movement, given as

$$\mathbf{u} = \frac{\varepsilon_0 \varepsilon_f \zeta_p}{\mu} (\mathbf{I} - \mathbf{n}_f \mathbf{n}_f) \bullet \nabla\phi + \frac{\partial \mathbf{w}}{\partial t} \qquad \text{on } \Gamma. \qquad (5.12)$$

In this equation, ζ_p is the zeta potential of the particle, and $\mathbf{w}$ is the displacement of the particle that is also governed by Equations (5.3) and (5.4). However, the DEP force is also induced due to the spatially nonuniform electric field around the particle. Thus, the traction force on the particle-fluid interface is written as

$$\sigma(\mathbf{w}) \bullet \mathbf{n}_p = \sigma_f \bullet \mathbf{n}_f + \sigma_E \bullet \mathbf{n}_f. \qquad (5.13)$$

where $\sigma_E = \varepsilon_f \left[\mathbf{E}\mathbf{E} - \frac{1}{2}(\mathbf{E} \bullet \mathbf{E})\mathbf{I} \right]$ is the Maxwell stress tensor. The initial displacement of the solid phase is zero.

5.3.2 NUMERICAL IMPLEMENTATION AND CODE VALIDATION

The implementation of the electrokinetics-induced particle deformation model is similar to that described in Section 5.2.2. The differences include the addition of the electric field, the fluid velocity boundary conditions on the rigid channel wall and the deformable particle surface by adding the Smoluchowski slip velocity, and the traction force on the particle surface. Table 5.3 lists all the constants used in the modeling. Table 5.4 shows the detailed instructions of implementation in the GUI of COMSOL Multiphysics 3.5a. The steady distributions of the flow field and the electric field are solved at the beginning as the initial solution for the time-dependent problem. The full COMSOL MATLAB script M-file capable of automated remeshing is as follows:

TABLE 5.3
Constant Table

Variable	Value or Expression	Description
rho	1e3 [kg/m^3]	Densities of fluid and particle
eta	1e-3 [Pa*s]	Fluid viscosity
a	1e-06 [m]	Particle radius
E0	10000 [V/m]	Applied electric field
Ca	eps_f*zeta_p*E0/G0/a	Capillary number
zeta_p	50e-3 [V]	Particle zeta potential
eps_f	80*8.854187817e-12 [F/m]	Fluid permittivity
G0	25 [Pa]	Shear modulus
Bulk	1e3 [Pa]	Bulk modulus
zetar	0	Zeta potential ratio
Re	rho*U0*a/eta	Reynolds number
U0	eps_f*zeta_p*E0/eta	Electrophoretic velocity

TABLE 5.4
Model Setup in the GUI of COMSOL Multiphysics 3.5a

Model Navigator	Select **2D** in space dimension and click **Multiphysics** button. Select **Structural Mechanics Module|Plane Strain|Viscoelastic transient initialization**. Remove the predefined variables in the **Dependent variables** and enter u2 v2 p2. Click **Add** button. Select **COMSOL Multiphysics|Deformed Mesh|Moving Mesh (ALE)|Transient analysis**. Click **Add** button. Select **MEMS Module|Microfluidics|Stokes Flow|Transient analysis**. Click **Add** button. Select **COMSOL Multiphysics|Electromagnetics|Conductive Media DC**. Click **Add** button.
Option Menu|**Constants**	Define variables in Table 5.3.
Physics Menu|**Subdomain** **Setting**	smpn mode
	Deactivate Subdomain 1 (Fluid)
	Subdomain 2 (Solid)
	Material model: Hyperelastic Hyperelastic model: Neo-hookean
	$\mu = G0$; $\kappa = $ Bulk; $\alpha = 1.2\text{e-}5$; $\rho = $ rho; thickness = 1
	ale model
	Subdomain 1
	Free displacement
	Subdomain 2
	Physics induced displacement
	dx = u2; dy = v2
	mmglf mode
	Subdomain 1
	$\rho = $ rho; $\eta = $ eta; Fx = 0; Fy = 0; Thickness = 1
	Deactivate Subdomain 2
	emdc mode
	$J_e = 0$; $Q_j = 0$; d = 1; $\sigma = 1$
Physics Menu|**Boundary** **Setting**	smpn mode
	Particle surface
	Tab Constraint
	Constrain condition: Free
	Tab Load
	Type of load: Distributed load
	Fx = $-(lm3+Fdep\_x)*dvol\_ale/dvol$; Fy = $-(lm4+Fdep\_y)*dvol\_$ ale/dvol
	Other boundaries are inactive
	ale mode
	Particle surface

continued

TABLE 5.4 (CONTINUED)
Model Setup in the GUI of COMSOL Multiphysics 3.5a

	dx = u2; dy = v2
	Other boundaries
	dx = 0; dy = 0
	mmglf mode
	Left boundary: Inlet|Pressure, no viscous stress $P_0 = 0$
	Right boundary: Outlet| Pressure, no viscous stress $P_0 = 0$
	Surface of particle I: Wall|Moving/leaking wall Uw = u2t+ueofp
	Vw = v2t+veofp
	Lower and upper boundaries: Wall|Moving/leaking wall
	Uw = ueof
	Vw = veof
	emdc mode
	Left and right boundaries: V0 = E0*x
	Other boundaries: Electric insulation
Other Settings	Properties of ale model (**Physics** Menu|**Properties**)
	Smoothing method: Winslow
	Allow remeshing: On
	Properties of mmglf model (**Physics** Menu|**Properties**)
	Weak constraints: On
	Constraint type: Non-ideal
Options Menu|**Integration** **Coupling** **Variables**|**Subdomain** **Variables**|	Subdomain selection: subdomain 2 (Particle)
	Name: area Expression: 1; Integration order: 4; Frame: Frame(mesh)
	Name: vel Expression: u2t; Integration order: 4; Frame: Frame(mesh)
	Name: disx Expression: u2; Integration order: 4; Frame: Frame(mesh)
	Name: disy Expression: v2; Integration order: 4; Frame: Frame(mesh)
Options Menu| **Expressions**|**Boundary** **Expressions**	Boundary selection: 4, 5, 7, 8, 9, 10 (Particle surface)
	Name: ueofp Expression: zeta_p*eps_f* tEx_emdc /eta
	Name: veofp Expression: zeta_p*eps_f* tEy_emdc /eta
	Name: Fdep_x Expression: 0.5*eps_f*(tEx_emdc^2+tEy_emdc^2)*nx_emdc
	Name: Fdep_y Expression: 0.5*eps_f*(tEx_emdc^2+tEy_emdc^2)*ny_emdc
	Boundary selection: 2, 3 (Channel wall)
	Name: ueof Expression: zetar*zeta_p*eps_f* tEx_emdc /eta
	Name: veof Expression: zetar*zeta_p*eps_f* tEy_emdc /eta
Options Menu| **Expressions**|**Global** **Expressions**	Name: up Expression: vel/area
	Name: disx_p Expression: disx/area
	Name: disy_p Expression: disy/area

TABLE 5.4 (CONTINUED)
Model Setup in the GUI of COMSOL Multiphysics 3.5a

Mesh|Free Mesh Parameters	Tab subdomain
	Subdomain 1|Maximum element size: 1.2e-6
	Subdomain 2|Maximum element size: 0.15e-6
	Tab boundary
	Particle surface: 4, 5, 7, 8, 9, 10|**Maximum element size**: 0.15e-6
	Channel wall: 2, 3|**Maximum element size**: 0.5e-6
Solve Menu|	**Solver Parameters**
	Tab General
	Times: Range(0, 1e-4, 0.1)
	Relative tolerance: 1e-4
	Absolute tolerance: 1e-6
	Tab Timing Stepping
	Activate Use stop condition
	minqual1_ale-0.6
	Click = **Solve Problem**
Postprocessing Menu|	Result check: Particle shape
	Domain Plot Parameters|Line/Extrusion
	Boundary select: 4, 5, 7, 8, 9, 10
	y-axis data: y; x-axis data: x
	Click Apply
	Result check: up
	Global Variables Plot|Select up, click >
	Click Apply

```
%%%%%%%%%%%%Electrokinetics-induced Particle
  Deformation.m%%%%%%%%%%%%%%

flclear fem

% COMSOL version
clear vrsn
vrsn.name = 'COMSOL 3.5';
vrsn.ext = 'a';
vrsn.major = 0;
vrsn.build = 608;
vrsn.rcs = '$Name: v35ap $';
vrsn.date = '$Date: 2009/05/11 07:38:49 $';
fem.version = vrsn;

% New: Define the particle radius
ap=1e-6;
% New: Define the aspect ratio of the particle
```

```
ratio=12;

% New: Define the electric field

Electric=[1e4,2e4,3e4];

% NEW: FOR loop to go through all the electric fields
for i=1:length(Electric)

% Constants
fem.const = {'rho','1e3[kg/m^3]', ...
             'eta','1e-3[Pa*s]', ...
             'a',num2str(ap), ...
             'E0',num2str(Electric(i)), ...
             'Ca','eps_f*zeta_p*E0/G0/a', ...
             'zeta_p','50e-3[V]', ...
             'eps_f','80*8.854187817e-12[F/m]', ...
             'G0','25[Pa]', ...
             'Bulk','1e3[Pa]', ...
             'zetar','0', ...
             'Re','rho*U0*a/eta', ...
             'U0','eps_f*zeta_p*E0/eta'};

% ********************
% Geometry
g1=rect2('100e-6','50e-6','base','center','pos',{'0','0'},
   'rot','0');
g2=rect2(num2str(2*ap),num2str((ratio-2)*ap),'base',
   'center','pos',{'0','0'},'rot','0');
g3=circ2(num2str(ap),'base','center','pos',{'0',num2str
   ((ratio-2)*ap/2)},'rot','0');
g4=circ2(num2str(ap),'base','center','pos',{'0',num2str
   (-(ratio-2)*ap/2)},'rot','0');
g5=geomcomp({g2,g3,g4},'ns',{'g2','g3','g4'},'sf',
   'g2+g3+g4','edge','none');
g6=geomdel(g5);
g6=move(g6,[-20e-6,0]);

% Analyzed geometry
clear s
s.objs={g1,g6};
s.name={'R1','CO2'};
s.tags={'g1','g6'};

fem.draw=struct('s',s);
fem.geom=geomcsg(fem);

% NEW: Display geometry
geomplot(fem, 'Labelcolor','r','Edgelabels','on','submode',
   'off');
```

```
% NEW: Solve plane strain of neo-hookean material
% Application mode 1
clear appl
appl.mode.class = 'SmePlaneStrain';
appl.dim = {'u2','v2','p2'};
appl.sdim = {'X','Y','Z'};
appl.module = 'SME';
appl.gporder = 4;
appl.cporder = 2;
appl.assignsuffix = '_smpn';
clear prop
prop.analysis='time';
prop.deformframe='ref';
prop.frame='ref';
clear weakconstr
weakconstr.value = 'off';
weakconstr.dim = {'lm9','lm10'};
prop.weakconstr = weakconstr;
appl.prop = prop;
clear bnd
bnd.name = {'fluid','solid'};
bnd.Fx = {0,'(-lm3+Fdep_x)*dvol_ale/dvol'};
bnd.Fy = {0,'(-lm4+Fdep_y)*dvol_ale/dvol'};
bnd.ind = [1,1,1,2,2,1,2,2,2,2];
appl.bnd = bnd;
clear equ
equ.kappa = {10000000000,'Bulk'};
equ.rho = {7850,'rho'};
equ.mu = {800000,'G0'};
equ.materialmodel = {'iso','hyper'};
equ.name = {'fluid','solid'};
equ.usage = {0,1};
equ.ind = [1,2];
appl.equ = equ;
fem.appl{1} = appl;

% NEW: Solve moving mesh
% Application mode 2
clear appl
appl.mode.class = 'MovingMesh';
appl.sdim = {'Xm','Ym','Zm'};
appl.shape = {'shlag(2,''lm1'')','shlag(2,''lm2'')','shlag
   (2,''x'')','shlag(2,''y'')'};
appl.gporder = {30,4};
appl.cporder = 2;
appl.assignsuffix = '_ale';
clear prop
prop.smoothing='winslow';
prop.analysis='transient';
prop.allowremesh='on';
```

```
prop.origrefframe='ref';
appl.prop = prop;
clear bnd
bnd.defflag = {{1;1}};
bnd.wcshape = [1;2];
bnd.name = {'fluid','solid'};
bnd.deform = {{0;0},{'u2';'v2'}};
bnd.ind = [1,1,1,2,2,1,2,2,2,2];
appl.bnd = bnd;
clear equ
equ.gporder = 2;
equ.physexpr = {{0;0},{'u2';'v2'}};
equ.name = {'fluid','solid'};
equ.shape = [3;4];
equ.type = {'free','phys'};
equ.ind = [1,2];
appl.equ = equ;
fem.appl{2} = appl;

% NEW: Solve flow field using Stokes equations
% Application mode 3
clear appl
appl.mode.class = 'GeneralLaminarFlow';
appl.module = 'MEMS';
appl.shape = {'shlag(2,''lm3'')','shlag(2,''lm4'')','shlag
   (1,''lm5'')','shlag(2,''u'')','shlag(2,''v'')','shlag(1,
   ''p'')'};
appl.gporder = {30,4,2};
appl.cporder = {2,1};
appl.assignsuffix = '_mmglf';
clear prop
prop.weakcompflow='Off';
prop.inerterm='Off';
clear weakconstr
weakconstr.value = 'on';
weakconstr.dim = {'lm3','lm4','lm5','lm6','lm7'};
prop.weakconstr = weakconstr;
prop.constrtype='non-ideal';
appl.prop = prop;
clear bnd
bnd.type = {'inlet','walltype','walltype','outlet'};
bnd.weakconstr = {0,1,1,0};
bnd.vwall = {0,'veof','v2t+veofp',0};
bnd.wcshape = [1;2;3];
bnd.wcgporder = 1;
bnd.intype = {'p','uv','uv','uv'};
bnd.walltype = {'noslip','lwall','lwall','noslip'};
bnd.uwall = {0,'ueof','u2t+ueofp',0};
bnd.name = {'inlet','wall','solid','outlet'};
bnd.ind = [1,2,2,3,3,4,3,3,3,3];
```

```
appl.bnd = bnd;
clear equ
equ.eta = {'eta',0.001};
equ.gporder = {{2;2;3}};
equ.rho = {'rho',1000};
equ.cporder = {{1;1;2}};
equ.shape = [4;5;6];
equ.name = {'fluid','solid'};
equ.usage = {1,0};
equ.ind = [1,2];
appl.equ = equ;
fem.appl{3} = appl;

% NEW: Solve electric field
% Application mode 4
clear appl
appl.mode.class = 'EmConductiveMediaDC';
appl.module = 'MEMS';
appl.assignsuffix = '_emdc';
clear prop
clear weakconstr
weakconstr.value = 'off';
weakconstr.dim = {'lm8'};
prop.weakconstr = weakconstr;
appl.prop = prop;
clear bnd
bnd.V0 = {0,'E0*x'};
bnd.type = {'nJ0','V'};
bnd.name = {'Insulation','Electrode'};
bnd.ind = [2,1,1,1,1,2,1,1,1,1];
appl.bnd = bnd;
clear equ
equ.sigma = {1,{1;1;1}};
equ.name = {'fluid','solid'};
equ.usage = {1,0};
equ.ind = [1,2];
appl.equ = equ;
fem.appl{4} = appl;
fem.sdim = {{'Xm','Ym'},{'X','Y'},{'x','y'}};
fem.frame = {'mesh','ref','ale'};
fem.border = 1;
clear units;
units.basesystem = 'SI';
fem.units = units;

% Boundary settings
clear bnd
bnd.ind = [1,2,2,3,3,1,3,3,3,3];
bnd.dim = {'u2','v2','p2','x','y','lm1','lm2','u','v','p',
  'lmx_mmglf', ...
```

```
'lmy_mmglf','lm3','lm4','lm5','V'};

% Boundary expressions
bnd.expr = {'ueofp',{'','','zeta_p*eps_f*tEx_emdc/eta'}, ...
  'veofp',{'','','zeta_p*eps_f*tEy_emdc/eta'}, ...
  'ueof',{'','zetar*zeta_p*eps_f*tEx_emdc/eta', ...
  ''}, ...
  'veof',{'','zetar*zeta_p*eps_f*tEy_emdc/eta', ...
  ''}, ...
  'Fdep_x',{'','','0.5*eps_f*(tEx_emdc^2+tEy_emdc^2)*nx_
  emdc'}, ...
  'Fdep_y',{'','','0.5*eps_f*(tEx_emdc^2+tEy_emdc^2)*ny_
  emdc'}};
fem.bnd = bnd;

% Coupling variable elements
clear elemcpl
% Integration coupling variables
clear elem
elem.elem = 'elcplscalar';
elem.g = {'1'};
src = cell(1,1);
clear equ
equ.expr = {{{},'1'},{{},'u2t'},{{},'u2'},{{},'v2'}};
equ.ipoints = {{{},'4'},{{},'4'},{{},'4'},{{},'4'}};
equ.frame = {{{},'mesh'},{{},'mesh'},{{},'mesh'},{{},
  'mesh'}};
equ.ind = {{'1'},{'2'}};
src{1} = {{},{},equ};
elem.src = src;
geomdim = cell(1,1);
geomdim{1} = {};
elem.geomdim = geomdim;
elem.var = {'area','vel','disx','disy'};
elem.global = {'1','2','3','4'};
elem.maxvars = {};
elemcpl{1} = elem;
fem.elemcpl = elemcpl;

% Global expressions
fem.globalexpr = {'up','vel/area', ...
                  'disx_p','disx/area', ...
                  'disy_p','disy/area'};

% Multiphysics
fem=multiphysics(fem);

% NEW: Loop for continuous particle tracking
% NEW: Remeshing index
j=1;
```

```
% NEW: Define initial time
time_e=0;

% NEW: Define time_step
time_step=1e-4/i;

% NEW: Define data storage index
num=1;

% NEW: Define matrix for data storage
particle=zeros(10,2);

% NEW: Define end condition of the computation (Loop con-
   trol)
while time_e<0.15/i

% Initialize mesh
fem.mesh=meshinit(fem, ...
                  'hauto',5, ...
'hmaxedg',[2,0.5e-6,3,0.5e-6,4,0.15e-6,5,0.15e-6,7,0.15e-
   6,8,0.15e-6,9,0.15e-6,10,0.15e-6], ...
                  'hmaxsub',[1,1.2e-6,2,0.15e-6]);

% Extend mesh
fem.xmesh=meshextend(fem);

if j==1

% NEW: Solve the steady flow field and electric field as the
   initial condition
% Solve problem
fem.sol=femstatic(fem, ...
  'solcomp',{'lm4','v','lm3','u','V','p'}, ...
  'outcomp',{'lm4','lm3','lm2','lm1','u2','v2','v','u','V',
  'p','Y','X','y','x'}, ...
            'blocksize','auto', ...
            'callback','postcallback', ...

  'callbparam',{'tridata',{'normE_emdc','cont','internal',
  'unit','V/m'},'trimap','Rainbow','solnum','end','title',
  'Surface: Electric field, norm [V/m]','axis',
  [-1.469611004279722E-4,1.469611004279722E-4,
  -1.20999969432829E-4,1.20999969432829E-4]});

% Save current fem structure for restart purposes
fem1=fem;

% Solve problem
```

```
fem.sol=femtime(fem, ...
                'init',fem1.sol, ...
'solcomp',{'lm4','lm3','v','u','lm2','lm1','V','u2','v2',
 'p','y','x'}, ...
'outcomp',{'lm4','lm3','lm2','lm1','u2','v2','v','u','V',
 'p','Y','X','y','x'}, ...
                'blocksize','auto', ...
                'tlist',[colon(0,time_step,0.1)], ...
                'tout','tlist', ...
                'nlsolver','manual', ...
                'ntolfact',1, ...
                'maxiter',4, ...
                'rtol',1e-4, ...
                'atol',1e-6, ...
                'dtech','const', ...
                'damp',1.0, ...
                'jtech','minimal', ...
                'callback','postcallback', ...

  'callbparam',{'tridata',{'normE_emdc','cont','internal',
  'unit','V/m'},'trimap','Rainbow','title','Surface:
  Electric field, norm [V/m]','axis',[-30E-6,30E-6,
  -30E-6,30E-6]}, ...
                'stopcond','minqual1_ale-0.6');

else

% Mapping current solution to extended mesh
init = asseminit(fem,'init',fem0.sol,'xmesh',fem0.xmesh,
  'blocksize','auto','framesrc','ale','domwise','on');

% Solve problem
fem.sol=femtime(fem, ...
                'init',init, ...
'solcomp',{'lm4','lm3','v','lm2','u','V','lm1','u2','p',
 'v2','y','x'}, ...
'outcomp',{'lm4','lm3','lm2','lm1','u2','v2','v','u','V',
 'p','Y','X','y','x'}, ...
            'blocksize','auto', ...
            'tlist',[colon(time_e,time_step,0.1)], ...
            'tout','tlist', ...
            'nlsolver','manual', ...
            'ntolfact',1, ...
            'maxiter',4, ...
            'rtol',1e-4, ...
            'atol',1e-6, ...
            'dtech','const', ...
            'damp',1.0, ...
            'jtech','minimal', ...
            'callback','postcallback', ...
```

```
    'callbparam',{'tridata',{'normE_emdc','cont','internal',
    'unit','V/m'},'trimap','Rainbow','title','Surface:
    Electric field, norm [V/m]','axis',[-30E-6,30E-6,-30E-
    6,30E-6]}, ...
    'stopcond','minqual1_ale-0.6');

end

% Save current fem structure for restart purposes
fem0=fem;

% NEW: Get the last time from solution
time_e = fem.sol.tlist(end);

% NEW: Get the length of the time steps
Leng =length(fem.sol.tlist);

% Global variables plot
data=postglobalplot(fem,{'up/U0','disx_p','disy_p'}, ...
                'Outtype','postdata', ...
                'title','up/U0', ...
                'axislabel',{'Time','up/U0'});

% NEW: Store the interested data
for n=5:Leng
    particle(num,1)=data.p(1,n);
    particle(num,2)=data.p(2,n);
    particle(num,3)=data.p(2,n+Leng);
    particle(num,4)=data.p(2,n+2*Leng);
    num=num+1;

% NEW: Write the data into a file
dlmwrite(strcat('DEP_G0_25Pa_ratio_12_zetar_0_ang_90_center_
    bulk=1000_E=',num2str(Electric(i)),'.dat'),particle,',');

% NEW: Output the COMSOL Multiphysics GUI .mph file if
    necessary
flsave(strcat('DEP_G0_25Pa_ratio_12_zetar_0_ang_90_center_
    bulk=1000_E=',num2str(Electric(i)),'_',num2str(j),
    '.mph'),fem);

end

j=j+1;

% Generate geom from mesh
fem = mesh2geom(fem, ...
                'frame','ale', ...
                'srcdata','deformed', ...
                'destfield',{'geom','mesh'}, ...
```

```
'srcfem',1,  ...
'destfem',1);

end

end
```

The developed numerical model is verified by simulating electrophoresis of a circular particle translating along the centerline of a slit channel. The particle can be considered as a rigid particle by assigning a very high shear modulus (G_0 = 337 Pa). Figure 5.16 shows the particle velocity normalized by $\varepsilon_0 \varepsilon_f \zeta_p E/\mu$ as a function of the ratio of the particle diameter to the height of the microchannel. Obviously, the numerical results obtained by the developed model for a deformable particle quantitatively agree with the predictions obtained by the previous numerical model for rigid particles.

5.3.3 Results and Discussion

The channel is assumed to be uncharged without further specification. The particle of radius a = 1 μm and length L_p = 12 μm is initially positioned at

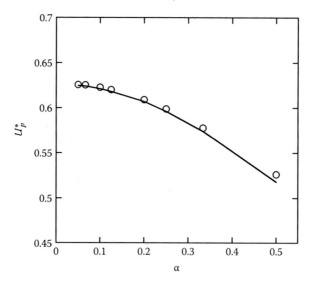

FIGURE 5.16 Dimensionless translational velocity normalized by $\varepsilon_0 \varepsilon_f \zeta_p E/\mu$ of a circular particle translating along the centerline of a slit microchannel as a function of the ratio of the particle diameter to the height of the microchannel. Solid line and circles represent, respectively, the results obtained by the model for a rigid particle and the present model for a deformable particle with a very high shear modulus. The ratio of the zeta potential of the microchannel to that of the particle is 0.375. (From Ai, Y., M. Benjamin, A. Sharma, and S. Qian S. 2010. Electrokinetic motion of a deformable particle: Dielectrophoretic effect. *Electrophoresis* 32:2282–2291 with permission from Wiley-VCH.)

$(x_{p0}, y_{p0}) = (-20 \ \mu\text{m}, 0)$ and electrokinetically moves from left to right in all the following simulations.

5.3.3.1 DEP Effect

We first investigate the DEP effect on particle motion and deformation when $E = 10$ kV/m, $\zeta_p = -50$ mV, and $d = 50 \ \mu\text{m}$. The particle is initially presenting $\theta_{p0} = 60°$ with respect to the centerline of the microchannel. Figure 5.17 shows the trajectory and deformation of a hyperelastic particle when $G_0 = 25$ Pa (Figures 5.17a and 5.17d), $G_0 = 50$ Pa (Figures 5.17b and 5.17e), and $G_0 = 500$ Pa (Figures 5.17c and 5.17f) at seven different time steps (from left to right): $t = 0$, 12, 24, 36, 48, 72, and 100 ms. The results on the left (a, b, and c) and right (d, e, and f) columns are, respectively, obtained with and without considering the DEP effect. The second term, $\sigma_E \bullet \mathbf{n}_f$, on the right-hand-side of Equation (5.13) is removed when the DEP effect is not considered. When $G_0 = 25$ Pa (Figures 5.17a

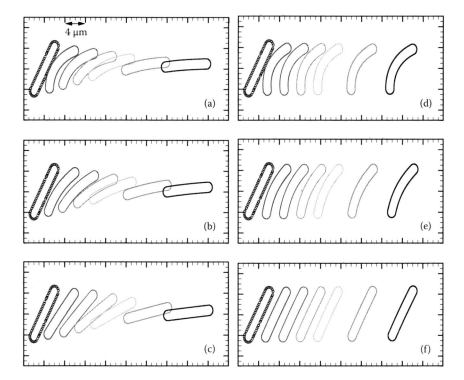

FIGURE 5.17 Dynamics of a hyperelastic particle when $G_0 = 25$ Pa (a and d), $G_0 = 50$ Pa (b and e), and $G_0 = 500$ Pa (c and f) at seven different time steps (from left to right): $t = 0$, 12, 24, 36, 48, 72, and 100 ms. The results in (a)–(c) and (d)–(f) are, respectively, obtained by the models with and without considering the DEP effect. $E = 10$ KV/m, $\theta_{p0} = 60°$, $\zeta_p = -50$ mV, and $d = 50 \ \mu\text{m}$. (From Ai, Y., M. Benjamin, A. Sharma, and S. Qian S. 2010. Electrokinetic motion of a deformable particle: Dielectrophoretic effect. *Electrophoresis* 32:2282–2291 with permission from Wiley-VCH.)

and 5.17d), the particle deforms dramatically at the beginning (t = 12 ms). If the DEP effect is not considered, the particle remains deformed while further translating. However, if the DEP effect is taken into account, the particle rotates clockwise and gradually aligns with its longest axis parallel to the applied electric field. The particle alignment phenomenon owing to the DEP effect has also been predicted using the numerical model for rigid particles in Chapter 4. The nonuniformity of the electric field on the particle surface decreases as the particle becomes aligned to the applied electric field. Therefore, the particle deformation decreases during the alignment process. The particle becomes more rigid as the shear modulus increases, which accordingly decreases the degree of deformation, as shown in Figure 5.17b. When G_0 = 500 Pa (Figure 5.17c), the particle can be regarded as a rigid particle.

Although the particle deformation under a different shear modulus varies, the tendency of the particle alignment shows an obvious similarity. When the DEP effect is not taken into account, the particle alignment to the applied electric field is not predicted in Figures 5.17d, 5.17e, or 5.17f. It is confirmed that the DEP effect is responsible for the particle alignment phenomenon.

When the particle is initially perpendicular to the applied electric field, θ_{p0} = 90°, the net torque is zero because of a symmetric distribution of the DEP force on the particle. Therefore, the particle alignment is eliminated in this singular orientation. The deformed particle at its equilibrium state when E = 20 kV/m, G_0 = 25 Pa, ζ_p = −50 mV, and d = 50 µm is in a C shape symmetric with respect to the centerline of the microchannel when the particle moves from left to right, as shown in Figure 5.18a. This prediction is similar to the numerical results obtained by Swaminathan, Gao, and Hu (2010) using their own ALE code without considering the DEP effect. However, we further found that the particle deformation considering the DEP effect is more significant than that ignoring the DEP effect. At the beginning, the DEP force acting on the particle is symmetric with respect to the center of the undeformed particle, which thus has no contribution to particle deformation. However, the electric field near the two ends of the particle is higher than that in the other regions; the nonuniform Smoluchowski slip velocity on the particle surface thus generates a larger shear force at the two ends of the particle, which deforms the particle to a C shape as it moves from left to right. As a result, the electric field around the particle becomes asymmetric, which leads to an asymmetric distribution of the DEP force on the particle surface, as shown in Figure 5.18b. Obviously, the asymmetric DEP distribution further enhances particle deformation.

The averaged particle velocity in the x direction is defined as

$$U_p = \frac{\int \frac{\partial w_x}{\partial t} d\Omega_p}{\int 1 d\Omega_p}, \tag{5.14}$$

where w_x is the particle displacement in the x direction. Figure 5.19 shows the averaged particle velocity at its equilibrium state as a function of the applied electric field when $G_0 = 25$ Pa, $\theta_{p0} = 90°$, $\zeta_p = -50$ mV, and $d = 50$ μm. The DEP effect is proportional to the square of the applied electric field. Therefore, the DEP effect is relatively weak under a relatively low electric field. The steady averaged particle velocity considering DEP is very close to that without considering DEP when the applied electric field is $E = 5$ kV/m. As the applied electric field increases, the averaged particle velocity considering DEP significantly deviates from that without considering DEP. Figure 5.18 shows that the net DEP force is pointed toward the negative x direction, which in turn slows the particle motion. The DEP effect can significantly affect both particle deformation and particle mobility, which is thus considered in all the following simulations.

5.3.3.2 Effect of Particle's Shear Modulus

Figure 5.17 revealed that the particle deformation increases as its shear modulus decreases. However, the particle deformation is also reduced as the particle is aligned to the electric field. Here, we set the initial particle orientation $\theta_{p0} = 90°$ to get rid of the particle alignment and investigate the effect of the shear modulus of the particle on its deformation. Figure 5.20 shows the deformation evolution of a hyperelastic particle when $E = 10$ kV/m, $\zeta_p = -50$ mV, $d = 50$ μm, $G_0 = 12.5$ Pa (Figure 5.20a), $G_0 = 25$ Pa (Figure 5.20b), $G_0 = 50$ Pa (Figure 5.20c), and $G_0 = 100$ Pa (Figure 5.20d). As previously explained, the particle deforms because of a higher Smoluchowski slip velocity at the two ends of the particle. The DEP effect also enhances particle deformation. A smaller shear modulus leads to more significant deformation, which implies a softer particle, as shown in Figure 5.20.

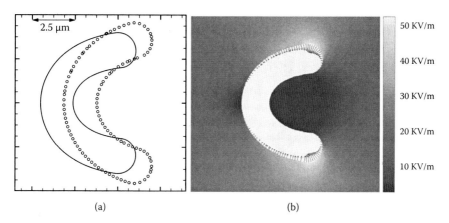

(a)   (b)

FIGURE 5.18 (a) Equilibrium shape of a hyperelastic particle when $E = 20$ kV/m, $G_0 = 25$ Pa, $\theta_{p0} = 90°$, $\zeta_p = -50$ mV, and $d = 50$ μm. Solid line and circles represent, respectively, the results obtained by the models with and without considering the DEP effect. (b) Distribution of the electric field around the deformed particle. The gray levels denote the magnitude of the electric field strength, and the arrows on the particle surface indicate the distribution of the DEP force. (From Ai, Y., M. Benjamin, A. Sharma, and S. Qian S. 2010. Electrokinetic motion of a deformable particle: Dielectrophoretic effect. *Electrophoresis* 32:2282–2291 with permission from Wiley-VCH.)

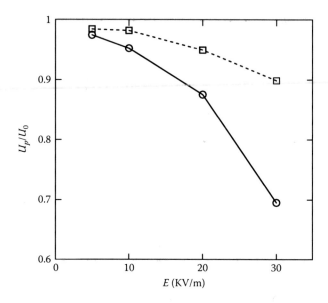

FIGURE 5.19 Steady averaged particle velocity in the x direction normalized by $U_0 = \varepsilon_0 \varepsilon_f \zeta_p E / \mu$ as a function of the applied electric field when $G_0 = 25$ Pa, $\theta_{p0} = 90°$, $\zeta_p = -50$ mV, and $d = 50$ μm. Solid line with circles and dashed line with squares represent, respectively, the results obtained by the models with and without considering the DEP effect. (From Ai, Y., M. Benjamin, A. Sharma, and S. Qian S. 2010. Electrokinetic motion of a deformable particle: Dielectrophoretic effect. *Electrophoresis* 32:2282–2291 with permission from Wiley-VCH.)

5.3.3.3 Effect of Electric Field

The intensity of the imposed electric field affects both the Smoluchowski slip velocity on the particle surface and the DEP force acting on the particle, which are the main factors for particle deformation. Figure 5.21 shows the deformation evolution of a hyperelastic particle under two different electric fields, $E = 20$ kV/m (Figure 5.21a) and $E = 30$ kV/m (Figure 5.21b), when $G_0 = 25$ Pa, $\theta_{p0} = 90°$, $\zeta_p = -50$ mV, and $d = 50$ μm. The equilibrium particle shape when $E = 30$ kV/m (Figure 5.21b) is more deformed than the case of $E = 20$ kV/m (Figure 5.21a), which is also more deformed than the case of $E = 10$ kV/m (Figure 5.20b). Thus, particle deformation can be effectively controlled by adjusting the external electric field.

As stated, the nonuniform EOF near the particle induces a nonuniform shear force, which in turn deforms the particle. Because the fluid-structure interaction is a two-way coupling problem, particle deformation can also affect the flow field around the particle. Figure 5.22 shows the evolution of the flow field near the particle when $E = 20$ kV/m, $G_0 = 25$ Pa, $\theta_{p0} = 90°$, $\zeta_p = -50$ mV, and $d = 50$ μm. A vortex is generated at either end of the particle due to the combined effects of the particle movement and the nearby EOF, as shown in Figure 5.22a. As the particle curls up, the two vortices at the ends of the particle are brought closer to each other, as indicated in Figures 5.22b, 5.22c, and 5.22d.

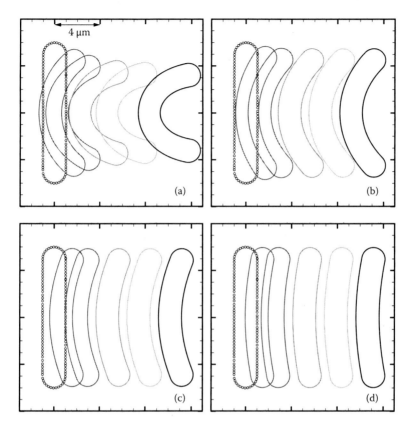

FIGURE 5.20 Deformation of a hyperelastic particle when $G_0 = 12.5$ Pa (a), $G_0 = 25$ Pa (b), $G_0 = 50$ Pa (c), and $G_0 = 100$ Pa (d) at six different time steps (from left to right): $t = 0$, 4, 8, 16, 24, and 32 ms. $E = 10$ kV/m, $\theta_{p0} = 90°$, $\zeta_p = -50$ mV, and $d = 50$ μm. (From Ai, Y., M. Benjamin, A. Sharma, and S. Qian S. 2010. Electrokinetic motion of a deformable particle: Dielectrophoretic effect. *Electrophoresis* 32:2282–2291 with permission from Wiley-VCH.)

We also investigated the effect of the applied electric field on particle motion when it is not initially perpendicular to the applied electric field. Therefore, the particle alignment and the resultant relaxation of particle deformation come into play. Figure 5.23 shows the transient behavior of a hyperelastic particle under two different electric fields, $E = 20$ kV/m (Figure 5.23a) and $E = 30$ kV/m (Figure 5.23b), when $G_0 = 50$ Pa, $\theta_{p0} = 60°$, $\zeta_p = -50$ mV, and $d = 50$ μm. The case of $E = 10$ kV/m is shown in Figure 5.17b. The particle deformation at the beginning under a higher electric field is more significant due to a higher Smoluchowski slip velocity at the two ends of the particle. Later, the particle is aligned to the applied electric field due to the DEP effect. In addition, a higher electric field leads to faster alignment, which also was predicted in Chapter 4. Once the particle becomes parallel to the applied electric field, the particle deformation is almost eliminated.

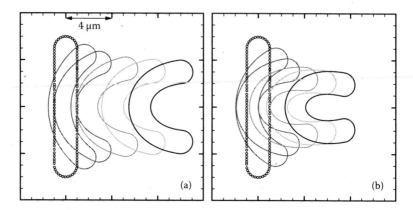

FIGURE 5.21 Deformation of a hyperelastic particle when $E = 20$ kV/m (a) and $E = 30$ kV/m (b) at six different time steps (from left to right): $t = 0, 2, 4, 8, 12,$ and 16 ms (a) and $t = 0, 1.33, 2.67, 5.33, 8,$ and 10.67 ms (b). $G_0 = 25$ Pa, $\theta_{p0} = 90°$, $\zeta_p = -50$ mV, and $d = 50$ μm. (From Ai, Y., M. Benjamin, A. Sharma, and S. Qian S. 2010. Electrokinetic motion of a deformable particle: Dielectrophoretic effect. *Electrophoresis* 32:2282–2291 with permission from Wiley-VCH.)

5.3.3.4 Effect of Particle's Zeta Potential

Equation (5.12) shows that the Smoluchowski slip velocity is proportional to the applied electric field and the particle's zeta potential. The effect of the particle's zeta potential on particle deformation is shown in Figure 5.24 when $E = 20$ kV/m, $G_0 = 25$ Pa, $\theta_{p0} = 90°$, and $d = 50$ μm. The time step is scaled by the magnitude of the particle's zeta potential to compare the equilibrium particle shape at approximately the same location. A higher zeta potential implies a higher Smoluchowski slip velocity at the ends of the particle, which in turn increases particle deformation, as shown in Figure 5.24. Previous experimental studies showed that the particle subjected to a pressure-driven flow deforms as a horizontally reversed C shape when it moves from left to right (Abkarian, Faivre, and Stone 2006; Abkarian et al. 2008). Obviously, the electrokinetics-induced particle deformation is totally different from that subjected to a pressure-driven flow.

5.3.3.5 Effect of Rigid Channel Boundary

Figure 5.25 shows the equilibrium particle shapes between two parallel walls with different distances when $E = 10$ kV/m, $G_0 = 25$ Pa, $\theta_{p0} = 90°$, $\zeta_p = -50$ mV, and $t = 50$ ms. When the distance between the two parallel walls decreases, the electric field between the particle and the rigid wall is enhanced. Therefore, particle deformation is slightly increased. In addition, the retardation effect arising from the stationary rigid wall becomes more pronounced when the distance between the two parallel walls decreases. Accordingly, particle velocity is also slightly decreased.

In all of these results, the rigid walls are assumed uncharged. To investigate the effect of the charged wall on particle deformation, we assume the zeta potential of the particle and the rigid wall are, respectively, $\zeta_p = 0$ and $\zeta_w = 50$ mV and

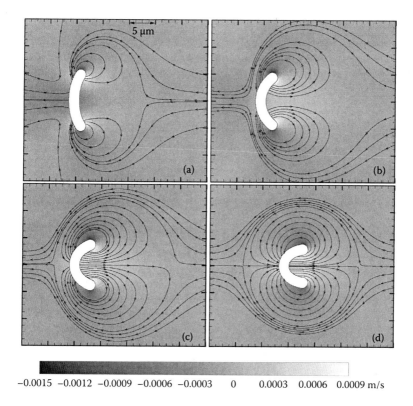

FIGURE 5.22   Flow field around a hyperelastic particle at $t = 1$ ms (a), 2 ms (b), 4 ms (c), and 12 ms (d) when $E = 20$ kV/m, $G_0 = 25$ Pa, $\theta_{p0} = 90°$, $\zeta_p = -50$ mV, and $d = 50$ μm. The gray levels denote the x component flow velocity, and the streamlines with arrows indicate the flow field. (From Ai, Y., M. Benjamin, A. Sharma, and S. Qian S. 2010. Electrokinetic motion of a deformable particle: Dielectrophoretic effect. *Electrophoresis* 32:2282–2291 with permission from Wiley-VCH.)

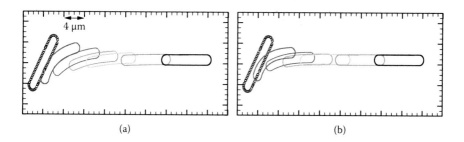

FIGURE 5.23   Dynamics of a hyperelastic particle when $E = 20$ kV/m (a) and $E = 30$ kV/m (b) at six different time steps (from left to right): $t = 0, 6, 12, 18, 24, 36,$ and 50 ms (a) $t = 0, 4, 8, 12, 16, 24,$ and 33.3 ms. $G_0 = 50$ Pa, $\theta_{p0} = 60°$, $\zeta_p = -50$ mV, and $d = 50$ μm. (From Ai, Y., M. Benjamin, A. Sharma, and S. Qian S. 2010. Electrokinetic motion of a deformable particle: Dielectrophoretic effect. *Electrophoresis* 32:2282–2291 with permission from Wiley-VCH.)

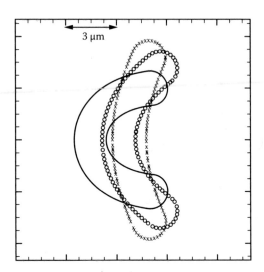

FIGURE 5.24 Equilibrium shape of a hyperelastic particle when $E = 20$ kV/m, $G_0 = 25$ Pa, $\theta_{p0} = 90°$, and $d = 50$ µm. Solid line, circles, and crosses denote, respectively, the equilibrium particle shape when $\zeta_p = -50$ mV ($t = 16$ ms), -25 mV ($t = 32$ ms), and -10 mV ($t = 80$ ms). (From Ai, Y., M. Benjamin, A. Sharma, and S. Qian S. 2010. Electrokinetic motion of a deformable particle: Dielectrophoretic effect. *Electrophoresis* 32:2282–2291 with permission from Wiley-VCH.)

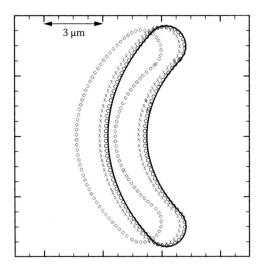

FIGURE 5.25 Equilibrium shape of a hyperelastic particle when $E = 10$ kV/m, $G_0 = 25$ Pa, $\theta_{p0} = 90°$, and $\zeta_p = -50$ mV. Solid line, circles, crosses, and diamonds denote, respectively, the equilibrium particle shape when $d = 50$, 40, 30, and 20 µm. (From Ai, Y., M. Benjamin, A. Sharma, and S. Qian S. 2010. Electrokinetic motion of a deformable particle: Dielectrophoretic effect. *Electrophoresis* 32:2282–2291 with permission from Wiley-VCH.)

keep all the other conditions in Figure 5.25 unchanged. Therefore, the particle motion is driven by the EOF stemming from the charged wall. Because the EOF is plug-like flow, the flow around the particle in an unbounded medium is uniform and unable to induce particle deformation. When the distance between the two parallel walls is $d = 50$ μm and 40 μm, the boundary effect is negligible. Therefore, it is found that the particle does not deform at all. As the boundary effect increases, the electric field between the particle and the rigid wall is enhanced, which could induce a slightly nonuniform EOF near the particle. Hence, the particle slightly deforms as the boundary effect becomes important. However, the particle deformation is still much smaller than those shown in Figure 5.25. Thus, the particle deformation due to the EOF effect arising from the charged wall could be largely ignored compared to that due to particle electrophoresis.

5.4 CONCLUDING REMARKS

In this chapter, we have demonstrated how to use COMSOL Multiphysics 3.5a to study shear-induced and electrokinetics-induced particle deformations in a slit channel numerically. The flow field and structural deformation were solved using the ALE algorithm in the simulation of shear-induced particle deformation. The circular particle located at the centerline of the channel is deformed as a perfect ellipse when the inertial effect is negligible. The numerical predictions obtained in the present study were in good agreement with the existing results obtained by other researchers' ALE code.

The electrokinetics-induced particle deformation is mainly due to the nonuniform Smoluchowski slip velocity on the particle surface. The asymmetric DEP force with respect to the center of the deformed particle could further enhance the particle deformation when the particle is initially perpendicular to the applied electric field. The deformation direction undergoing electrokinetics is opposite to that subjected to a pressure-driven flow when the particle moves in the same direction. A smaller shear modulus implies a softer particle, which leads to more significant deformation. The particle deformation increases as the applied electric field and the particle's zeta potential increase. However, the EOF flow arising from the charged wall has limited effect on particle deformation. If the particle is not initially perpendicular to the applied electric field, the particle is aligned with its longest axis parallel to the applied electric field because of the DEP effect. This numerical prediction was in good agreement with our previous experimental study presented in Chapter 4.

REFERENCES

Abkarian, M., M. Faivre, R. Horton, K. Smistrup, C. A. Best-Popescu, and H. A. Stone. 2008. Cellular-scale hydrodynamics. *Biomedical Materials* 3 (3):034011.
Abkarian, M., M. Faivre, and H. A. Stone. 2006. High-speed microfluidic differential manometer for cellular-scale hydrodynamics. *Proceedings of the National Academy of Sciences of the United States of America* 103 (3):538–542.

Ai, Y., M. Benjamin, A. Sharma, and S. Qian S. 2010. Electrokinetic motion of a deformable particle: Dielectrophoretic effect. *Electrophoresis* 32:2282–2291.

Chen, Y. C., G. Y. Chen, Y. C. Lin, and G. J. Wang. 2010. A lab-on-a-chip capillary network for red blood cell hydrodynamics. *Microfluidics and Nanofluidics* 9 (2–3):585–591.

Doddi, S. K., and P. Bagchi. 2009. Three-dimensional computational modeling of multiple deformable cells flowing in microvessels. *Physical Review E* 79 (4):046318.

Dondorp, A. M., B. J. Angus, K. Chotivanich, K. Silamut, R. Ruangveerayuth, M. R. Hardeman, P. A. Kager, J. Vreeken, and N. J. White. 1999. Red blood cell deformability as a predictor of anemia in severe falciparum malaria. *American Journal of Tropical Medicine and Hygiene* 60 (5):733–737.

Eggleton, C. D., and A. S. Popel. 1998. Large deformation of red blood cell ghosts in a simple shear flow. *Physics of Fluids* 10 (8):1834–1845.

Gao, T., and H. H. Hu. 2009. Deformation of elastic particles in viscous shear flow. *Journal of Computational Physics* 228 (6):2132–2151.

Hu, H. H., D. D. Joseph, and M. J. Crochet. 1992. Direct simulation of fluid particle motions. *Theoretical and Computational Fluid Dynamics* 3 (5):285–306.

Hu, H. H., N. A. Patankar, and M. Y. Zhu. 2001. Direct numerical simulations of fluid-solid systems using the arbitrary Lagrangian-Eulerian technique. *Journal of Computational Physics* 169 (2):427–462.

Korin, N., A. Bransky, and U. Dinnar. 2007. Theoretical model and experimental study of red blood cell (RBC) deformation in microchannels. *Journal of Biomechanics* 40 (9):2088–2095.

MacMeccan, R. M., J. R. Clausen, G. P. Neitzel, and C. K. Aidun. 2009. Simulating deformable particle suspensions using a coupled lattice-Boltzmann and finite-element method. *Journal of Fluid Mechanics* 618:13–39.

Secomb, T. W., B. Styp-Rekowska, and A. R. Pries. 2007. Two-dimensional simulation of red blood cell deformation and lateral migration in microvessels. *Annals of Biomedical Engineering* 35 (5):755–765.

Sugiyama, K., S. Ii, S. Takeuchi, S. Takagi, and Y. Matsumoto. 2011. A full Eulerian finite difference approach for solving fluid-structure coupling problems. *Journal of Computational Physics* 230 (3):596–627.

Swaminathan, T. N., T. Gao, and H. H. Hu. 2010. Deformation of a long elastic particle undergoing electrophoresis. *Journal of Colloid and Interface Science* 346 (1):270–276.

Tomaiuolo, G., M. Barra, V. Preziosi, A. Cassinese, B. Rotoli, and S. Guido. 2011. Microfluidics analysis of red blood cell membrane viscoelasticity. *Lab on a Chip* 11:449–454.

Tomaiuolo, G., M. Simeone, V. Martinelli, B. Rotoli, and S. Guido. 2009. Red blood cell deformation in microconfined flow. *Soft Matter* 5 (19):3736–3740.

van den Heuvel, M. G. L., R. Bondesan, M. C. Lagomarsino, and C. Dekker. 2008. Single-molecule observation of anomalous electrohydrodynamic orientation of microtubules. *Physical Review Letters* 101 (11):118301.

6 Pair Interaction between Two Colloidal Particles under DC Electric Field

In this chapter, pair interaction between two circular uncharged colloidal particles immersed in an aqueous solution subjected to an external direct current (DC) electric field is numerically investigated taking into account the particle-fluid-electric field interactions under the thin electrical double layer (EDL) assumption. Two nearby particles interact with the external electric field as well as each other when they are close enough, inducing a spatially nonuniform electric field surrounding them. The interaction between the induced nonuniform electric field and the induced electric dipole in each particle arising from the difference in the electrical properties of the particle and the background liquid medium induces a mutual dielectrophoretic (DEP) force exerted on each particle, leading to particle chaining, which has been experimentally observed in the DEP-directed particle assembly process. Particle assembly via DEP force has been recognized as one of the major fabrication tools to assemble novel nanomaterials with superior electric, magnetic, and optical properties for various applications, including photonics, photovoltaic solar cells, electronics, as well as data storage.

A transient, two-dimensional (2D) multiphysics model, which includes the Stokes and continuity equations for the fluid flow, the Laplace equation for the electric field externally imposed, and Newton's equations for the particles' translation and rotation, is developed to investigate the pair interaction between two particles and their relative motions under various conditions. The resulting DEP force is calculated by integrating the Maxwell stress tensor (MST) over the particle surface. Prior to the study of the DEP particle–particle interaction, comparison between the DEP particle–particle interaction and Brownian motion is performed. When the DEP particle–particle interaction dominates over the random Brownian motion, the induced electric dipole leads to the formation of particle chain aligned along the direction of the imposed electric field, independent of the particle's initial orientation. The numerical predictions are in qualitative agreement with the experimental observations available from the literature. During the particle chaining, their velocities tend to decrease dramatically due to the rapid increase in the repulsive hydrodynamic pressure force when the particle distance decreases to a certain value. One exclusive exception of the particle chaining occurs when the initial connecting line of the particles is perpendicular to the electric field imposed, which is extremely unstable owing to the inevitable Brownian motion.

6.1 INTRODUCTION

A particle, submerged in a liquid medium under an external electric field, polarizes and acquires a dipole moment. In the presence of a spatially nonuniform electric field surrounding the particle, the poles of the dipole experience different electric field strengths, giving rise to a net force on the particle, which in turn drives the particle motion, called dielectrophoresis. When the particle is more (less) polarizable than the background medium, the particle is driven toward a location with higher (lower) electric field intensity, which is referred to as positive (negative) DEP. DEP has become one of the most promising tools for various particle manipulations (e.g., particle separation, focusing, and trapping) in microfluidics (Kang and Li 2009; Li et al. 2007; Lapizco-Encinas and Rito-Palomares 2007). Further, DEP has been successfully used to assemble colloidal particles or biological entities to construct microscopic functional structures (Gangwal, Cayre, and Velev 2008; Gupta et al. 2008; Hermanson et al. 2001; Velev and Bhatt 2006; Velev, Gangwal, and Petsev 2009; Juarez and Bevan 2009) and exhibits many unique advantages over other particle assembly techniques: strong forces, speed, tunable parameters, simplicity, and robustness (Gupta et al. 2008). Particle chaining is typically observed in the particle assembly via DEP, which has been comprehensively reviewed (Velev, Gangwal, and Petsev 2009). Particles interact with the electric field imposed as well as each other when they are close enough. The presence of particles distorts the local electric field surrounding them. For example, the tangential electric field varies over the surface of a spherical particle and is not simply equal to the uniform electric field imposed far away from the particle. The tangential electric field is zero at the "poles" of the sphere and attains a maximum at the "equator," defined based on the direction of the field imposed. The induced nonuniform electric field causes mutual DEP particle–particle interaction force on each other, which plays an important role in particle chaining.

The induced DEP force depends on three characteristic length scales: the particle size, the characteristic length of the local electric field's variation, and the distance between two particles (Kadaksham, Singh, and Aubry 2005). The second length scale is related to the DEP force arising from the interaction between the nonuniform electric field and an individual particle; while the last length scale is responsible for the DEP particle–particle interaction force. If the latter DEP force dominates over the former one, the particles are aligned along the direction of the electric field imposed. Otherwise, the DEP particle–particle interaction force is negligible, and particles do not form a chain. Hwang, Kim, and Park (2008) experimentally observed the chaining and alignment of a pair of spherical particles initially presenting an angle with respect to the electric field imposed as a result of the DEP particle–particle interaction force.

Swaminathan and Hu (2004) and Yariv (2004) studied the particle–particle interaction in electrophoresis and denominated this "inertia-induced particle interaction." Under the assumption of thin EDL, Swaminathan and Hu (2004) found that the stable orientation of a pair of particles occurs when their connecting line is perpendicular to the electric field imposed. Yariv (2004) derived approximation solutions of the inertia force and the particles' trajectories, which are valid when the distance between the two particles is larger than the particle's radius. Furthermore, it is predicted that a pair of particles with an arbitrary initial orientation tends to rotate toward the aforementioned stable orientation. However, the DEP particle–particle interaction is neglected in these studies.

To elucidate the mechanism of particle assembly by an electric field, theoretical understanding of the dynamics of both fluid and particles in a DEP-directed particle assembly process is desired. A numerical Lagrange multiplier method has been developed to study the particle motion arising from the DEP particle–particle interaction (Kadaksham, Singh, and Aubry 2004a, 2004b, 2006; Aubry and Singh 2006a). However, the DEP force is obtained by using the point dipole moment approximation, in which the electric field is evaluated in the absence of the particles, and it is valid only when the distance between the particles is larger than the particle size (Aubry and Singh 2006b).

Kang and Li (2006) adopted an approximation solution of the DEP particle–particle interaction force and used a constant Stokes drag coefficient to derive an approximation solution of the particle trajectory. They found that a pair of particles with an arbitrary initial orientation tended to attract each other and align to the external electric field as the stable orientation, which is totally different from the results obtained by Swaminathan and Hu (2004) and Yariv (2004), in which the DEP particle–particle interaction was not considered. The approximation solution of the DEP particle–particle interaction force is valid only when the distance between the particles is larger than the particle size. The assumption of a constant Stokes drag coefficient is also not appropriate due to the presence of the hydrodynamic particle–particle interaction.

In this chapter, the DEP and hydrodynamic particle–particle interactions are investigated using a transient multiphysics model in which the fluid flow field, electric field, and particle motion are simultaneously solved using an arbitrary Lagrangian–Eulerian (ALE) method. The DEP force is obtained by integrating the MST over the particle surface, considered as the most rigorous approach for DEP calculation (Al-Jarro et al. 2007; Rosales and Lim 2005; Wang, Wang, and Gascoyne 1997). The current model takes into account the full interactions of fluid-particle-electric field and is valid under the required conditions of thin EDL approximation.

6.2 MATHEMATICAL MODEL

Let us consider two identical circular particles of radius a suspended in an incompressible and Newtonian aqueous solution of density ρ, dynamic viscosity μ, and permittivity ε_f confined in a square of length L, as shown in Figure 6.1. The origin of the Cartesian coordinate system (x, y) is located at the middle point of the connecting line of the two particles. An electric field $\mathbf{E}$ is imposed in the x direction across the computational domain, which is enclosed by segments ABCD and the

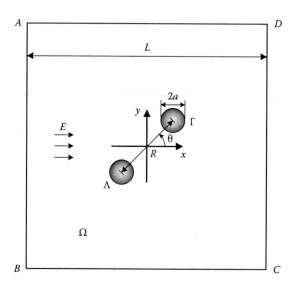

FIGURE 6.1 A pair of identical particles suspended in electrolyte confined in a square (ABCD) under an externally applied electric field **E**. The origin of the Cartesian coordinate systems (x, y) is located at the center point of the connecting line of the two particles and also the center of the square. The distance between the two particles and the angle between the connecting line of the two particles and the external electric field are, respectively, R and θ. (From Ai, Y., and S. Qian. 2010. DC dielectrophoretic particle–particle interactions and their relative motions. *Journal of Colloid and Interface Science* 346:448–454 with permission from Elsevier.)

particle surfaces Λ and Γ. The two particles with a center-to-center distance of R present an orientation θ with respect to the direction of the applied electric field. As discussed in the previous chapters, thin EDL approximation is valid in microscale electrokinetics as the Debye length is much smaller than the characteristic length in the present study. As a result, the EDL interaction force and van der Waals force, described by a well-established Derjaguin–Landau–Verwey–Overbeek (DLVO) theory (Das and Bhattacharjee 2004; Malysheva, Tang, and Schiavone 2008; Young and Li 2005), are neglected in the present study.

The net charge density in the computational domain is zero due to the thin EDL approximation. Accordingly, the distribution of the electric potential is governed by the Laplace equation:

$$\nabla^2 \phi = 0 \quad \text{in } \Omega. \tag{6.1}$$

The electric potentials on segments AB and CD are, respectively, $\phi = \phi_0$ and $\phi = 0$. All the other boundaries are assumed to be insulating, $\mathbf{n} \bullet \nabla \phi = 0$.

The flow field is governed by the modified Stokes equations because of a negligible inertial effect due to extremely low Reynolds number:

$$\rho \frac{\partial \mathbf{u}}{\partial t} - \mu \nabla^2 \mathbf{u} + \nabla p = 0, \tag{6.2}$$

$$\nabla \bullet \mathbf{u} = 0. \tag{6.3}$$

A symmetrical boundary condition is imposed on segments BC and AD, which are far away from the particles. Segments AB and CD are assumed to be a normal flow with zero pressure. The flow boundary condition on the ith particle surface includes the particle's translational velocity $\mathbf{U}_{pi}$ and rotational velocity $\boldsymbol{\omega}_{pi}$,

$$\mathbf{u}_i = \mathbf{U}_{pi} + \boldsymbol{\omega}_{pi} \times (\mathbf{x}_{si} - \mathbf{x}_{pi}), \tag{6.4}$$

where $\mathbf{x}_{si}$ and $\mathbf{x}_{pi}$ are, respectively, the position vector of the surface and center of the ith particle.

The translational velocity and rotational velocity of the ith particle are, respectively, determined by

$$m_{pi} \frac{d\mathbf{U}_{pi}}{dt} = \mathbf{F}_{Hi} + \mathbf{F}_{DEPi}, \tag{6.5}$$

and

$$I_{pi} \frac{d\boldsymbol{\omega}_{pi}}{dt} = \int (\mathbf{x}_{si} - \mathbf{x}_{pi}) \times (\mathbf{T}^H \bullet \mathbf{n} + \mathbf{T}^E \bullet \mathbf{n}) dS_i. \tag{6.6}$$

In these equations, $\mathbf{F}_{Hi}$ and $\mathbf{F}_{DEPi}$ are, respectively, the hydrodynamic force and the DEP force acting on the ith particle:

$$\mathbf{F}_{Hi} = \int \mathbf{T}^H \bullet \mathbf{n} dS_i = \int \left[-p\mathbf{I} + \mu \left(\nabla \mathbf{u} + \nabla \mathbf{u}^T \right) \right] \bullet \mathbf{n} dS_i, \tag{6.7}$$

$$\mathbf{F}_{DEPi} = \int \mathbf{T}^E \bullet \mathbf{n} dS_i = \int \varepsilon_0 \varepsilon_f \left[\mathbf{EE} - \frac{1}{2} (\mathbf{E} \bullet \mathbf{E}) \mathbf{I} \right] \bullet \mathbf{n} dS_i. \tag{6.8}$$

In these equations, $\mathbf{T}^H$ and $\mathbf{T}^E$ are, respectively, the hydrodynamic and Maxwell stress tensors; ε_0 is the permittivity of the vacuum.

The particle radius a as the length scale; the electric potential applied on segment AB ϕ_0 as the potential scale; $U_0 = \varepsilon_0 \varepsilon_f \phi_0^2 / (\mu a)$ as the velocity scale; and $\mu U_0 / a$ as the pressure scale are selected to normalize all the previous governing equations:

$$\nabla^{*2} \phi^* = 0, \tag{6.9}$$

$$\mathrm{Re} \frac{\partial \mathbf{u}^*}{\partial t^*} - \nabla^{*2} \mathbf{u}^* + \nabla^* p^* = 0, \tag{6.10}$$

$$\nabla^* \bullet \mathbf{u}^* = 0. \tag{6.11}$$

The force and torque are, respectively, normalized by μU_0 and $a\mu U_0$, yielding the dimensionless equations of particle motion:

$$m_{pi}^* \frac{d\mathbf{U}_{pi}^*}{dt^*} = \int (\mathbf{T}^{H*} \bullet \mathbf{n} + \mathbf{T}^{E*} \bullet \mathbf{n}) dS_i^*, \tag{6.12}$$

$$I_{pi}^* \frac{d\omega_{pi}^*}{dt^*} = \int \left(\mathbf{x}_{si}^* - \mathbf{x}_{pi}^*\right) \times \left(\mathbf{T}^{H*} \bullet \mathbf{n} + \mathbf{T}^{E*} \bullet \mathbf{n}\right) dS_i^*, \tag{6.13}$$

where the mass and the moment of inertia are, respectively, normalized by $a\mu/U_0$ and $a^3\mu/U_0$,

$$\mathbf{T}^{H*} = -p^*\mathbf{I} + \left(\nabla^*\mathbf{u}^* + \nabla^*\mathbf{u}^{*T}\right),$$

and

$$\mathbf{T}^{E*} = \mathbf{E}^*\mathbf{E}^* - \frac{1}{2}\left(\mathbf{E}^* \bullet \mathbf{E}^*\right)\mathbf{I}.$$

6.3   NUMERICAL IMPLEMENTATION IN COMSOL

To neglect the gravity effect, we assume that the densities of fluid and particle are the same, $\rho_f = \rho_p = 1.0 \times 10^3$ kg/m^3. The relative permittivity and viscosity of fluid are, respectively, $\varepsilon_f = 80$ and $\mu = 1.0 \times 10^{-3}$ kg/(m•s). The side length of the square is 20 times the particle radius, which ensures that the boundary effect is negligible on particle motion. Table 6.1 shows all the constants defined in COMSOL Multiphysics® 3.5a. Table 6.2 shows detailed instructions for setting up the dimensionless model in the graphical user interface (GUI) of COMSOL Multiphysics. The full MATLAB script M-file capable of automated remeshing is as follows:

```
%%%%%%%%%%%DEP particle-particle interaction.m%%%%%%%%%%%
%%%%%%%%%%%%%%%%E=10KV/m

flclear fem

% NEW: Define the center to center distance between two
  particles
dis=3;

% NEW: Define angle between particle connection line and x
  axis
ang=85*3.14159265359/180;
```

(continued on page 210)

TABLE 6.1
Constant Table

Variable	Value or Expression	Description
eps0	8.854187817e-12 [F/m]	Permittivity of vacuum
eps_r	80	Relative permittivity of fluid
eta	1e-3 [pa*s]	Fluid viscosity
rho_f	1e3 [kg/m^3]	Fluid density
rho	1e3 [kg/m^3]	Particle density
a	10e-6 [m]	Particle radius
XpI0	−2	Dimensionless initial x location of particle I
YpI0	0	Dimensionless initial y location of particle I
Mass	rho*pi*a^2	Dimensional particle mass
Massd	Mass*Uc/eta/a	Dimensionless particle mass
Ird	1/2*Mass*a^2*Uc/eta/a^3	Dimensionless moment of inertia
XpII0	2	Dimensionless initial x location of particle II
YpII0	0	Dimensionless initial y location of particle II
Uc	eps_r*eps0*v0^2/(eta*a)	Velocity scale
v0	2 [V]	Applied voltage
Re	rho_f*a*Uc/eta	Reynolds number

TABLE 6.2
Model Setup in the GUI of COMSOL Multiphysics 3.5a

Model Navigator	Select **2D** in space dimension and click **Multiphysics** button.
	Select **COMSOL Multiphysics|Deformed Mesh|Moving Mesh (ALE)|Transient analysis**. Click **Add** button.
	Select **COMSOL Multiphysics|Electromagnetics|Conductive Media DC**. Click **Add** button.
	Select **MEMS Module|Microfluidics|Stokes Flow|Transient analysis**. Click **Add** button.
Option Menu| **Constants**	Define variables in Table 6.1.
Physics Menu| **Subdomain** **Setting**	ale model
	Subdomain 1
	Free displacement
	emdc mode (external electric field)
	Subdomain 1
	$J_e = 0$; $Q_j = 0$; $d = 1$; $\sigma = 1$
	mmglf mode
	Subdomain 1
	Tab Physics
	$\rho = Re$; $\eta = 1$

continued

TABLE 6.2 (CONTINUED)
Model Setup in the GUI of COMSOL Multiphysics 3.5a

Physics Menu|	ale mode
Boundary	Surface of particle I: Mesh velocity vx = upI−omegaI*(y-YpI)
Setting	vy = vpI+omegaI*(x-XpI)
	Surface of particle II: Mesh velocity vx = upII−omegaII*(y-YpII)
	vy = vpII+omegaII*(x-XpII)
	Other boundaries: Mesh displacement dx = 0; dy = 0
	emdc mode
	Left boundary: V_0 = 1
	Right boundary: Ground
	Other boundaries: Electric insulation
	ns mode
	Left boundary: Inlet|Pressure, no viscous stress P_0 = 0
	Right boundary: Outlet|Pressure, no viscous stress P_0 = 0
	Surface of particle I: Wall|Moving/leaking wall Uw = upI−omegaI*(y-YpI)
	Vw = vpI+omegaI*(x-XpI)
	Surface of particle II: Wall|Moving/leaking wall Uw= upII−omegaII*(y-YpII)
	Vw = vpII+omegaII*(x-XpII)
	Other boundaries: Symmetry boundary
Other Settings	Properties of ale model (**Physics Menu|Properties**)
	Smoothing method: Winslow
	Allow remeshing: On
	Properties of mmglf model (**Physics Menu|Properties**)
	Weak constraints: On
	Constraint type: Nonideal
	Deactivate the unit system (**Physics Menu|Model Setting**)
	Base unit system: None
Options Menu|	Boundary selection: 5, 6, 9, 10 (Surface of Particle I)
Integration	Name: Fh_xI Expression: -lm3; Integration order: 4; Frame: Frame(mesh)
Coupling	Name: Fh_yI Expression: -lm4; Integration order: 4; Frame: Frame(mesh)
Variables|	Name: Fe_xI Expression: (Ex_emdc^2+Ey_emdc^2)*nx/2; Integration order:
Boundary	4; Frame: Frame(mesh)
Variables	Name: Fe_yI Expression: (Ex_emdc^2+Ey_emdc^2)*ny/2; Integration order:
	4; Frame: Frame(mesh)
	Name: TrI Expression: (x-XpI)*((Ex_emdc^2+Ey_emdc^2)*ny/2-lm4)
	−(y-YpI)*((Ex_emdc^2+Ey_emdc^2)*nx/2-lm3); Integration order: 4;
	Frame: Frame(mesh)
	Boundary selection: 7, 8, 11, 12 (Surface of Particle II)
	Name: Fh_xII Expression: -lm3; Integration order: 4; Frame: Frame(mesh)
	Name: Fh_yII Expression: -lm4; Integration order: 4; Frame: Frame(mesh)
	Name: Fe_xII Expression: (Ex_emdc^2+Ey_emdc^2)*nx/2; Integration
	order: 4; Frame: Frame(mesh)

TABLE 6.2 (CONTINUED)
Model Setup in the GUI of COMSOL Multiphysics 3.5a

	Name: Fe_yII Expression: (Ex_emdc^2+Ey_emdc^2)*ny/2; Integration order: 4; Frame: Frame(mesh)
	Name: TrII Expression: (x-XpII)*((Ex_emdc^2+Ey_emdc^2)*ny/2-lm4) −(y-YpII)*((Ex_emdc^2+Ey_emdc^2)*nx/2-lm3); Integration order: 4; Frame: Frame(mesh)
Physics Menu\	Name: upI; Expression: Massd*upIt-(Fh_xI+Fe_xI); Init (u): 0; Init (ut): 0
Global Equations	Name: XpI; Expression: XpIt-upI; Init (u): XpI0; Init (ut): 0
	Name: vpI; Expression: Massd*vpIt-(Fh_yI+Fe_yI); Init (u): 0; Init (ut): 0
	Name: YpI; Expression: YpIt-vpI; Init (u): YpI0; Init (ut): 0
	Name: omegaI; Expression: Ird*omegaIt-TrI; Init (u): 0; Init (ut): 0
	Name: angI; Expression: angIt-omegaI; Init (u): 0; Init (ut): 0
	Name: upII; Expression: Massd*upIIt-(Fh_xII+Fe_xII); Init (u): 0; Init (ut): 0
	Name: XpII; Expression: XpIIt-upII; Init (u): XpII0; Init (ut): 0
	Name: vpII; Expression: Massd*vpIIt-(Fh_yII+Fe_yII); Init (u): 0; Init (ut): 0
	Name: YpII; Expression: YpIIt-vpII; Init (u): YpII0; Init (ut): 0
	Name: omegaII; Expression: Ird*omegaIIt-TrII; Init (u): 0; Init (ut): 0
	Name: angII; Expression: angIIt-omegaII; Init (u): 0; Init (ut): 0
Mesh|Free Mesh	Tab subdomain
Parameters	Subdomain 1|Maximum element size: 0.4
	Tab boundary
	Particle surface: 5, 6, 7, 8, 9, 10, 11, 12|**Maximum element size**: 0.03
	Refine mesh in [-5 5 -5 5]
Solve Menu\	**Solver Parameters**
	Tab General
	Times: Range(0, 200, 80000)
	Relative tolerance: 1e-3
	Absolute tolerance: 1e-7
	Tab Timing Stepping
	Activate Use stop condition
	minqual1_ale-0.4
	Click = **Solve Problem**
Postprocessing	Result check: upI, XpI, vpI, YpI, omegaI, angI, upII, XpII, vpII, YpII,
Menu\	omegaII, angII
	Global Variables Plot|Select all the predefined quantities, click >>
	Click Apply

```
% NEW: Initial location of Particle I (bottom)
XpI=-dis*cos(ang)/2;
YpI=-dis*sin(ang)/2;

% NEW: Initial location of Particle II (bottom)
XpII=dis*cos(ang)/2;
YpII=dis*sin(ang)/2;

% NEW: Define particle radius
radius=10e-6;

% Constants
fem.const = {'eps0','8.854187817e-12', ...
             'eps_r','80', ...
             'eta','1e-3', ...
             'rho_f','1e3', ...
             'rho','1e3', ...
             'a','10e-6', ...
             'XpI0',num2str(XpI), ...
             'YpI0',num2str(YpI), ...
             'Mass','rho*pi*a^2', ...
             'Massd','Mass*Uc/eta/a', ...
             'Ird','1/2*Mass*a^2*Uc/eta/a^3', ...
             'XpII0',num2str(XpII), ...
             'YpII0',num2str(YpII), ...
             'Uc','eps_r*eps0*v0^2/(eta*a)', ...
             'v0','2', ...
             'Re','rho_f*a*Uc/eta'};

% Geometry
% NEW: Generate Particle II
g4=circ2(num2str(radius*1e6),'base','center','pos',{num2str(
  XpII*10),num2str(YpII*10)},'rot','0');

% NEW: Generate Particle I
g5=circ2(num2str(radius*1e6),'base','center','pos',{num2str(
  XpI*10),num2str(YpI*10)},'rot','0');

g7=square2('200','base','center','pos',{'0','0'},'rot','0');
g8=geomcomp({g7,g5,g4},'ns',{'g7','g5','g4'},'sf',
  'g7-g5-g4','edge','none');

% Analyzed geometry
clear s
s.objs={g8};
s.name={'CO1'};
s.tags={'g8'};

fem.draw=struct('s',s);
fem.geom=geomcsg(fem);
```

```
% NEW: Scale the geometry by particle radius
g8=scale(g8,0.1,0.1,0,0);

% Analyzed geometry
clear s
s.objs={g8};
s.name={'CO1'};
s.tags={'g8'};

fem.draw=struct('s',s);
fem.geom=geomcsg(fem);

% NEW: Display geometry with the sequence of the boundary
 geomplot(fem, 'Labelcolor','r','Edgelabels','on');

% (Default values are not included)
% NEW: Solve moving mesh
% Application mode 1
clear appl
appl.mode.class = 'MovingMesh';
appl.sdim = {'Xm','Ym','Zm'};
appl.shape = {'shlag(2,''lm1'')','shlag(2,''lm2'')',
  'shlag(2,''x'')','shlag(2,''y'')'};
appl.gporder = {30,4};
appl.cporder = 2;
appl.assignsuffix = '_ale';
clear prop
prop.smoothing='winslow';
prop.analysis='transient';
prop.allowremesh='on';
prop.origrefframe='ref';
appl.prop = prop;
clear bnd
bnd.defflag = {{1;1},{0;0},{0;0}};
bnd.veldeform = {{0;0},{'upI+UrI';'vpI+VrI'},{'upII+UrII';
  'vpII+VrII'}};
bnd.wcshape = [1;2];
bnd.name = {'fixed','Particle I','Particle II'};
bnd.type = {'def','vel','vel'};
bnd.veldefflag = {{0;0},{1;1},{1;1}};
bnd.ind = [1,1,1,1,2,2,3,3,2,2,3,3];
appl.bnd = bnd;
clear equ
equ.gporder = 2;

equ.shape = [3;4];
equ.ind = [1];
appl.equ = equ;
fem.appl{1} = appl;
```

```
% NEW: Solve flow field using Navier-Stokes equations
% Application mode 2
clear appl
appl.mode.class = 'GeneralLaminarFlow';
appl.module = 'MEMS';
appl.shape = {'shlag(2,''lm3'')','shlag(2,''lm4'')',
  'shlag(1,''lm5'')','shlag(2,''u'')','shlag(2,''v'')',
  'shlag(1,''p'')'};
appl.gporder = {30,4,2};
appl.cporder = {2,1};
appl.assignsuffix = '_mmglf';
clear prop
prop.weakcompflow='Off';
prop.inerterm='Off';
clear weakconstr
weakconstr.value = 'on';
weakconstr.dim = {'lm3','lm4','lm5','lm6','lm7'};
prop.weakconstr = weakconstr;
prop.constrtype='non-ideal';
appl.prop = prop;
clear bnd
bnd.type = {'inlet','sym','outlet','walltype','walltype'};
bnd.vwall = {0,'Veof',0,'vpI+VrI','vpII+VrII'};
bnd.wcshape = [1;2;3];
bnd.wcgporder = 1;
bnd.intype = {'p','uv','uv','uv','uv'};
bnd.U0 = {0.001,0,0,0,0};
bnd.walltype = {'noslip','slip','noslip','lwall','lwall'};
bnd.uwall = {0,'Ueof',0,'upI+UrI','upII+UrII'};
bnd.name = {'Inlet','wall','Outlet','Particle I','Particle
  II'};
bnd.ind = [1,2,2,3,4,4,5,5,4,4,5,5];
appl.bnd = bnd;
clear equ
equ.eta = 1;
equ.gporder = {{2;2;3}};
equ.rho = 'Re';
equ.cporder = {{1;1;2}};
equ.shape = [4;5;6];
equ.ind = [1];
appl.equ = equ;
fem.appl{2} = appl;

% NEW: Solve electrostatics using Laplace equation
% Application mode 3
clear appl
appl.mode.class = 'EmConductiveMediaDC';
appl.module = 'MEMS';
appl.sshape = 2;
appl.assignsuffix = '_emdc';
```

```
clear prop
clear weakconstr
weakconstr.value = 'off';
weakconstr.dim = {'lm8'};
prop.weakconstr = weakconstr;
appl.prop = prop;
clear bnd
bnd.V0 = {1,0,0};
bnd.type = {'V','V0','nJ0'};

bnd.name = {'Electrode','Ground','insulator'};
bnd.ind = [1,3,3,2,3,3,3,3,3,3,3,3];
appl.bnd = bnd;
clear equ
equ.sigma = 1;
equ.name = 'default';
equ.ind = [1];
appl.equ = equ;
fem.appl{3} = appl;
fem.sdim = {{'Xm','Ym'},{'X','Y'},{'x','y'}};
fem.frame = {'mesh','ref','ale'};
fem.border = 1;

% Boundary settings
clear bnd
bnd.ind = [1,1,1,1,2,2,3,3,2,2,3,3];
bnd.dim = {'x','y','lm1','lm2','u','v','p','lmx_mmglf',  ·
   'lmy_mmglf', ...
   'lm3','lm4','lm5','V'};
% Boundary expressions
bnd.expr = {'UrI',{'','-omegaI*(y-YpI)',''}, ...
   'VrI',{'','omegaI*(x-XpI)',''}, ...
   'UrII',{'','','-omegaII*(y-YpII)'}, ...
   'VrII',{'','','omegaII*(x-XpII)'}};
fem.bnd = bnd;

% NEW: Boundary integration to calculate the hydrodynamic
  force
% NEW: and DEP force on the particle
% Coupling variable elements
clear elemcpl
% Integration coupling variables
clear elem
elem.elem = 'elcplscalar';
elem.g = {'1'};
src = cell(1,1);
clear bnd
bnd.expr = {{{},{},'-lm3'},{{},{},'-lm4'},{{},{}, ...

   '(x-XpII)*((Ex_emdc^2+Ey_emdc^2)*ny/2-lm4)-(y-YpII)*((Ex_
   emdc^2+Ey_emdc^2)*nx/2-lm3)'}, ...
```

```
    {{},{},'(Ex_emdc^2+Ey_emdc^2)*nx/2'},{{},{}, ...
    '(Ex_emdc^2+Ey_emdc^2)*ny/2'},{{},'-
    lm3',{}}},{{},'-lm4',{}},{{}, ...
    '(x-XpI)*((Ex_emdc^2+Ey_emdc^2)*ny/2-lm4)-(y-YpI)*((Ex_
    emdc^2+Ey_emdc^2)*nx/2-lm3)', ...
    {}}},{{},'(Ex_emdc^2+Ey_emdc^2)*nx/2',{}}},{{}, ...
    '(Ex_emdc^2+Ey_emdc^2)*ny/2',{}}}};
bnd.ipoints = {{{},{},'4'},{{},{},'4'},{{},{},'4'},{{},{},
    '4'},{{},{},'4'},{ ...
    {},'4',{}},{{},'4',{}},{{},'4',{}},{{},'4',{}},{{},'4',{}}
    };
bnd.frame = {{{},{},'mesh'},{{},{},'mesh'},{{},{},'mesh'},
    ...
    {{},{},'mesh'},{{},{},'mesh'},{{},'mesh',{}},{{},
    'mesh',{}},{{}, ...
    'mesh',{}},{{},'mesh',{}},{{},'mesh',{}}};
bnd.ind = {{'1','2','3','4'},{'5','6','9','10'},{'7','8',
    '11','12'}};
src{1} = {{},bnd,{}};
elem.src = src;
geomdim = cell(1,1);
geomdim{1} = {};
elem.geomdim = geomdim;
elem.var = {'F_xII','F_yII','TrII','Fd_xII','Fd_yII','F_xI',
    'F_yI','TrI', ...
    'Fd_xI','Fd_yI'};
elem.global = {'1','2','3','4','5','6','7','8','9','10'};
elem.maxvars = {};
elemcpl{1} = elem;

fem.elemcpl = elemcpl;

% NEW: Solve six ODEs to determine the particle's transla-
  tional velocity and rotational velocity
% ODE Settings
clear ode
ode.dim={'upI','XpI','vpI','YpI','omegaI','angI','upII','XpI
   I','vpII','YpII','omegaII','angII'};
ode.f={'Massd*upIt-(F_xI+Fd_xI)','XpIt-upI','Massd*vpIt-(F_
   yI+Fd_yI)','YpIt-vpI','Ird*omegaIt-TrI','angIt-
   omegaI','Massd*upIIt-(F_xII+Fd_xII)','XpIIt-
   upII','Massd*vpIIt-(F_yII+Fd_yII)','YpIIt-
   vpII','Ird*omegaIIt-TrII','angIIt-omegaII'};
ode.init={'0','XpI0','0','YpI0','0','0','0','XpII0','0',
   'YpII0','0','0'};
ode.dinit={'0','0','0','0','0','0','0','0','0','0','0','0'};
fem.ode=ode;

% Multiphysics
fem=multiphysics(fem);
```

```
% NEW: Loop for continuous particle tracking
% NEW: Remeshing index
j=1;

% NEW: Define matrix for data storage
particle=zeros(10,17);

% NEW: Define data storage index
num=1;

% NEW: Define initial time
time_e=0;

% NEW: Define time_step
time_step=400;

% NEW: Define end condition of the computation (Loop
  control)
while sqrt((XpI-XpII)^2+(YpI-YpII)^2)>2.1

% Initialize mesh

mesh=0.03;

% Initialize mesh
fem.mesh=meshinit(fem, ...
                  'hauto',5, ...

   'hmaxedg', [5,mesh,6,mesh,7,mesh,8,mesh,9,mesh,10,mesh,11,
   mesh,12,mesh], ...
                  'hmaxsub',[1,0.4]);

% Refine mesh
fem.mesh=meshrefine(fem, ...
                  'mcase',0, ...
                  'boxcoord',[-5 5 -5 5], ...
                  'rmethod','regular');

% Extend mesh
fem.xmesh=meshextend(fem);

if j==1
% Solve problem
fem.sol=femtime(fem, ...
'solcomp',{'upI','vpI','omegaI','upII','vpII','omegaII',
  'angI','lm4','YpI','lm3','lm2','lm1','YpII','lm5','XpI',
  'XpII','v','u','V','angII','p','y','x'}, ...

'outcomp',{'upI','vpI','omegaI','upII','vpII','omegaII',
  'angI','lm4','YpI','lm3','lm2','lm1','YpII','lm5','XpI',
  'XpII','v','u','V','angII','p','Y','X','y','x'}, ...
```

```
                        'blocksize','auto', ...
                        'tlist',[0:time_step:80000], ...
                        'tout','tlist', ...
                        'rtol',1e-3, ...
                        'atol',1e-7, ...
                        'stopcond','minquall_ale-0.4');

else
% NEW: Map the solution in the deformed mesh into the new
  geometry with undeformed mesh
init = asseminit(fem,'init',fem0.sol,'xmesh',fem0.xmesh,
  'framesrc','ale','domwise','on');

fem.sol=femtime(fem, ...
                'init',init,...

'solcomp',{'upI','vpI','omegaI','upII','vpII','omegaII',
  'angI','lm4','YpI','lm3','lm2','lm1','YpII','lm5','XpI',
  'XpII','v','u','V','angII','p','y','x'}, ...

'outcomp',{'upI','vpI','omegaI','upII','vpII','omegaII',
  'angI','lm4','YpI','lm3','lm2','lm1','YpII','lm5','XpI',
  'XpII','v','u','V','angII','p','Y','X','y','x'}, ...
                'blocksize','auto', ...
                'tlist',[time_e:time_step:80000], ...
                'tout','tlist', ...
                'rtol',1e-3, ...
                'atol',1e-7, ...
                'stopcond','minquall_ale-0.4');
end

% Save current fem structure for restart purposes

fem0=fem;

% NEW: Get the last time from solution
time_e = fem.sol.tlist(end);

% NEW: Get the length of the time steps
Leng =length(fem.sol.tlist);

% Global variables plot
data=postglobalplot(fem,{'angI','omegaI','XpI','YpI','upI',
  'vpI','angII','omegaII','XpII','YpII','upII','vpII'}, ...
                'linlegend','on', ...
                'title','Particle', ...
                'Outtype','postdata', ...
                'axislabel',{'Time','Parameters'});

% NEW: Store interested data
```

```
for n=2:Leng
    particle(num,1)= data.p(1,n);
    particle(num,2)= data.p(2,n);
    particle(num,3)= data.p(2,n+Leng);
    particle(num,4)= data.p(2,n+2*Leng);
    particle(num,5)= data.p(2,n+3*Leng);
    particle(num,6)= data.p(2,n+4*Leng);
    particle(num,7)= data.p(2,n+5*Leng);
    particle(num,8)= data.p(2,n+6*Leng);
    particle(num,9)= data.p(2,n+7*Leng);
    particle(num,10)=data.p(2,n+8*Leng);
    particle(num,11)=data.p(2,n+9*Leng);
    particle(num,12)=data.p(2,n+10*Leng);
    particle(num,13)=data.p(2,n+11*Leng);

    % Integrate
    I1=postint(fem,'(Ex_emdc^2+Ey_emdc^2)*nx/2', ...
                    'dl',[5,6,9,10], ...
                    'edim',1, ...
                    'solnum',n);

    % Integrate
    I2=postint(fem,'(Ex_emdc^2+Ey_emdc^2)*ny/2', ...
                    'dl',[5,6,9,10], ...
                    'edim',1, ...
                    'solnum',n);

    % Integrate
    I3=postint(fem,'(Ex_emdc^2+Ey_emdc^2)*nx/2', ...
                    'dl',[7,8,11,12], ...
                    'edim',1, ...
                    'solnum',n);
    % Integrate
    I4=postint(fem,'(Ex_emdc^2+Ey_emdc^2)*ny/2', ...
                    'dl',[7,8,11,12], ...
                    'edim',1, ...
                    'solnum',n);
    particle(num,14)=I1;
    particle(num,15)=I2;
    particle(num,16)=I3;
    particle(num,17)=I4;
    num=num+1;
end

% NEW: Write the data into a file
dlmwrite('dis_3_ang_85_10KV_New.dat',particle,',');

% NEW: update the particle's location and orientation
% NEW: Output them in the MATLAB
XpI=data.p(2,3*Leng)
```

```
XpII=data.p(2,9*Leng)
YpI=data.p(2,4*Leng)
YpII=data.p(2,10*Leng)
split=' '

% NEW: Output the COMSOL Multiphysics GUI .mph file if
  necessary
flsave(strcat('dis_3_ang_85_10KV_New_',num2str(j),'.
  mph'),fem);

% Plot solution
postplot(fem, ...
        'tridata',{'normE_emdc','cont','internal'}, ...
        'trimap','jet(1024)', ...
        'tridlim',[0 0.15], ...
        'solnum','end', ...
        'title','Surface: Electric field', ...
        'geom','off', ...
        'axis',[-12,12,-11,11]);

% Geometry
% Generate geom from mesh
fem = mesh2geom(fem, ...
              'frame','ale', ...
              'srcdata','deformed', ...
              'destfield',{'geom','mesh'}, ...
              'srcfem',1, ...
              'destfem',1);

j=j+1;
end
```

6.4 RESULTS AND DISCUSSION

6.4.1 Comparison between DEP Particle–Particle Interaction and Brownian Motion

Brownian motion is a random thermal motion that gives rise to collisions between suspended particles and surrounding liquid molecules. The importance of Brownian motion in the present study can be evaluated by Péclet number, defined as

$$Pe = \frac{aU_p}{D_B}.$$

(6.14)

In this equation, a, U_p, and D_B are, respectively, the particle radius, particle velocity due to the DEP particle–particle interaction, and Brownian diffusion (Wilson, Pietraszewski, and Davis 2000). It has been found that the DEP particle–particle interaction force decays very fast as the particle distance increases

(Kang and Li 2006). As a result, there is a possibility that the Brownian motion could become comparable to the DEP particle–particle interaction when the particle distance is large enough. In addition, the maximum DEP particle–particle interaction exists when two particles are parallel to the externally applied electric field. An approximation solution of the DEP particle–particle interaction force in the parallel configuration is (Kang and Li 2006)

$$F_{DEP} = \frac{3\pi\varepsilon_0\varepsilon_f E^2 a^2}{(R/a)^4}, \tag{6.15}$$

which is valid when $R > 4a$. Assume the Stokes drag acting on the particle is a constant; the particle velocity due to the DEP particle–particle interaction is given as

$$U_P = \frac{F_{DEP}}{6\pi\mu a}. \tag{6.16}$$

The Brownian diffusion is given as (Wilson, Pietraszewski, and Davis 2000)

$$D_B = \frac{KT}{3\pi\mu a}, \tag{6.17}$$

where K and T are, respectively, the Boltzmann constant and the absolute temperature. By substituting Equations (6.15)–(6.17) into Equation (6.14), the Péclet number can be rewritten as

$$Pe = \frac{3\pi\varepsilon_0\varepsilon_f E^2 a^3}{2KTR^{*4}}. \tag{6.18}$$

Apparently, the DEP particle–particle interaction is on the same order of the Brownian motion when $Pe \approx 1$. Therefore, the critical particle distance beyond which Brownian motion could dramatically disturb the particle motion due to the DEP particle–particle interaction is given as

$$R^* = \left(\frac{3\pi\varepsilon_0\varepsilon_f E^2 a^3}{2KT}\right)^{\frac{1}{4}}. \tag{6.19}$$

For example, the critical particle distance is approximately $R = 16a$ when $E = 10$ kV/m, $a = 10$ µm, $\varepsilon_f = 80$, and $T = 300$ K. To minimize the disturbance due to the Brownian motion, the particle distance in the present study is restricted to being no larger than $R = 5a$. As a result, the Péclet number is larger than 130, under which the Brownian motion becomes negligible compared to the DEP particle–particle interaction and thus is ignored in the present study.

6.4.2 PARALLEL ORIENTATION, θ = 0°

All the following results are presented in dimensionless form and are organized on the basis of the particle's initial orientation with respect to the applied electric field. In this section, we consider two particles initially located at $(x_0^*, y_0^*) = (\pm2.5, 0)$, whose center-to-center connecting lines are parallel to an external electric field $E^* = 0.05$ ($E = 10$ kV/m). The numerical prediction shows that the two particles move toward each other at an identical translational velocity. The predicted attractive motion is in good agreement with the previous study (Kang and Li 2006).

Figure 6.2a shows the electric field around the two particles located at $(x^*, y^*) = (\pm1.56, 0)$. As the electric field between the two particles is reduced, the negative DEP thus leads to the attractive motion. In addition, the nonuniform electric field stems from the coexistence of the two particles, which is thus called DEP particle–particle interaction. The attractive DEP particle–particle interaction is responsible for the particle chaining, which has been widely used in particle self-assembly (Velev, Gangwal, and Petsev 2009). Because of the symmetric attractive motion, the fluids between the two particles are pushed out, which generates a symmetric flow pattern with respect to the center of the computational domain, as shown in Figure 6.2b. In addition, the pressure between the two particles is enhanced; however, this resists the attractive motion. Therefore, it is not appropriate to ignore the resistant pressure force in the previous study (Kang and Li 2006) when the two particles are very close to each other.

Figure 6.3 shows the translational velocities of the two particles along their corresponding travelling paths. It is confirmed that their velocities are identical in magnitude but opposite in direction. At the beginning, the particle velocity increases as the two particles approach each other due to an increased DEP force. The resistant pressure force also increases accordingly; however, this is slower than the attractive DEP force. When the center-to-center particle distance decreases to a critical value (e.g., $R^* \approx 3$ for this case), the resistant pressure force increases faster than the attractive DEP force. As a result, the particle velocity begins to decrease. To maintain enough finite elements between the two particles, they actually do not contact each other in the present study. In addition, when the distance between the two particles is on the order of the Debye length, the EDL interaction force must be considered (Das and Bhattacharjee 2004; Malysheva, Tang, and Schiavone 2008; Young and Li 2005); however, this is beyond the scope of this study. Nevertheless, it is still predictable that the particle velocity is almost zero when they nearly contact each other.

6.4.3 PERPENDICULAR ORIENTATION, θ = 90°

Here, we consider two particles initially located at $(x_0^*, y_0^*) = (0, \pm1.5)$ whose center-to-center connecting lines are perpendicular to the external electric field $E^* = 0.05$ ($E = 10$ kV/m). The inset in Figure 6.4 shows that the electric field

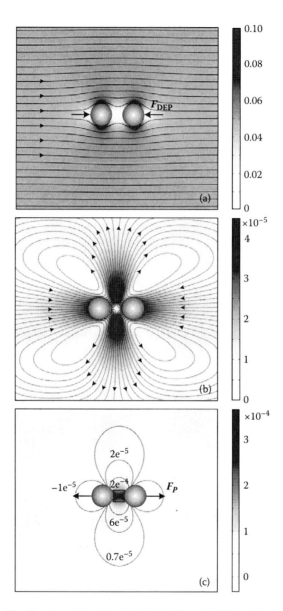

FIGURE 6.2   Distributions of the electric field (a), flow field (b), and pressure (c) around a pair of particles located at $(x^*, y^*) = (\pm 1.56, 0)$ subjected to an external electric field $\mathbf{E}^* = 0.05$ ($\mathbf{E} = 10$ KV/m). Lines in (a), (b), and (c) represent, respectively, the streamlines of the electric field and flow field and the contours of the pressure. Darkness represents the magnitude of the corresponding parameters. The DEP particle–particle interaction force shown in (a), $\mathbf{F}_{DEP}$, tends to attract the two particles; the pressure force denoted in (c), $\mathbf{F}_P$, retards the attraction motion. (From Ai, Y., and S. Qian. 2010. DC dielectrophoretic particle–particle interactions and their relative motions. *Journal of Colloid and Interface Science* 346:448–454 with permission from Elsevier.)

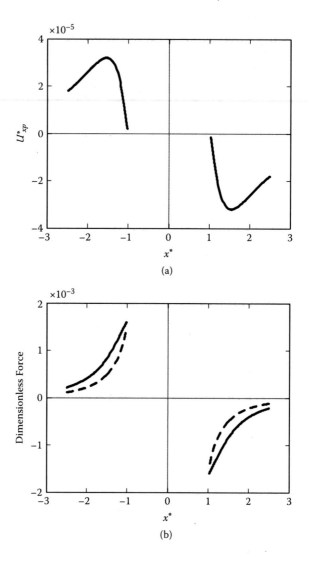

FIGURE 6.3 Velocity (a) and force (b) variations of a pair of particles initially located at $(x^*, y^*) = (\pm 2.5, 0)$ subjected to an external electric field $\mathbf{E}^* = 0.05$. The solid line and dashed line in (b) represent, respectively, the magnitude of the DEP particle–particle interaction force and the hydrodynamic pressure force in the x direction. (From Ai, Y., and S. Qian. 2010. DC dielectrophoretic particle–particle interactions and their relative motions. *Journal of Colloid and Interface Science* 346:448–454 with permission from Elsevier.)

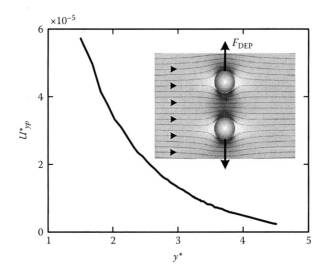

FIGURE 6.4   Velocity variation of a pair of particles initially located at $(x_0^*, y_0^*) =$ (0, ±1.5) subjected to an external electric field $\mathbf{E}^* = 0.05$. The inset denotes the distribution and streamlines of the local electric field around the two particles. The darkness represents the magnitude of the electric field strength. The arrows represent the direction of the DEP particle–particle interaction force. (From Ai, Y., and S. Qian. 2010. DC dielectrophoretic particle–particle interactions and their relative motions. *Journal of Colloid and Interface Science* 346:448–454 with permission from Elsevier.)

between the two particles is enhanced. The resultant DEP particle–particle interaction thus acts as a repulsive force, which pushes the two particles away from each other. As the DEP particle–particle interaction decays as the particle distance increases, the particle velocity accordingly decreases until the DEP particle–particle interaction becomes negligible, as shown in Figure 6.4. Meanwhile, the effect of Brownian motion becomes increasingly significant as the particle distance increases; a perfect perpendicular orientation is thus extremely unstable (Kang and Li 2006).

6.4.4   INTERMEDIATE ORIENTATION, $0° < \theta < 90°$

In general, the initial particle orientation is mostly between the two previous critical configurations. Therefore, it is more reasonable to study the DEP particle–particle interaction with an arbitrary orientation. Figure 6.5 shows the trajectories of two particles initially located at $(x_0^*, y_0^*) = [\pm 1.5 \cos(85°), \pm 1.5 \sin(85°)]$ under the electric field $\mathbf{E}^* = 0.05$. As the connecting line of the two particles is nearly perpendicular to the applied electric field, the DEP particle–particle interaction force pushes the two particles away from each other at the beginning. In addition,

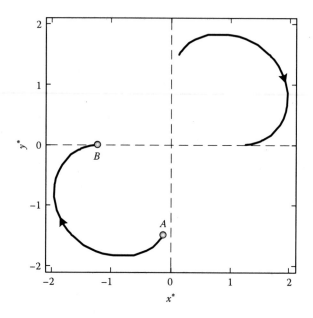

FIGURE 6.5 Trajectories of a pair of particles initially located with $R^* = 3$ and $\theta = 85°$ subjected to an external electric field $\mathbf{E}^* = 0.05$. Point A and B represent, respectively, the starting and ending positions of the particle. (From Ai, Y., and S. Qian. 2010. DC dielectrophoretic particle–particle interactions and their relative motions. *Journal of Colloid and Interface Science* 346:448–454 with permission from Elsevier.)

the two particles rotate with respect to each other owing to the x component of the DEP particle–particle interaction force. As a result, the nearly perpendicular configuration gradually becomes a nearly parallel configuration. The DEP particle–particle interaction force thus becomes attractive and pulls the two particles moving toward each other, similar to the attractive motion described in Section 6.4.2. The attractive motion of two particles initially presenting a large angle with respect to the external electric field has been experimentally observed (Hwang, Kim, and Park 2008), which also verifies our numerical model. It can be concluded that the DEP particle–particle interaction always tends to attract and align particles parallel to the external electric field except for the unstable perfect perpendicular configuration.

Figure 6.6 shows the effect of the applied electric field on the trajectories of two particles initially located at $(x_0^*, y_0^*) = [\pm 1.5 \cos(85°), \pm 1.5 \sin(85°)]$. Because of the antisymmetrical nature of the particle trajectories, we only show the trajectories of the particle located at $(x_0^*, y_0^*) = [1.5 \cos(85°), 1.5 \sin(85°)]$ under three different electric fields. The particle distance during the attractive motion under a higher electric field is slightly larger than that under a lower electric field. However, the traveling time required to achieve particle chaining under $\mathbf{E}^* = 0.05$ is approximately four times that under $\mathbf{E}^* = 0.1$, which indicates that the required

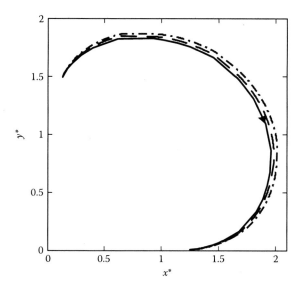

FIGURE 6.6   Trajectories of the upper one in a pair of particles initially located with $R = 3$ and $\theta = 85°$ under different electric fields, $E^* = 0.05$ (solid line), 0.1 (dashed line), and 0.15 (dash-dotted line). (From Ai, Y., and S. Qian. 2010. DC dielectrophoretic particle–particle interactions and their relative motions. *Journal of Colloid and Interface Science* 346:448–454 with permission from Elsevier.)

traveling time is proportional to $(1/E^*)^2$. Thus, it is efficient to reduce the time required for particle assembly by increasing the electric field.

Keeping the initial particle distance $R^* = 3$ unchanged and reducing the initial orientation to $\theta = 45°$, the particles also end up with an attractive motion, as shown in Figure 6.7. The attractive motion is even faster than the case when $R^* = 3$ and $\theta = 85°$. If the initial particle distance is increased to $R^* = 4$ while keeping the initial orientation $\theta = 45°$, the two particles also eventually form a chain parallel to the applied electric field. As the DEP particle–particle interaction becomes weaker at a larger initial particle distance, the time required for particle chaining increases to 1.6 times that when $R^* = 3$ and $\theta = 45°$.

Figure 6.8 shows the velocity of the lower particle shown in Figure 6.7. As the DEP particle–particle interaction force for $R^* = 3$ is stronger than that for $R^* = 4$, both x and y component velocities for $R^* = 3$ are higher than those for $R^* = 4$. Because the velocity variation for $R^* = 3$ is very similar to that for $R^* = 4$, we only discuss the latter case. At the beginning, the x component velocity (solid line) is negative, and its magnitude decreases until it becomes positive. After that, the two particles are approaching each other, resulting in an increased DEP particle–particle interaction. Therefore, the magnitude of the x component velocity increases and maximizes at a certain value and then decreases toward zero, as discussed in Figure 6.3a. The y component velocity (dashed line) is positive during the entire movement. It varies slightly before the x component velocity

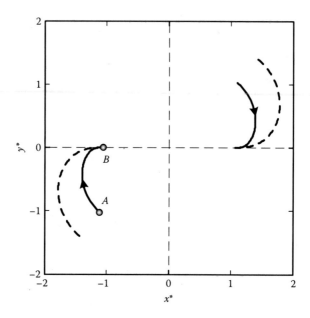

FIGURE 6.7 Trajectories of a pair of particles initially located with $\theta = 45°$ and $R^* = 3$ (solid lines) and $R^* = 4$ (dashed lines) subjected to an external electric field $\mathbf{E}^* = 0.05$. Point A and B represent, respectively, the starting and ending positions of the particle. (From Ai, Y., and S. Qian. 2010. DC dielectrophoretic particle–particle interactions and their relative motions. *Journal of Colloid and Interface Science* 346:448–454 with permission from Elsevier.)

becomes positive. Subsequently, it begins to decrease toward zero as the attractive motion mainly happens along the x axis.

6.5 CONCLUDING REMARKS

DEP particle–particle interaction and the particles' relative motions were numerically studied under the thin EDL approximation. The critical particle distance beyond which Brownian motion could significantly disturb the DEP particle–particle interaction is proportional to $(E^2 a^3)^{1/4}$. When the center-to-center particle distance is much shorter than the critical distance mentioned, the DEP particle–particle interaction always dominates Brownian motion. It was further found that the DEP particle–particle interaction always ends up with an attractive motion independent of their initial orientation except for the unstable perpendicular configuration. This numerical prediction is in qualitative agreement with the existing experimental observation. When two particles approach each other, their velocities increase at the beginning and maximize at a certain particle distance. After that, the repulsive pressure force increases faster and thus reduces the particle velocity toward zero in a near-contact mode. The DEP particle–particle interaction has huge applications in particle self-assembly.

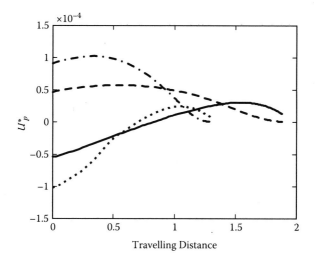

FIGURE 6.8 Velocity variations of one of the particles in Figure 6.7. The dotted line and solid line denote, respectively, the x component velocities of the particle with initial distance $R^* = 3$ and 4. The dash-dotted line and dashed line denote, respectively, the y component velocities of the particle with initial distance $R^* = 3$ and 4. (From Ai, Y., and S. Qian. 2010. DC dielectrophoretic particle–particle interactions and their relative motions. *Journal of Colloid and Interface Science* 346:448–454 with permission from Elsevier.)

REFERENCES

Ai, Y., and S. Qian. 2010. DC dielectrophoretic particle–particle interactions and their relative motions. *Journal of Colloid and Interface Science* 346:448–454.

Al-Jarro, A., J. Paul, D. W. P. Thomas, J. Crowe, N. Sawyer, F. R. A. Rose, and K. M. Shakesheff. 2007. Direct calculation of Maxwell stress tensor for accurate trajectory prediction during DEP for 2D and 3D structures. *Journal of Physics D–Applied Physics* 40 (1):71–77.

Aubry, N., and P. Singh. 2006a. Control of electrostatic particle–particle interactions in dielectrophoresis. *Europhysics Letters* 74 (4):623–629.

Aubry, N., and P. Singh. 2006b. Influence of particle–particle interactions and particles rotational motion in traveling wave dielectrophoresis. *Electrophoresis* 27:703–715.

Das, P. K., and S. Bhattacharjee. 2004. Electrostatic double-layer interaction between spherical particles inside a rough capillary. *Journal of Colloid and Interface Science* 273 (1):278–290.

Gangwal, S., O. J. Cayre, and O. D. Velev. 2008. Dielectrophoretic assembly of metallodielectric janus particles in AC electric fields. *Langmuir* 24 (23):13312–13320.

Gupta, S., R. G. Alargova, P. K. Kilpatrick, and O. D. Velev. 2008. On-chip electric field driven assembly of biocomposites from live cells and functionalized particles. *Soft Matter* 4 (4):726–730.

Hermanson, K. D., S. O. Lumsdon, J. P. Williams, E. W. Kaler, and O. D. Velev. 2001. Dielectrophoretic assembly of electrically functional microwires from nanoparticle suspensions. *Science* 294 (5544):1082–1086.

Hwang, H., J. J. Kim, and J. K. Park. 2008. Experimental investigation of electrostatic particle–particle interactions in optoelectronic tweezers. *Journal of Physical Chemistry B* 112 (32):9903–9908.

Juarez, J. J., and M. A. Bevan. 2009. Interactions and microstructures in electric field mediated colloidal assembly. *Journal of Chemical Physics* 131 (13):134704.

Kadaksham, A. T. J., P. Singh, and N. Aubry. 2004a. Dielectrophoresis of nanoparticles. *Electrophoresis* 25 (21–22):3625–3632.

Kadaksham, J., P. Singh, and N. Aubry. 2004b. Dynamics of electrorheological suspensions subjected to spatially nonuniform electric fields. *Journal of Fluids Engineering–Transactions of the ASME* 126 (2):170–179.

Kadaksham, J., P. Singh, and N. Aubry. 2005. Dielectrophoresis induced clustering regimes of viable yeast cells. *Electrophoresis* 26 (19):3738–3744.

Kadaksham, J., P. Singh, and N. Aubry. 2006. Manipulation of particles using dielectrophoresis. *Mechanics Research Communications* 33 (1):108–122.

Kang, K. H., and D. Q. Li. 2006. Dielectric force and relative motion between two spherical particles in electrophoresis. *Langmuir* 22 (4):1602–1608.

Kang, Y. J., and D. Q. Li. 2009. Electrokinetic motion of particles and cells in microchannels. *Microfluidics and Nanofluidics* 6 (4):431–460.

Lapizco-Encinas, B. H., and M. Rito-Palomares. 2007. Dielectrophoresis for the manipulation of nanobioparticles. *Electrophoresis* 28 (24):4521–4538.

Li, Y. L., C. Dalton, H. J. Crabtree, G. Nilsson, and K. Kaler. 2007. Continuous dielectrophoretic cell separation microfluidic device. *Lab on a Chip* 7 (2):239–248.

Malysheva, O., T. Tang, and P. Schiavone. 2008. Adhesion between a charged particle in an electrolyte solution and a charged substrate: Electrostatic and van der Waals interactions. *Journal of Colloid and Interface Science* 327 (1):251–260.

Rosales, C., and K. M. Lim. 2005. Numerical comparison between Maxwell stress method and equivalent multipole approach for calculation of the dielectrophoretic force in single-cell traps. *Electrophoresis* 26 (11):2057–2065.

Swaminathan, T. N., and H. H. Hu. 2004. Particle interactions in electrophoresis due to inertia. *Journal of Colloid and Interface Science* 273 (1):324–330.

Velev, O. D., and K. H. Bhatt. 2006. On-chip micromanipulation and assembly of colloidal particles by electric fields. *Soft Matter* 2 (9):738–750.

Velev, O. D., S. Gangwal, and D. N. Petsev. 2009. Particle-localized AC and DC manipulation and electrokinetics. *Annual Reports Section "C" (Physical Chemistry)* 105:213–246.

Wang, X. J., X. B. Wang, and P. R. C. Gascoyne. 1997. General expressions for dielectrophoretic force and electrorotational torque derived using the Maxwell stress tensor method. *Journal of Electrostatics* 39 (4):277–295.

Wilson, H. J., L. A. Pietraszewski, and R. H. Davis. 2000. Aggregation of charged particles under electrophoresis or gravity at arbitrary Peclet numbers. *Journal of Colloid and Interface Science* 221 (1):87–103.

Yariv, E. 2004. Inertia-induced electrophoretic interactions. *Physics of Fluids* 16 (4):L24-L27.

Young, E. W. K., and D. Q. Li. 2005. Dielectrophoretic force on a sphere near a planar boundary. *Langmuir* 21 (25):12037–12046.

7 Electrokinetic Translocation of a Cylindrical Particle through a Nanopore *Poisson–Boltzmann Approach*

In this chapter, the electric field-induced translocation of a cylindrical nanoparticle through a nanopore is theoretically investigated with full consideration of the particle-fluid-electric field-ionic concentration field interactions. The coupled Poisson–Boltzmann (PB) equation for the ionic concentrations and the electric field stemming from the charged surfaces, the Laplace equation for the electric field externally imposed, the modified Stokes equations for the flow field, and the Newton equations for particle translation and rotation are simultaneously solved using the arbitrary Lagrangian–Eulerian (ALE) finite element method. The dynamic electrokinetic translocation of a cylindrical nanoparticle through a nanopore and the corresponding ionic current response are investigated as functions of the electric field intensity imposed, the electrical double layer (EDL) thickness, the nanopore's surface charge density, and the particle's initial orientation and lateral offset from the centerline of the nanopore. The translocating particle blocks the ionic current when the EDLs of the particle and the nanopore are not overlapped, and the predictions are in qualitative agreement with the existing experimental observations.

Under a relatively low imposed electric field, the particle experiences a significant rotation and lateral movement if the particle's axis is not coincident with the centerline of the nanopore. The particle is aligned with its longest axis parallel to the local electric field very quickly, arising from the dielectrophoretic (DEP) effect when the imposed electric field is relatively high. The PB model is valid only when the EDL is relatively thin (the EDL thickness is smaller than the particle size) and is not disturbed by the external electric field and nearby EDLs of solid boundaries.

7.1 INTRODUCTION

We consider two identical fluid reservoirs separated by an electrically insulating membrane equipped with a single nanopore, as schematically shown in Figure 7.1. The nanofluidic device is filled with an electrolyte solution. One of the fluid reservoirs (say the bottom one in Figure 7.1) contains charged nanoparticles. An electric field is externally imposed between the two reservoirs by applying a potential difference between two electrodes, each positioned in one fluid reservoir, inducing ionic current through the nanopore. The charged particle electrophoretically translocates from one reservoir to the other and simultaneously affects the ionic current through the pore. By monitoring the change in the ionic current through the pore, one hopes to characterize and sense individual nanoscale particles, especially DNA molecules, proteins, and organic polymers, for various bioanalytical applications (Choi et al. 2006; Martin and Siwy 2007; Howorka and Siwy 2009; Gu and Shim 2010). For example, such nanopore-based technology may be used to read and sequence human DNA directly and offers significant advantages in cost, throughput, and speed compared to other DNA sequencing technologies (Bayley 2006; Rhee and Burns 2006; Mukhopadhyay 2009).

The encouraging benefits of nanopore-based DNA sequencing have stimulated a fast-growing body of research on electrokinetic particle translocation through a nanopore and the corresponding ionic current response (Chang et al. 2004; Storm et al. 2005; Kim et al. 2007; Lathrop et al. 2010). Due to the small size of the

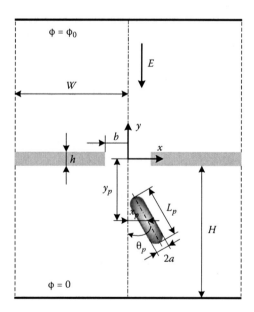

FIGURE 7.1 Schematics of the translocation of a nanoparticle through a nanopore. (From Ai, Y., and S. Qian. 2011. Direct numerical simulation of electrokinetic translocation of a cylindrical particle through a nanopore using a Poisson–Boltzmann approach. *Electrophoresis* 32:996–1005 with permission from Wiley-VCH.)

particles, it is not trivial to observe the dynamics of particle translocation through a nanopore experimentally. Therefore, mathematical modeling of electrokinetic particle translocation through a nanopore is highly desired to explore the underlying mechanism of particle translocation and provide a knowledge base for rational design of experiments.

Molecular dynamics (MD) simulations have been used to model the electrokinetic translocation and even conformational change of DNA molecules through a nanopore (Aksimentiev et al. 2004; Sigalov et al. 2007; Zhao, Payne, and Cummings 2008; Comer et al. 2009). However, the time scale of MD simulations is restricted to about 100 ns and is unable to capture a practical DNA translocation through a nanopore, which takes a much longer time (on the order of microseconds to milliseconds) than the capacity of MD simulations. To reduce the time of simulations, the MD simulations were carried out at a much higher electric field than those typically used in the actual experiments. The predictions of a continuum-based model, including the Poisson equation for the electrostatics, Nernst–Planck equations for the transport of ions in solution, and Navier–Stokes and continuity equations for the flow field, are in good qualitative agreement with the predictions obtained from the MD simulations and the experimental data for DNA translocation (Qian, Wang, and Afonien 2006; Qian and Joo 2008; Qian et al. 2008, 2009; Ai et al. 2010). Therefore, a continuum-based model still successfully captures the essential physics of the translocation process.

In the continuum-based models, the fully coupled Poisson–Nernst–Planck (PNP) equations are the most rigorous method to determine the distributions of the ionic concentration field and the electric field within the EDL in the vicinity of a charged surface (Qian, Wang, and Afonien 2006; Qian and Joo 2008; Qian et al. 2008, 2009; Ai et al. 2010). When the formed EDL is not disturbed or distorted by an external field, including the electric or concentration field imposed and the fluid motion induced, the ionic concentrations obey Boltzmann distribution. Accordingly, the fully coupled PNP equations can be degraded to the PB equation, which significantly reduces the computational complexity. It is found that the PB-based continuum model is still valid for electrokinetic particle translocation through a nanopore when the particle size is not smaller than the Debye length as the characteristic length of the EDL (Corry, Kuyucak, and Chung 2000; Moy et al. 2000; van Dorp et al. 2009). Liu, Qian, and Bau (2007) also confirmed that the predictions from the PB model and the PNP model are in good agreement when the EDL is thin.

Based on the PB model, Henry (1931) derived the famous Henry's function to account for the finite EDL effect on electrophoresis of a sphere in an unbounded medium when the surface potential (or zeta potential) of the particle is small. Later, Ennis and Anderson (1997) used the PB model and the method of reflection and successfully derived analytical approximation solutions for the electrophoretic velocity of a charged sphere near a single flat wall, within a slit and cylindrical tube when the zeta potentials are relatively small under the conditions mentioned as necessary for the PB-based model. For complicated geometries of nanopores and particles, the particle's translational velocity can be determined by balancing the total force exerted on the particle, assuming all the physical fields are at their equilibrium states (Hsu, Kuo, and Ku 2006, 2008; Qian, Wang, and Afonien 2006; Liu, Qian, and Bau 2007; Qian and Joo 2008; Qian et al. 2008, 2009; Chen and Conlisk 2010).

As a large number of nanoparticles (e.g., DNA molecules and synthetic nanowires) can be approximated as a cylindrical shape, Hsu's group (Hsu and Kao 2002; Hsu and Kuo 2006; Hsu, Chen, et al. 2008), Liu et al. (Liu, Bau, and Hu 2004; Liu, Qian, and Bau 2007), and Chen and Conlisk (Chen and Conlisk 2010) used the quasi-static method and PB model mentioned to investigate the electrokinetic translocation of cylindrical particles through pores under various conditions. However, the quasi-static method is unable to capture the particle translocation dynamics and especially the particle rotation during the translocation process, which could play important roles in particle translocation and the corresponding change in the ionic current.

In this chapter, a transient continuum-based model, consisting of the PB equation for the ionic concentration field and the electric field stemming from the surface charges of the nanoparticle and the nanopore, the Laplace equation for the electric field externally imposed, the modified Stokes equations for the hydrodynamic field, and the Newton equations for the particle translation and rotation, is formulated to investigate the dynamic translocation of a cylindrical particle through a nanopore. The ALE method is used to track the particle's translation and rotation. Detailed implementation of direct numerical simulation of the model is introduced in Section 7.3. Section 7.4 introduces some representative simulation results on the dynamic particle translocation and the ionic current through the nanopore as functions of the electric field intensity imposed, the EDL thickness, the nanopore's surface charge density, and the particle's initial orientation and lateral offset from the centerline of the nanopore.

7.2 MATHEMATICAL MODEL

Consider a negatively charged cylindrical particle of length L_p and radius a having two hemispherical caps of radius a at both ends; it is submerged in a binary symmetric electrolyte solution (such as KCl) with density ρ, dynamic viscosity μ, and relative permittivity ε_f, as shown in Figure 7.1. The solution is confined in a vessel that is separated by an electrically insulating membrane of thickness h into two identical fluid reservoirs, of width $2W$ and height H. The membrane is equipped with a single pore of radius b and has a uniformly distributed surface charge density σ_w. We assume the particle's surface is uniformly charged with surface charge density σ_p. Far away from the charged nanopore and the particle, the electrolyte solution is neutral and has its bulk concentration C_0.

To drastically reduce computational effort, only a two-dimensional (2D) model is considered here, and all variables are defined in a 2D Cartesian coordinate system (x, y) with the origin fixed at the center of the nanopore. The particle, with its center of mass initially located at (x_{p0}, y_{p0}), presents an initial angle θ_{p0} with respect to the centerline of the nanopore. The initial angle θ_{p0} is defined as positive when the particle rotates counterclockwise with respect to the centerline of the nanopore. An electric field $\mathbf{E}$ is imposed by applying a potential difference ϕ_0 across two electrodes positioned inside the two reservoirs, which generates ionic current through the pore and causes the particle to translocate through the nanopore due to electrophoresis.

7.2.1 Dimensional Form of Mathematical Model

When the externally applied electric field is relatively weaker than the electric field arising from the surface charges, it is appropriate to decompose the overall electric field linearly into the two electric fields mentioned. The external electric field associated with the electric potential ϕ is governed by the Laplace equation:

$$\nabla^2 \phi = 0 \tag{7.1}$$

The electric potentials applied at the two ends of the reservoirs are, respectively, $\phi(x, -(H+h/2)) = 0$ and $\phi(x, (H+h/2)) = \phi_0$. All the other boundaries are assumed to be insulation, $\mathbf{n} \bullet \nabla \phi = 0$, where $\mathbf{n}$ is the unit normal vector pointing from the boundary into the fluid.

As described in Chapter 1, co-ions and counterions near a charged surface both obey the Boltzmann distribution when its EDL is not disturbed by the external fields, nearby boundaries, and EDLs:

$$c_i = C_0 \exp(-z_i \frac{F\psi}{RT}). \tag{7.2}$$

In this equation, c_i and z_i are, respectively, the molar concentration and valence of the ith ionic species; F is the Faraday constant; R is the universal gas constant; T is the absolute temperature of the aqueous solution; ψ is the electric potential arising from the surface charge, which is governed by the Poisson equation:

$$-\varepsilon_0 \varepsilon_f \nabla^2 \psi = \sum_{i=1}^{n} F z_i c_i. \tag{7.3}$$

The KCl aqueous solution used in the present study is a binary and symmetric ionic solution ($z_1 = 1$ for K⁺ and $z_2 = -1$ for Cl⁻). By substituting Equation (7.2) into Equation (7.3), the PB equation is obtained:

$$\nabla^2 \frac{\psi}{RT/F} = \frac{1}{\lambda_D^2} \sinh(\frac{\psi}{RT/F}), \tag{7.4}$$

where $\lambda_D = \kappa^{-1} = \sqrt{\varepsilon_0 \varepsilon_f RT/(2F^2 C_0)}$ is the Debye length. The boundary conditions for the potential ψ on surfaces of the particle and the nanopore are, respectively, $-\varepsilon_0 \varepsilon_f \mathbf{n} \bullet \nabla \psi = \sigma_p$ and $-\varepsilon_0 \varepsilon_f \mathbf{n} \bullet \nabla \psi = \sigma_w$. The electric potentials at the two ends of the reservoirs are both zero. All the other boundaries are considered electric insulation.

Since the Reynolds numbers of electrokinetic flows in nanofluidics are very small (i.e., $Re \ll 1$), the inertial terms in the Navier–Stokes equations can be neglected. Thus, the flow field is modeled by the Stokes equations,

$$\rho \frac{\partial \mathbf{u}}{\partial t} = -\nabla p + \mu \nabla^2 \mathbf{u} - F(z_1 c_1 + z_2 c_2)\nabla(\phi + \psi), \qquad (7.5)$$

and the continuity equation for the incompressible fluid,

$$\nabla \bullet \mathbf{u} = 0. \qquad (7.6)$$

In these equations, p is the pressure. The last term on the right-hand side of Equation (7.5) is the electrostatic force through the interaction between the net charge and the overall electric field, generating the electroosmotic flow (EOF).

There is no external pressure gradient imposed at the two ends of the fluid reservoirs. The flow boundary condition on the surfaces of the nanopore and the membrane is nonslip. A slip boundary condition is applied at the side boundaries of the two reservoirs, which are far away from the nanopore. Thus, normal fluid velocity and tangential viscous stress on these boundaries are both zero, $\mathbf{n} \bullet \mathbf{u} = 0$ and $\tau \bullet \{[-p\mathbf{I} + \mu(\nabla \mathbf{u} + \nabla \mathbf{u}^T)] \bullet \mathbf{n}\} = 0$, where τ is the unit tangential vector on the boundary. The particle translates and rotates due to the electrokinetic effects; the flow boundary condition on the particle surface is thus given as

$$\mathbf{u} = \mathbf{U}_p + \boldsymbol{\omega}_p \times (\mathbf{x}_s - \mathbf{x}_p), \qquad (7.7)$$

where $\mathbf{U}_p$ is the particle translational velocity; $\boldsymbol{\omega}_p$ is the rotational velocity; $\mathbf{x}_s$ and $\mathbf{x}_p$ are, respectively, the position vector of the surface and center of mass of the particle.

The particle's translational velocity is governed by Newton's second law,

$$m_p \frac{d\mathbf{U}_p}{dt} = \mathbf{F}_H + \mathbf{F}_E, \qquad (7.8)$$

where m_p is the mass of the particle. $\mathbf{F}_H$ and $\mathbf{F}_E$ are, respectively, the hydrodynamic force and the electrical force acting on the particle, which are obtained, respectively, by integrating the hydrodynamic stress tensor $\mathbf{T}^H$ and the Maxwell stress tensor $\mathbf{T}^E$ over the particle surface,

$$\mathbf{F}_H = \int \mathbf{T}^H \bullet \mathbf{n}d\Gamma = \int \left[-p\mathbf{I} + \mu\left(\nabla \mathbf{u} + \nabla \mathbf{u}^T\right) \right] \bullet \mathbf{n}d\Gamma, \qquad (7.9)$$

$$\mathbf{F}_E = \int \mathbf{T}^E \bullet \mathbf{n}d\Gamma = \int \varepsilon_0 \varepsilon_f \left[\mathbf{E}_t \mathbf{E}_t - \frac{1}{2}(\mathbf{E}_t \bullet \mathbf{E}_t)\mathbf{I} \right] \bullet \mathbf{n}d\Gamma. \qquad (7.10)$$

In these equations, $\mathbf{E}_t = -\nabla(\phi + \psi)$ is the overall electric field intensity, and Γ is the particle surface.

The rotational velocity of the particle is determined by

$$I_p \frac{d\omega_p}{dt} = \int (\mathbf{x}_s - \mathbf{x}_p) \times (\mathbf{T}^H \bullet \mathbf{n} + \mathbf{T}^E \bullet \mathbf{n}) d\Gamma, \qquad (7.11)$$

where I_p is the particle's moment of inertia. The right-hand side of Equation (7.11) is the total torque acting on the particle. The center of mass $\mathbf{x}_p$ and the orientation θ_p of the particle are expressed, respectively, by

$$\mathbf{x}_p = \mathbf{x}_{p0} + \int_0^t \mathbf{U}_p \, dt, \qquad (7.12)$$

$$\theta_p = \theta_{p0} + \int_0^t \omega_p \, dt, \qquad (7.13)$$

where $\mathbf{x}_{p0}$ and θ_{p0} denote, respectively, the initial position and orientation of the particle.

The induced ionic current through the nanopore is given as

$$I = \int F(z_1 \mathbf{N}_1 + z_2 \mathbf{N}_2) \bullet \mathbf{n} dS, \qquad (7.14)$$

where $\mathbf{N}_i = \mathbf{u}c_i - D_i \nabla c_i - z_i \dfrac{D_i}{RT} F c_i \nabla(\psi + \phi)$ is the ionic flux density of the ith ionic species; D_i is the diffusivity of the ith ionic species; and S denotes the opening of either reservoir due to current conservation.

7.2.2 DIMENSIONLESS FORM OF MATHEMATICAL MODEL

We use the particle radius a as the length scale, RT/F as the potential scale, $U_0 = C_0 RTa/\mu$ as the velocity scale, and $\mu U_0/a$ as the pressure scale to normalize the previous governing equations:

$$\nabla^{*2} \phi^* = 0, \qquad (7.15)$$

$$\nabla^{*2} \psi^* = (\kappa a)^2 \sinh \psi^*, \qquad (7.16)$$

$$\text{Re} \frac{\partial \mathbf{u}^*}{\partial t^*} = -\nabla^* p^* + \nabla^{*2} \mathbf{u}^* + 2 \sinh \psi^* \nabla^* (\phi^* + \psi^*), \qquad (7.17)$$

$$\nabla^* \bullet \mathbf{u}^* = 0, \qquad (7.18)$$

with the following dimensionless boundary conditions:

$$\phi^*(x^*, -(H^* + h^*/2)) = \phi^*(x^*, (H^* + h^*/2)) - \phi_0^* = 0, \qquad (7.19)$$

$$-\mathbf{n} \bullet \nabla^* \phi^* = 0, \qquad (7.20)$$

$$-\mathbf{n} \bullet \nabla^* \psi^* = \sigma_p^*.$$ (7.21)

$$-\mathbf{n} \bullet \nabla^* \psi^* = \sigma_w^*.$$ (7.22)

$$\mathbf{u}^* = \mathbf{U}_p^* + \omega_p^* \times (\mathbf{x}_s^* - \mathbf{x}_p^*).$$ (7.23)

Here, the surface charge densities are normalized by $\varepsilon_0 \varepsilon_f RT /(Fa)$. The force and torque are, respectively, normalized by μU_0 and $\mu U_0 a$, yielding the dimensionless equations of particle motion:

$$m_p^* \frac{d\mathbf{U}_p^*}{dt^*} = \int \mathbf{T}^{H*} \bullet \mathbf{n} d\Gamma^* + 2(\kappa a)^{-2} \int \mathbf{T}^{E*} \bullet \mathbf{n} d\Gamma^*,$$ (7.24)

$$I_p^* \frac{d\omega_p^*}{dt^*} = \int \left(\mathbf{x}_s^* - \mathbf{x}_p^* \right) \times \left(\mathbf{T}^{H*} \bullet \mathbf{n} + 2(\kappa a)^{-2} \mathbf{T}^{E*} \bullet \mathbf{n} \right) d\Gamma^*,$$ (7.25)

where the mass and the moment of inertia are, respectively, normalized by $a\mu/U_0$ and $a^3 \mu/U_0$ in 2D:

$$\mathbf{T}^{H*} = -p^* \mathbf{I} + \left(\nabla^* \mathbf{u}^* + \nabla^* \mathbf{u}^{*T} \right)$$

and

$$\mathbf{T}^{E*} = \mathbf{E}_t^* \mathbf{E}_t^* - \frac{1}{2} \left(\mathbf{E}_t^* \bullet \mathbf{E}_t^* \right) \mathbf{I}.$$

The dimensionless ionic current through the nanopore normalized by $FU_0 C_0 a^2$ is

$$I^* = \int (z_1 \mathbf{N}_1^* + z_2 \mathbf{N}_2^*) \bullet \mathbf{n} \, dS^*,$$ (7.26)

where $\mathbf{N}_i^* = \exp(-z_i \psi^*)(\mathbf{u}^* - \dfrac{z_i}{Pe_i} \nabla^* \phi^*)$ is the dimensionless ionic flux of the ith ionic species normalized by $U_0 C_0$, and $Pe_i = aU_0/D_i$ is the Peclet number of the ith ionic species.

7.3 NUMERICAL IMPLEMENTATION IN COMSOL AND CODE VALIDATION

In this section, we describe the implementation of the developed mathematical model in COMSOL Multiphysics® 3.5a. The computational domain is shown in Figure 7.1. The dimensions used in the present study are $b = 5$ nm, $h = 5$ nm, $W = 25$ nm, $H = 40$ nm, $a = 1$ nm, and $L_p = 10$ nm. The physical parameters used in the simulation are $\varepsilon_f = 80$; $\rho = 1 \times 10^3$ kg/m^3; $\mu = 1 \times 10^{-3}$ Pa·s; D_1 (diffusivity

TABLE 7.1
Constant Table Defined in COMSOL

Variable	Value or Expression	Description
D1	1.95e-9 [m^2/s]	Diffusivity of K^+
D2	2.03e-9 [m^2/s]	Diffusivity of Cl^-
z1	1	Valence of K^+
z2	−1	Valence of Cl^-
F	96485.3415 [C/mol]	Faraday constant
R0	8.314472 [J/K/mol]	Gas constant
T	300 [K]	Temperature
eps0	8.854187817e-12 [F/m]	Permittivity of vacuum
eps_r	80	Relative permittivity of fluid
eta	1.0e-3 [pa*s]	Fluid viscosity
rho	1e3 [kg/m^3]	Density of the fluid and the particle
V0	2000000*85e-9[V]	Applied voltage to achieve electric field 2000 kV/m
cp	−0.01 [C/m^2]	Surface charge density of the particle
cw	0 [C/m^2]	Surface charge density of the nanopore
c0	100 [mM]	Bulk concentration
a	1e-9 [m]	Particle radius
lamda	sqrt(eps_r*eps0*R0*T/2/F^2/c0)	Debye length
Lp	10*a	Length of the particle
Mass1	rho*pi*a^2	Mass of the two semicircles
Mass2	rho*2*a*(Lp-2*a)	Mass of the middle region of the particle
Massd	(Mass1+Mass2)*U0/eta/a	Dimensionless particle mass
Ird	(Mass1*(0.5*a^2+(Lp-2*a)^2/4+ (Lp-2*a)*4*a/(3*pi))+Mass2*1/12* ((Lp-2*a)^2+4*a^2))*U0/eta/a^3	Dimensionless moment of inertia
xpin	0	Initial x location of the particle
ypin	−15	Initial y location of the particle
angin	60*pi/180	Initial orientation of the particle
ai	a/lamda	κa
ef	2*(lamda/a)^2	Coefficient in dimensionless electrical force
U0	c0*R0*T*a/eta	Velocity scale
charges	F*a/R0/T/eps_r/eps0	Surface charge density scale
Re	rho*U0*a/eta	Reynolds number
cpd	cp*charges	Dimensionless particle's surface charge density
cwd	cw*charges	Dimensionless nanopore's surface charge density
V0d	V0*F/R0/T	Dimensionless applied voltage

of K+) = 1.95 × 10⁻⁹ m²/s; D_2 (diffusivity of Cl⁻) = 2.03 × 10⁻⁹ m²/s; σ_p = −0.01 C/m²; and T = 300 K. Table 7.1 lists all the constants defined in COMSOL Multiphysics 3.5a. Table 7.2 shows detailed instructions for setting up the dimensionless model in the graphical user interface (GUI) of COMSOL Multiphysics. The GUI model can be saved as an M-file, which could be further modified to achieve an automated remeshing for the tracking of long-term particle motion. The full COMSOL MATLAB script M-file is as follows:

TABLE 7.2
Model Setup in the GUI of COMSOL Multiphysics 3.5a

Model Navigator	Select **2D** in space dimension and click **Multiphysics** button.
	Select **COMSOL Multiphysics|Deformed Mesh|Moving Mesh (ALE)| Transient analysis**. Click **Add** button
	Select **COMSOL Multiphysics|Electromagnetics|Electrostatics**. Remove the predefined variables in the **Dependent variables** and enter $V1$. Click **Add** button.
	Select **COMSOL Multiphysics|Electromagnetics|Electrostatics**. Remove the predefined variables in the **Dependent variables** and enter $V2$. Click **Add** button.
	Select **COMSOL Multiphysics|Fluid Dynamics|Incompressible Navier–Stokes|Transient analysis**. Click **Add** button.
Option Menu| **Constants**	Define constants in Table 7.1.
Physics Menu| **Subdomain Setting**	ale model
	Subdomain 1
	Free displacement
	es mode (external electric field)
	Subdomain 1
	d = 1; ρ = 0; ε_r = 1/eps0
	es2 mode (electric field due to the surface charges)
	Subdomain 1
	d = 1; ρ = -(ai)^2*sinh(V2); ε_r = 1/eps0
	ns mode
	Subdomain 1
	Tab Physics
	ρ = Re; η = 1; F_x = 2*sinh(V2)*(V1x+V2x); F_y = 2*sinh(V2)*(V1y+V2y)
	Tab Stabilization
	Deactivate Streamline diffusion and Crosswind diffusion
Physics Menu| **Boundary Setting**	ale mode
	Particle surface: Mesh velocity
	vx = up-omega*(y-Yp)
	vy = vp+omega*(x-Xp)
	Other boundaries: Mesh displacement dx = 0; dy = 0

TABLE 7.2 (CONTINUED)
Model Setup in the GUI of COMSOL Multiphysics 3.5a

	es mode
	Lower end of the nanopore: $V_0 = 0$
	Upper end of the nanopore: $V_0 = V0d$
	Others: Zero charge/Symmetry
	es2 mode
	Lower and upper ends of the nanopore: $V_0 = 0$
	Particle surface: Surface charge $\rho_s = cpd$
	Nanopore wall: Surface charge $\rho_s = cwd$
	Others: Zero charge/Symmetry
	ns mode
	Two ends of the nanopore: Outlet|Pressure, no viscous stress $P_0 = 0$
	Particle surface: Inlet|Velocity $u_0 = up-omega*(y-Yp)$
	$v_0 = vp+omega*(x-Xp)$
	Nanopore wall: Wall|No slip
	Reservoir wall: Wall|Symmetry boundary
Other Settings	Properties of ale model (**Physics** Menu|**Properties**)
	Smoothing method: Winslow
	Allow remeshing: On
	Properties of ns model (**Physics** Menu|**Properties**)
	Weak constraints: On
	Constraint type: Nonideal
	Deactivate the unit system (**Physics** Menu|**Model Setting**)
	Base unit system: None
Options Menu|	Boundary selection: 8, 9, 17,18, 19, 20 (Particle surface)
Integration	Name: Fh_x Expression: -lm5; Integration order: 4; Frame: Frame(mesh)
Coupling	Name: Fh_y Expression: -lm6; Integration order: 4; Frame: Frame(mesh)
Variables|Boundary	Name: Fe_x Expression: ef*(-(V1Ty+V2Ty+cpd*ny)*(V1Tx+V2Tx+cpd*
Variables	nx)*ny
	$-0.5*(V1Tx+V2Tx+cpd*nx)^2*nx+0.5*(V1Ty+V2Ty+cpd*ny)^2*nx);$
	Integration order: 4; Frame: Frame(mesh)
	Name: Fe_y Expression: ef*(-(V1Ty+V2Ty+cpd*ny)*(V1Tx+V2Tx+cpd*
	nx)*nx
	$+0.5*(V1Tx+V2Tx+cpd*nx)^2*ny-0.5*(V1Ty+V2Ty+cpd*ny)^2*ny);$
	Integration order: 4; Frame: Frame(mesh)
	Name: Tr Expression: (x-Xp)*(ef*(-(V1Ty+V2Ty+cpd*ny)*(V1Tx+V2Tx
	+cpd*nx)*nx
	$+0.5*(V1Tx+V2Tx+cpd*nx)^2*ny-0.5*(V1Ty+V2Ty+cpd*ny)^2*ny)$
	-lm6)
	$-(y-Yp)*(ef*(-(V1Ty+V2Ty+cpd*ny)*(V1Tx+V2Tx+cpd*nx)*ny$

continued

TABLE 7.2 (CONTINUED)
Model Setup in the GUI of COMSOL Multiphysics 3.5a

	$-0.5*(V1Tx+V2Tx+cpd*nx)^2*nx+0.5*(V1Ty+V2Ty+cpd*ny)^2$ $*nx)-lm5);$ Integration order: 4; Frame: Frame(mesh)
Options Menu| **Expressions**| **Boundary** **Expressions**	Boundary selection: 2, 6 (Ends of the two reservoirs) Name: N1; Expression: $c0*exp(-z1*V2)*(U0*v-z1*D1*V1y/a)$ Name: N2; Expression: $c0*exp(-z2*V2)*(U0*v-z2*D2*V1y/a)$
Physics Menu| **Global Equations**	Name: up; Expression: Massd*upt-(Fh_x+Fe_x); Init (u): 0; Init (ut): 0 Name: Xp; Expression: Xpt-up; Init (u): xpin; Init (ut): 0 Name: vp; Expression: Massd*vpt-(Fh_y+Fe_y); Init (u): 0; Init (ut): 0 Name: Yp; Expression: Ypt-vp; Init (u): ypin; Init (ut): 0 Name: omega; Expression: Ird*omegat-Tr; Init (u): 0; Init (ut): 0 Name: ang; Expression: angt-omega; Init (u): angin; Init (ut): 0
Mesh|**Free Mesh** **Parameters**	Tab subdomain Subdomain 1|**Maximum element size**: 0.7 Tab boundary Particle surface: 8, 9, 17, 18, 19, 20|**Maximum element size**: 0.2 Nanopore: 3, 5, 7, 10, 11, 12, 15, 16, 21, 22|**Maximum element size**: 0.2
Solve Menu|	**Solver Parameters** Tab General Times: Range(0, 2, 10000) Relative tolerance: 1e-3 Absolute tolerance: 1e-7 Tab Timing Stepping Activate Use stop condition minqual1_ale-0.5 Click = **Solve Problem**
Postprocessing Menu|	Result check: up, Xp, vp, Yp, omega, ang **Global Variables Plot**|Select all the predefined quantities, click >> Click Apply Result check: Ionic current through the nanopore Boundary selection: 2 or 6 (Either end of the reservoirs) **Boundary Integration**|**Expression**: $a^2*F*(N1-N2)$ Click **Apply**

```
%%%%%%%%%%%%%%%%%%%%%%%%%%PB model.m%%%%%%%%%%%%%%%%%%%%%%%%%%%
%%%%%%%%%%%%%%%%C0=400mM
%%%%%%%%%%%%%%%%ang=0
%%%%%%%%%%%%%%%%xp=5
%%%%%%%%%%%%%%%%yp=-15
%%%%%%%%%%%%%%%%E=2000KV/m
%%%%%%%%%%%%%%%%cw=0

flclear fem
```

```
% COMSOL version
clear vrsn
vrsn.name = 'COMSOL 3.5';
vrsn.ext = '';
vrsn.major = 0;
vrsn.build = 494;
vrsn.rcs = '$Name: $';
vrsn.date = '$Date: 2008/09/19 16:09:48 $';
fem.version = vrsn;

% NEW: Define particle's initial x, y location and orienta-
  tion
xp0=5;
yp0=-15;
angle0=0*pi/180;

% NEW: Length of the particle
Length_p=10;

% Constants
fem.const = {'D1','1.95e-9[m^2/s]', ...
  'D2','2.03e-9[m^2/s]', ...
  'F','96485.3415[C/mol]', ...
  'R0','8.314472[J/K/mol]', ...
  'T','300[K]', ...
  'z1','1', ...
  'z2','-1', ...
  'eps_r','80', ...
  'eta','1.0e-3[Pa*s]', ...
  'v0','2000000*85e-9[V]', ...
  'rho','1e3[kg/m^3]', ...
  'cp','-0.01[C/m^2]', ...
  'cw','0[C/m^2]', ...
  'c0','400[mol/m^3]', ...
  'a','1e-9[m]', ...
  'Mass1','rho*pi*a^2', ...
  'Mass2','rho*2*a*(Lp-2*a)', ...
  'Massd','(Mass1+Mass2)*U0/eta/a', ...
  'Lp',num2str(Length_p*1e-9), ...
  'Ir','(Mass1*(0.5*a^2+(Lp-2*a)^2/4+(Lp-2*a)*4*a/
(3*pi))+Mass2*1/12*((Lp-2*a)^2+4*a^2))', ...
  'Ird','Ir*U0/eta/a^3', ...
  'U0','c0*R0*T*a/eta', ...
  'lamda','sqrt(eps_r*eps0*R0*T/2/F^2/c0)', ...
  'ai','a/lamda', ...
  'ef','2*(lamda/a)^2', ...
  'v0d','v0*F/R0/T', ...
  'charges','F*a/R0/T/eps_r/eps0', ...
  'cpd','cp*charges', ...
  'cwd','cw*charges', ...
```

```
'Re','rho*U0*a/eta', ...
'eps0','8.854187817e-12[F/m]', ...
'xpin',num2str(xp0), ...
'ypin',num2str(yp0), ...
'angin',num2str(angle0)};

% Geometry
% NEW: Create the nanopore system
g1=rect2('10','5','base','center','pos',{'0','0'},'rot','0');
g2=rect2('50','40','base','center','pos',{'0','22.5'},'
  rot','0');
g3=rect2('50','40','base','center','pos',{'0','-
  22.5'},'rot','0');
g4=geomcomp({g1,g2,g3},'ns',{'g1','g2','g3'},'sf',
  'g1+g2+g3','edge','none');
g5=geomdel(g4);
g6=fillet(g5,'radii',1.5,'point',[5,6,7,8]);
% NEW: Create the nanoparticle
g7=rect2('2','8','base','center','pos',{num2str(xp0),num2str
  (yp0)},'rot','0');
g8=circ2('1','base','center','pos',{num2str(xp0),num2str(yp0
  +4)},'rot','0');
g9=circ2('1','base','center','pos',{num2str(xp0),num2str
  (yp0-4)},'rot','0');
g10=geomcomp({g7,g8,g9},'ns',{'g7','g8','g9'},'sf',
  'g7+g8+g9','edge','none');
g11=geomdel(g10);
g11=rotate(g11,angle0,[xp0,yp0]);
g12=geomcomp({g6,g11},'ns',{'g6','g11'},'sf',
  'g6-g11','edge','none');

% Geometry

% Analyzed geometry
clear s
s.objs={g12};
s.name={'CO2'};
s.tags={'g12'};

fem.draw=struct('s',s);
fem.geom=geomcsg(fem);

% NEW: Display geometry with the sequence of the boundary
geomplot(fem,
  'Labelcolor','r','Edgelabels','on','submode','off');

% (Default values are not included)

% NEW: Solve moving mesh
% Application mode 1
```

```
clear appl
appl.mode.class = 'MovingMesh';
appl.sdim = {'Xm','Ym','Zm'};
appl.shape = {'shlag(2,''lm1'')','shlag(2,''lm2'')',
  'shlag(2,''x'')','shlag(2,''y'')'};
appl.gporder = {30,4};
appl.cporder = 2;
appl.assignsuffix = '_ale';
clear prop
prop.smoothing='winslow';
prop.analysis='transient';
prop.allowremesh='on';
prop.origrefframe='xy';
appl.prop = prop;
clear bnd
bnd.defflag = {{1;1},{0;0}};
bnd.veldeform = {{0;0},{'up-omega*(y-Yp)';
  'vp+omega*(x-Xp)'}};
bnd.wcshape = [1;2];
bnd.name = {'Fix','particle'};
bnd.type = {'def','vel'};
bnd.veldefflag = {{0;0},{1;1}};
bnd.ind = [1,1,1,1,1,1,1,2,1,2,1,1,1,1,1,1,2,2,2,2,1,1];
appl.bnd = bnd;
clear equ
equ.gporder = 2;
equ.shape = [3;4];
equ.ind = [1];
appl.equ = equ;
fem.appl{1} = appl;

% NEW: Solve externally applied electric field
% Application mode 2
clear appl
appl.mode.class = 'Electrostatics';
appl.dim = {'V1'};
appl.sshape = 2;
appl.assignsuffix = '_es';
clear prop
clear weakconstr
weakconstr.value = 'off';
weakconstr.dim = {'lm3'};
prop.weakconstr = weakconstr;
appl.prop = prop;
clear bnd
bnd.V0 = {0,'v0d',0};
bnd.name = {'insulator','up','down'};
bnd.type = {'nD0','V','V'};
bnd.ind = [1,3,1,1,1,2,1,1,1,1,1,1,1,1,1,1,1,1,1,1,1,1,1];
appl.bnd = bnd;
```

```
clear equ
equ.epsilonr = '1/eps0';
equ.ind = [1];
appl.equ = equ;
fem.appl{2} = appl;

% NEW: Solve electric field due to the surface charges
% Application mode 3
clear appl
appl.mode.class = 'Electrostatics';
appl.dim = {'V2'};
appl.name = 'es2';
appl.sshape = 2;
appl.assignsuffix = '_es2';
clear prop
clear weakconstr
weakconstr.value = 'off';
weakconstr.dim = {'lm4'};
prop.weakconstr = weakconstr;
appl.prop = prop;
clear bnd
bnd.name = {'ground','particle','wall','insulator'};
bnd.rhos = {0,'cpd','cwd',0};
bnd.type = {'V0','r','r','nD0'};
bnd.ind = [4,1,3,4,3,1,3,2,3,2,3,3,4,4,3,3,2,2,2,2,3,3];
appl.bnd = bnd;
clear equ
equ.epsilonr = '1/eps0';
equ.rho = '-sinh(V2)/(lamda/a)^2';
equ.ind = [1];
appl.equ = equ;
fem.appl{3} = appl;

% NEW: Solve flow field using the modified Stokes equations
% Application mode 4
clear appl
appl.mode.class = 'FlNavierStokes';
appl.shape = {'shlag(2,''u'')','shlag(2,''v'')','shlag(1,
    ''p'')','shlag(2,''lm5'')','shlag(2,''lm6'')','shlag(1,
    ''lm7'')'};
appl.gporder = {4,2,30};
appl.cporder = {2,1};
appl.assignsuffix = '_ns';
clear prop
clear weakconstr
weakconstr.value = 'on';
weakconstr.dim = {'lm5','lm6','lm7'};
prop.weakconstr = weakconstr;
prop.constrtype='non-ideal';
appl.prop = prop;
```

```
clear bnd
bnd.v0 = {0,'vp+omega*(x-Xp)',0,0};
bnd.type = {'walltype','inlet','outlet','sym'};
bnd.walltype = {'noslip','lwall','noslip','noslip'};
bnd.name = {'wall','particle','outlet','sym'};
bnd.u0 = {0,'up-omega*(y-Yp)',0,0};
bnd.wcshape = [4;5;6];
bnd.velType = {'U0in','u0','U0in','U0in'};
bnd.wcgporder = 3;
bnd.ind = [4,3,1,4,1,3,1,2,1,2,1,1,4,4,1,1,2,2,2,2,1,1];
appl.bnd = bnd;
clear equ
equ.gporder = {{1;1;2}};
equ.F_y = '2*sinh(V2)*(V1y+V2y)';
equ.rho = 'Re';
equ.cporder = {{1;1;2}};
equ.F_x = '2*sinh(V2)*(V1x+V2x)';
equ.shape = [1;2;3];
equ.cdon = 0;
equ.sdon = 0;
equ.ind = [1];
appl.equ = equ;
fem.appl{4} = appl;
fem.sdim = {{'Xm','Ym'},{'X','Y'},{'x','y'}};
fem.frame = {'mesh','xy','ale'};
if isfield(fem,'sshape')
 fem=rmfield(fem,'sshape');
end
fem.border = 1;

% NEW: Boundary expression to describe the ionic flux
% Boundary settings
clear bnd
bnd.ind = [1,2,1,1,1,2,1,1,1,1,1,1,1,1,1,1,1,1,1,1,1,1];
bnd.dim = {'x','y','lm1','lm2','V1','V2','u','v','p','lm5',
  'lm6','lm7'};

% Boundary expressions
bnd.expr = {'N1',{'','c0*exp(-z1*V2)*(U0*v-1/
  a*z1*D1*V1y)'}, ...
  'N2',{'','c0*exp(-z2*V2)*(U0*v-1/a*z2*D2*V1y)'}};
fem.bnd = bnd;

% NEW: Boundary integration to calculate the hydrodynamic
  force
% NEW: and electrical force on the particle
% Coupling variable elements
clear elemcpl
% Integration coupling variables
clear elem
```

```
elem.elem = 'elcplscalar';
elem.g = {'1'};
src = cell(1,1);
clear bnd
bnd.expr = {{{},'-lm5'},{{},'-lm6'},{{}, ...

  'ef*(-(V1Ty+V2Ty+cpd*ny)*(V1Tx+V2Tx+cpd*nx)*ny-0.5*(V1Tx+V2
  Tx+cpd*nx)^2*nx+0.5*(V1Ty+V2Ty+cpd*ny)^2*nx)'},{ ...
  {}, ...

  'ef*(-(V1Ty+V2Ty+cpd*ny)*(V1Tx+V2Tx+cpd*nx)*nx+0.5*(V1Tx+V2
  Tx+cpd*nx)^2*ny-0.5*(V1Ty+V2Ty+cpd*ny)^2*ny)'},{ ...
  {}, ...

  '(x-Xp)*(ef*(-(V1Ty+V2Ty+cpd*ny)*(V1Tx+V2Tx+cpd*nx)*nx+0.5*
  (V1Tx+V2Tx+cpd*nx)^2*ny-0.5*(V1Ty+V2Ty+cpd*ny)^2*ny)-lm6)-
  (y-Yp)*(ef*(-(V1Ty+V2Ty+cpd*ny)*(V1Tx+V2Tx+cpd*nx)*ny-
  0.5*(V1Tx+V2Tx+cpd*nx)^2*nx+0.5*(V1Ty+-
  V2Ty+cpd*ny)^2*nx)-lm5)'}};
bnd.ipoints = {{{},'4'},{{},'4'},{{},'4'},{{},'4'},{{},'4'}};
bnd.frame = {{{},'mesh'},{{},'mesh'},{{},'mesh'},{{},'mesh'}
  ,{{},'mesh'}};
bnd.ind = {{'1','2','3','4','5','6','7','9','11','12','13','
  14','15', ...
  '16','21','22'},{'8','10','17','18','19','20'}};
src{1} = {{},bnd,{}};
elem.src = src;
geomdim = cell(1,1);
geomdim{1} = {};
elem.geomdim = geomdim;
elem.var = {'Fh_x','Fh_y','Fe_x','Fe_y','Tr'};
elem.global = {'1','2','3','4','5'};
elemcpl{1} = elem;
fem.elemcpl = elemcpl;


% NEW: Solve six ODEs to determine the particle's
% NEW: translational velocity and rotational velocity
% ODE Settings
clear ode
ode.dim={'up','Xp','vp','Yp','omega','ang'};
ode.f={'Massd*upt-(Fh_x+Fe_x)','Xpt-up','Massd*vpt-(Fh_
  y+Fe_y)','Ypt-vp','Ird*omegat-Tr','angt-omega'};
ode.init={'0','xpin','0','ypin','0','angin'};
ode.dinit={'0','0','0','0','0','0'};
clear units;
units.basesystem = 'SI';
ode.units = units;
fem.ode=ode;


% Multiphysics
```

```
fem=multiphysics(fem);

% NEW: Loop for continuous particle tracking
% NEW: Remeshing index
j=1;

% NEW: Define matrix for data storage
particle=zeros(10,7);

% NEW: Define data storage index
num=1;

% NEW: Define initial time
time_e=0;

% NEW: Define time_step
time_step=1.5;

% NEW: Update particle's location and orientation
xp=xp0;
yp=yp0;
angle=angle0;

% NEW: Define mesh size for particle surface and nanopore
mesh_p=0.2;
mesh_w=0.2;

% NEW: Define end condition of the computation (Loop control)
while yp<20

% Initialize mesh
fem.mesh=meshinit(fem, ...
                  'hauto',5, ...

  'hmaxedg',[3,mesh_w,5,mesh_w,7,mesh_w,8,mesh_p,9,mesh_p,10,
   mesh_w,11,mesh_w,12,mesh_w,15,mesh_w,16,mesh_w,17,mesh_p,1
   8,mesh_p,19,mesh_p,20,mesh_p,21,mesh_w,22,mesh_w], ...
  'hmaxsub',[1,1]);

% NEW: Refine mesh near the particle after each remeshing
% Refine mesh
fem.mesh=meshrefine(fem, ...
                    'mcase',0, ...
                    'boxcoord',[xp-2-Length_p*sin(angle)/2
  xp+2+Length_p*sin(angle)/2 yp-2-Length_p*cos(angle)/2
  yp+2+Length_p*cos(angle)/2], ...
                    'rmethod','regular');

% Extend mesh
fem.xmesh=meshextend(fem);
```

```
% NEW: Find out the mesh quality
quality=meshqual(fem.mesh);
minimum=min(quality)
used=minimum-0.1;


% NEW: Computation before first remeshing
if j==1

% Solve problem
fem.sol=femtime(fem, ...
'solcomp',{'V2','ang','up','vp','omega','lm2','lm1','lm7','l
   m6','V1','lm5','v','Yp','u','p','y','Xp' ,'x'}, ...
'outcomp',{'V2','ang','up','vp','omega','lm2','lm1','lm7','lm
   6','V1','lm5','v','Yp','u','p','Y','X','y','x','Xp'}, ...
            'blocksize','auto', ...
            'tlist',[colon(time_e,time_step,2e4)], ...
            'tout','tsteps', ...
            'tsteps','intermediate', ...
            'rtol',1e-3, ...
            'atol',1e-7, ...
            'stopcond',strcat('minqual1_ale-',num2str(used)));

else

% NEW: Map solution in the deformed mesh into the new geom-
   etry with undeformed mesh
init = asseminit(fem,'init',fem0.sol,'xmesh',fem0.xmesh,'blo
   cksize','auto','framesrc','ale','domwise','on');

% NEW: Computation after first remeshing
% Solve problem
fem.sol=femtime(fem, ...
                'init',init, ...
'solcomp',{'V2','ang','up','vp','omega','lm2','lm1' ,'lm7','
   lm6','V1','lm5','v','Yp','u','p','y' ,'Xp','x'}, ...
'outcomp',{'V2','ang','up','vp','omega','lm2','lm1' ,'lm7','
   lm6','V1','lm5','v','Yp','u','p','Y' ,'X','y','x','Xp'},
   ...
                'blocksize','auto', ...
                'tlist',[colon(time_e,time_step,2e4)], ...
                'tout','tsteps', ...
                'tsteps','intermediate', ...
                'rtol',1e-3, ...
                'atol',1e-7, ...
                'stopcond',strcat('minqual1_ale-',
   num2str(used)));

end


% NEW: Save current fem structure for restart purposes
```

```
fem0=fem;

% NEW: Get the last time from solution
time_e = fem.sol.tlist(end);

% NEW: Get the length of the time steps
Leng =length(fem.sol.tlist);

% NEW: Adaptive time step to ensure 100 outputs in one
  remeshing
time_step=Leng*time_step/100;

% Global variables plot
data=postglobalplot(fem,{'Xp','Yp','ang','up','vp','om
                    ega'}, ...
                  'linlegend','on', ...
                  'title','Particle', ...
                  'Outtype','postdata', ...
                  'axislabel',{'Time','Parameters'});

% NEW: Store the interested data
for n=20:Leng
        particle(num,1)=data.p(1,n);
        particle(num,2)=data.p(2,n);
        particle(num,3)=data.p(2,n+Leng);
        particle(num,4)=data.p(2,n+2*Leng);
        particle(num,5)=data.p(2,n+3*Leng);
        particle(num,6)=data.p(2,n+4*Leng);
        particle(num,7)=data.p(2,n+5*Leng);

% NEW: Current at anode
% NEW: Boundary integrate
I1=postint(fem,'a^2*F*(N1-N2)', ...
            'unit','mol/(m*s)', ...
            'recover','off', ...
            'dl',6, ...
            'edim',1, ...
            'solnum',n);

particle(num,8)=I1;

% NEW: Current at cathode
% NEW: Boundary integrate
I2=postint(fem,'a^2*F*(N1-N2)', ...
            'unit','mol/(m*s)', ...
            'recover','off', ...
            'dl',2, ...
            'edim',1, ...
            'solnum',n);
```

```
particle(num,9)=I2;

% NEW: current deviation
particle(num,10)=(abs(I1)-abs(I2))/abs(I1)*100;

    num=num+1;
end


% NEW: update the particle's location and orientation
% NEW: Output them in the MATLAB
xp=data.p(2,Leng)
yp=data.p(2,2*Leng)
angle=data.p(2,3*Leng)

% NEW: Output the COMSOL Multiphysics GUI .mph file if
  necessary
flsave(strcat('c400mM_E_2000KV_ang_0_xp_5_yp_n15_cw_0_
  yp_',num2str(yp),'_angle_',num2str(angle),'.mph'),fem);

% NEW: Write the data into a file
dlmwrite(strcat('c400mM_E_2000KV_ang_0_xp_5_yp_n15_cw_0','.
  dat'),particle,',');

j=j+1;

% Generate geom from mesh
fem = mesh2geom(fem, ...
                'frame','ale', ...
                'srcdata','deformed', ...
                'destfield',{'geom','mesh'}, ...
                'srcfem',1, ...
                'destfem',1);

end
% NEW: End of this script file
```

In the simulation, we found that it is not accurate enough to calculate the electrical force directly using Vx and Vy, in which V is the electric potential defined in COMSOL. The derivatives of V on the boundary in the tangential direction are more accurate compared to using Vx and Vy. As a result, we decompose Vx into the tangential derivative variable VTx and the normal derivative variable VNx, Vx = VTx + VNx. VTx is predefined in COMSOL, while VNx = (nx*Vx + ny*Vy)*nx. It is easy to obtain nx*Vx + ny*Vy = cpd from Equation (7.21), where cpd is the dimensionless surface charge density of the particle defined in COMSOL. Because V2 is the electric potential due to the surface charges, V2x is written as V2x = V2Tx + cpd*nx. For the external electric field associated with the electric potential V1 defined in COMSOL, the normal derivative variable 'V1Nx' on the particle surface is zero. As a result, V1x = V1Tx.

Thus, the overall Vx = V1x + V2x = V1Tx + V2Tx + cpd*nx. Accordingly, Vy = V1y + V2y = V1Ty + V2Ty + cpd*ny.

To validate the developed numerical model, we simulate the electrokinetic translocation of a sphere along the axis of an uncharged nanopore. Figure 7.2 shows the particle's steady axial velocity normalized by $\varepsilon_0\varepsilon_f\zeta E/\mu$ as a function of the ratio of the particle radius to the pore radius, a/b, when $a = 1$ nm, $\kappa a = 2.05$, $\zeta = 1$ mV, and $E = 100$ kV/m. The approximation solution of the particle velocity is available when the EDL is not disturbed by the external electric field, flow field, nearby solid boundaries, and EDLs, and the zeta potential of the particle is relatively low ($\zeta/(RT/F)<1$) (Ennis and Anderson 1997). Our numerical predictions are in good agreement with the approximation solution when the boundary effect is very small. The numerical results deviate from the approximation solution when the boundary effect becomes more significant.

7.4 RESULTS AND DISCUSSION

The effects due to the electric field intensity imposed E^*, the ratio of the particle radius to the Debye length κa, the particle's initial orientation θ^*_{p0}, the particle's

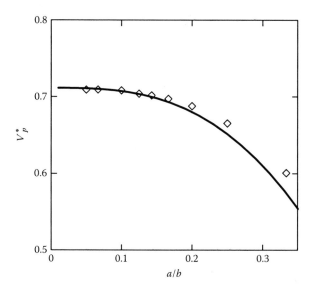

FIGURE 7.2 Effect of the ratio of the particle radius to the pore radius a/b on the axial electrophoretic velocity of a sphere translating along the axis of an uncharged cylindrical nanopore. Solid line and diamonds represent, respectively, the analytical approximation solution and our numerical results. The conditions are $a = 1$ nm; $\kappa a = 2.05$; the zeta potential of the particle $\zeta = 1$ mV; and the imposed axial electric field $E = 100$ kV/m. (From Ai, Y., and S. Qian. 2011. Direct numerical simulation of electrokinetic translocation of a cylindrical particle through a nanopore using a Poisson–Boltzmann approach. *Electrophoresis* 32:996–1005 with permission from Wiley-VCH.)

initial lateral offset from the nanopore's centerline x_{p0}^*, and the nanopore's surface charge density σ_w^*, on the electrokinetic particle translocation and the ionic current through a nanopore are thoughtfully investigated. The applied electric field intensity E is evaluated by dividing the electric potential difference ϕ_0 over the total height of the nanopore system $2H + h$.

7.4.1 EFFECT OF PARTICLE'S INITIAL ORIENTATION

Figures 7.3a and 7.3b show, respectively, the superposed particle trajectories under two different electric fields, $E^* = 7.7 \times 10^{-4}$ ($E = 20$ kV/m) and $E^* = 7.7 \times 10^{-2}$ ($E = 2,000$ kV/m) when $x_{p0}^* = 0$, $\theta_{p0}^* = 60°$, $\sigma_w^* = 0$, and $\kappa a = 1.03$. Obviously, the particle's nonzero initial orientation induces rotational motion as the particle translocates toward the nanopore. When the applied electric field is relatively low, $E^* = 7.7 \times 10^{-4}$, the particle rotates clockwise as it moves toward the nanopore. After passing through the nanopore, the particle slightly rotates counterclockwise; however, it cannot recover its initial orientation. Because of the nonzero initial orientation, the particle also experiences lateral movement, as shown in Figure 7.3a. When the particle's initial orientation is positive (negative), the particle laterally moves in the negative (positive) x direction. If the applied electric field is increased 100 times to $E^* = 7.7 \times 10^{-2}$, the particle's translocation through the nanopore is accordingly enhanced 100 times. In addition, the particle is aligned to the applied electric field quickly because of the DEP effect, as discussed in Chapter 4. However, the DEP effect is limited when the applied electric field is $E^* = 7.7 \times 10^{-4}$. As a result, DEP alignment is not observed in Figure 7.3a.

We further increase the bulk concentration to yield $\kappa a = 2.05$ and study the particle's translocation through the nanopore under the two electric fields. When the applied electric field is relatively low, $E^* = 7.7 \times 10^{-4}$, the particle experiences more significant lateral movement compared to Figure 7.3a. The zeta potential of a particle with a fixed surface charge decreases as κa increases (Ohshima 1998). Accordingly, the particle's y component translational velocity decreases, which in turn increases the duration of the particle's translocation process through the nanopore. In addition, the particle's x component translational velocity for $\kappa a = 2.05$ is larger than that for $\kappa a = 1.03$ after the particle exits the nanopore. Therefore, the particle experiences a more pronounced lateral movement as κa increases. When the applied electric field is relatively high, $E^* = 7.7 \times 10^{-2}$, the particle is aligned to the applied electric field as well.

Figures 7.4a and 7.4b show, respectively, the particle's y component translational velocity as a function of the particle's location y_p^* under $E^* = 7.7 \times 10^{-4}$ and $E^* = 7.7 \times 10^{-2}$ when $x_p^* = 0$ and $\sigma_w^* = 0$. When the particle's initial orientation is $\theta_{p0}^* = 0$, the particle's y component translational velocity is symmetric with respect to $y_p^* = 0$. It is also confirmed that the particle's y component translational velocity decreases as κa increases, which is attributed to the decrease in the zeta potential of the particle. In addition, the particle's y component translational velocity increases 100 times as the applied electric field increases 100 times. When the particle's initial orientation is

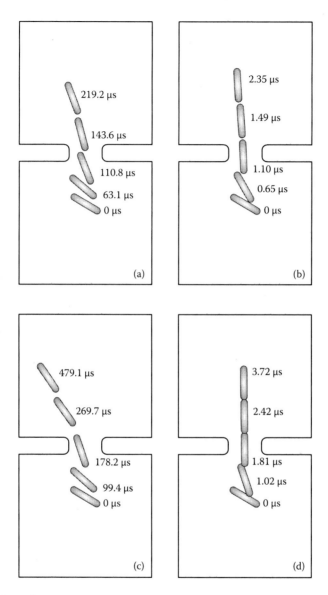

FIGURE 7.3 Superposed trajectories of the particle under $E^* = 7.7 \times 10^{-4}$ (a and c) and $E^* = 7.7 \times 10^{-2}$ (b and d). $x^*_{p0} = 0$, $\theta^*_{p0} = 60°$, $\sigma^*_w = 0$, $\kappa a = 1.03$ (a and b), and $\kappa a = 2.05$ (c and d). (From Ai, Y., and S. Qian. 2011. Direct numerical simulation of electrokinetic translocation of a cylindrical particle through a nanopore using a Poisson–Boltzmann approach. *Electrophoresis* 32:996–1005 with permission from Wiley-VCH.)

$\theta_{p0}^* = 60°$, the particle's y component translational velocity is significantly reduced before the particle enters the nanopore. When the applied electric field is relatively low, $E^* = 7.7 \times 10^{-4}$, the particle exits the nanopore with an obvious nonzero orientation, as shown in Figures 7.3a and 7.3c. As a result, the particle's y component translational velocity is slightly lower than the case of $\theta_{p0}^* = 0$ when the particle exits the nanopore. When the applied electric field is relatively high, $E^* = 7.7 \times 10^{-2}$, the particle is almost parallel to the applied electric field before entering the nanopore. Consequently, the particle's y component translational velocity coincides with the case of $\theta_{p0}^* = 0$ when the particle is completely aligned to the applied electric field.

Current deviation, defined as $\chi = (I^* - I_0^*)/I_0^*$, is used to quantify the change in the ionic current arising from the particle's translocation through the nanopore. In the equation, I_0^* refers to the base current when the particle is far away from the nanopore and is numerically obtained based on Equation (7.26) without including the particle in the simulation. Figure 7.5 shows the current deviations corresponding to the cases in Figure 7.4. When the particle's initial orientation is $\theta_{p0}^* = 0$, a symmetric current blockade with respect to $y_p^* = 0$ is predicted, which is in qualitative agreement with existing experimental observations (Storm et al. 2005; Kim et al. 2007) and numerical predictions by the PNP-based model using the quasi-static method (Liu, Qian, and Bau 2007).

When the applied electric field is relatively low, $E^* = 7.7 \times 10^{-2}$, the particle's nonzero initial orientation, $\theta_{p0}^* = 60°$, gives rise to a more significant current blockade compared to the case of $\theta_{p0}^* = 0$, as shown in Figure 7.5a. Because of

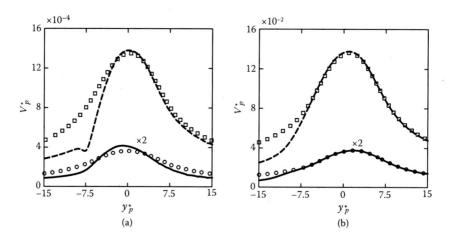

FIGURE 7.4 The y component translational velocity as a function of the particle's location y_p^* under $E^* = 7.7 \times 10^{-4}$ (a) and $E^* = 7.7 \times 10^{-2}$ (b). Symbols and lines represent, respectively, $\theta_{p0}^* = 0$ and 60°. $x_{p0}^* = 0$, $\sigma_w^* = 0$, $\kappa a = 1.03$ (dashed line and squares), and $\kappa a = 2.05$ (solid line and circles). A scale of 2 was applied to the solid line and circles for clear visualization. (From Ai, Y., and S. Qian. 2011. Direct numerical simulation of electrokinetic translocation of a cylindrical particle through a nanopore using a Poisson–Boltzmann approach. *Electrophoresis* 32:996–1005 with permission from Wiley-VCH.)

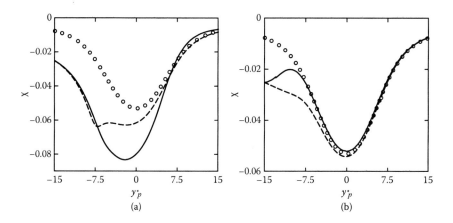

FIGURE 7.5 Current deviation as a function of the particle's location y_p^* under $E^* = 7.7 \times 10^{-4}$ (a) and $E^* = 7.7 \times 10^{-2}$ (b). Symbols and lines represent, respectively, $\theta_{p0}^* = 0$ and 60°. $x_{p0}^* = 0$, $\sigma_w^* = 0$, $\kappa a = 1.03$ (dashed line), and $\kappa a = 2.05$ (solid line and circles). (From Ai, Y., and S. Qian. 2011. Direct numerical simulation of electrokinetic translocation of a cylindrical particle through a nanopore using a Poisson–Boltzmann approach. *Electrophoresis* 32:996–1005 with permission from Wiley-VCH.)

the lateral movement, especially after the particle exits the nanopore, the current deviation when $\theta_{p0}^* = 60°$ begins to recover the case of $\theta_{p0}^* = 0$.

It is also concluded that the particle's initial orientation can significantly affect the resulting current deviation. When the applied electric field is relatively high, $E^* = 7.7 \times 10^{-2}$, the particle's nonzero initial orientation $\theta_{p0}^* = 60°$ can also induce a more significant current blockade at the beginning of the particle's translocation process. However, the DEP effect aligns the particle parallel to the applied electric field before the particle enters the nanopore. As a result, the current deviation when $\theta_{p0}^* = 60°$ gradually recovers the case of $\theta_{p0}^* = 0$ before the particle enters the nanopore, as shown in Figure 7.5b.

7.4.2 Effect of Particle's Initial Lateral Offset

In this section, we study the particle's initial lateral offset from the nanopore's centerline x_{p0}^* on the particle's translocation and the ionic current through the nanopore. Figure 7.6 shows the superposed particle trajectories under $E^* = 7.7 \times 10^{-4}$ (Figures 7.6a and 7.6b) and $E^* = 7.7 \times 10^{-2}$ (Figures 7.6c and 7.6d) when $\kappa a = 2.05$, $\theta_{p0}^* = 0$, and $\sigma_w^* = 0$. When the applied electric field is relatively low, $E^* = 7.7 \times 10^{-2}$, the nonzero initial lateral offset from the nanopore's centerline, $x_{p0}^* = 2.5$ (Figure 7.6a) and $x_{p0}^* = 5$ (Figure 7.6b), gives rise to a counterclockwise rotational motion as the particle moves toward the nanopore. Therefore, the particle exits the nanopore with a positive orientation with respect to the nanopore's centerline. It is also found that a large initial lateral offset induces a more significant rotation before entering the nanopore. As a result, the particle for $x_{p0}^* = 5$

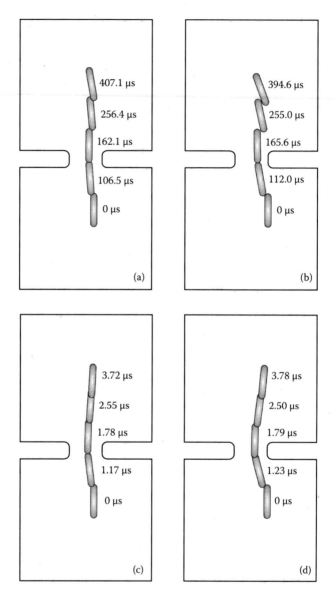

FIGURE 7.6 Superposed trajectories of the particle under $E^* = 7.7 \times 10^{-4}$ (a and b) and $E^* = 7.7 \times 10^{-2}$ (c and d). $\kappa a = 2.05$, $\sigma_w^* = 0$, $\theta_{p0}^* = 0$, $x_{p0}^* = 2.5$ (a and c), and $x_{p0}^* = 5$ (b and d). (From Ai, Y., and S. Qian. 2011. Direct numerical simulation of electrokinetic translocation of a cylindrical particle through a nanopore using a Poisson–Boltzmann approach. *Electrophoresis* 32:996–1005 with permission from Wiley-VCH.)

presents a larger angle than the case of $x_{p0}^* = 2.5$ when the particle exits the nanopore. When the applied electric field is relatively high, $E^* = 7.7 \times 10^{-2}$, the particle is aligned to the local electric field quickly. However, the electric field away from the nanopore's centerline is actually not parallel to the nanopore's centerline. As a result, the particle presents a negative orientation with respect to the nanopore's centerline when exiting the nanopore. Similarly, the particle for $x_{p0}^* = 5$ ends up with a larger angle than the case of $x_{p0}^* = 2.5$ when exiting the nanopore.

Furthermore, it is found that the particle is pushed toward the nanopore's centerline after passing through the nanopore due to the DEP effect, which has been utilized to focus particles in microfluidics (Zhu and Xuan 2009). However, the particle's initial offset from the nanopore's centerline indicates a minor effect on the particle's translocation and the ionic current through the nanopore.

A nonzero initial orientation, $\theta_{p0}^* = 60°$, is imposed, and all the other conditions in Figure 7.6 are unchanged to study the particle's translocation process, as shown in Figure 7.7. When the applied electric field is relatively low, $E^* = 7.7 \times 10^{-4}$, the particle rotates clockwise as it moves toward the nanopore and then rotates counterclockwise after it exits the nanopore, and the particle motion is similar to that shown in Figure 7.3c. As discussed, the effect of the particle's initial orientation overpowers the effect of the particle's initial lateral offset. Thus, the particle trajectory in Figure 7.7a is similar to that in Figure 7.7b. In addition, the particle experiences a significant lateral movement due to the nonzero initial orientation. When the applied electric field is relatively high, $E^* = 7.7 \times 10^{-2}$, the DEP effect aligns the particle parallel to the local electric field. Therefore, the particle trajectory shown in Figure 7.7c (7.7d) is similar to that shown in Figure 7.6c (7.6d), especially when the particle exits the nanopore.

Figures 7.8a and 7.8b show, respectively, the particle's y component translational velocity as a function of the particle's location y_p^* under $E^* = 7.7 \times 10^{-4}$ and $E^* = 7.7 \times 10^{-2}$ when $\kappa a = 2.05$, $\theta_{p0}^* = 60°$, and $\sigma_w^* = 0$. The particle's y component translational velocity when $\theta_{p0}^* = 0$ and $x_{p0}^* = 0$ is regarded as a reference. When the applied electric field is relatively low, $E^* = 7.7 \times 10^{-4}$, the particle's y component translational velocity is reduced compared to the reference when the particle is outside the nanopore. When the particle is inside the nanopore, the particle's nonzero orientation enhances the electric field, which slightly increases the particle's velocity. When the applied electric field is relatively high, $E^* = 7.7 \times 10^{-2}$, the particle's y component translational velocity is only reduced at the beginning of the particle's translocation process. As the particle is aligned to the applied electric field because of the DEP effect, the particle's y component translational velocity with a nonzero initial orientation recovers the reference before entering the nanopore. It is revealed that the particle's translocation is not very sensitive to its initial orientation and initial lateral offset when the applied electric field is very high.

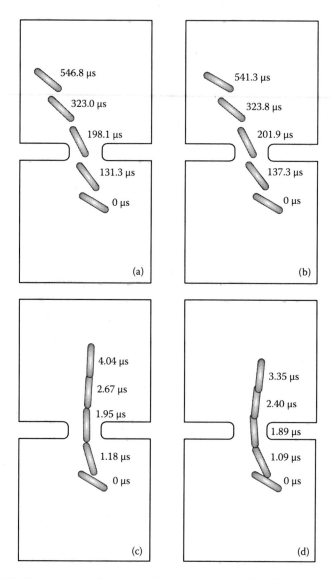

FIGURE 7.7 Superposed trajectories of the particle under $E^* = 7.7 \times 10^{-4}$ (a and b) and $E^* = 7.7 \times 10^{-2}$ (c and d). $\kappa a = 2.05$, $\sigma_w^* = 0$, $\theta_{p0}^* = 60°$, $x_{p0}^* = 2.5$ (a and c), and $x_{p0}^* = 5$ (b and d). (From Ai, Y., and S. Qian. 2011. Direct numerical simulation of electrokinetic translocation of a cylindrical particle through a nanopore using a Poisson–Boltzmann approach. *Electrophoresis* 32:996–1005 with permission from Wiley-VCH.)

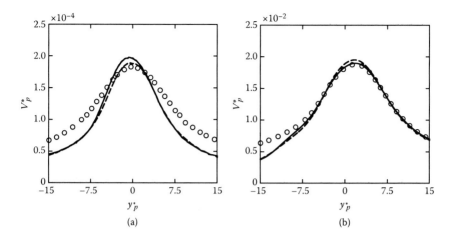

FIGURE 7.8 The y component translational velocity as a function of the particle's location y_p^* under $E^* = 7.7 \times 10^{-4}$ (a) and $E^* = 7.7 \times 10^{-2}$ (b). Symbols and lines represent, respectively, $\theta_{p0}^* = 0$ and 60°. $\kappa a = 2.05$, $\sigma_w^* = 0$, $x_{p0}^* = 0$ (circles), $x_{p0}^* = 2.5$ (solid line), and $x_{p0}^* = 5$ (dashed line). (From Ai, Y., and S. Qian. 2011. Direct numerical simulation of electrokinetic translocation of a cylindrical particle through a nanopore using a Poisson–Boltzmann approach. *Electrophoresis* 32:996–1005 with permission from Wiley-VCH.)

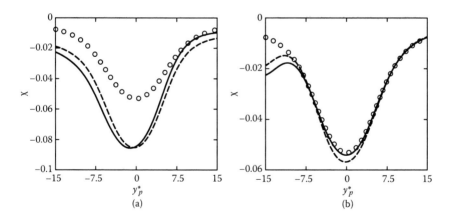

FIGURE 7.9 Current deviation as a function of the particle's location y_p^* under $E^* = 7.7 \times 10^{-4}$ (a) and $E^* = 7.7 \times 10^{-2}$ (b). Symbols and lines represent, respectively, $\theta_{p0}^* = 0$ and 60°. $\kappa a = 2.05$, $\sigma_w^* = 0$, $x_{p0}^* = 0$ (circles), $x_{p0}^* = 2.5$ (solid line), and $x_{p0}^* = 5$ (dashed line). (From Ai, Y., and S. Qian. 2011. Direct numerical simulation of electrokinetic transloca-tion of a cylindrical particle through a nanopore using a Poisson–Boltzmann approach. *Electrophoresis* 32:996–1005 with permission from Wiley-VCH.)

The current deviations corresponding to the cases in Figure 7.8 are shown in Figure 7.9, in which the current deviation when $\theta^*_{p0} = 0$ and $x^*_{p0} = 0$ is also included as a reference. When the applied electric field is relatively low, $E^* = 7.7 \times 10^{-4}$, the particle's initial orientation significantly increases the current blockade. When the particle exits the nanopore, the current deviation begins to recover the reference due to the lateral movement. The slight difference between the current deviations when $x^*_{p0} = 2.5$ and $x^*_{p0} = 5$ is due to the effect of the particle's initial lateral offset. When the applied electric field is relatively high, $E^* = 7.7 \times 10^{-2}$, the current deviation is only affected at the beginning of the particle's translocation process and later recovers the reference when the particle is well aligned to the applied electric field.

7.4.3 EFFECT OF NANOPORE'S SURFACE CHARGE DENSITY

In the studies discussed, the nanopore is assumed to be uncharged. Next, we further investigate the effect of the nanopore's surface charge. When the nanopore is also charged, an extra EOF is formed inside the nanopore, which in turn affects the particle's translocation through the viscous force acting on the particle. Figures 7.10a and 7.10b show, respectively, the particle's y component translational velocity as a function of the particle's location y^*_p under $E^* = 7.7 \times 10^{-4}$ and $E^* = 7.7 \times 10^{-2}$ when $\kappa a = 2.05$, $\theta^*_{p0} = 0$, and $x^*_{p0} = 0$. The particle's y component translational velocity for $\sigma^*_w = 0$ is included as a reference.

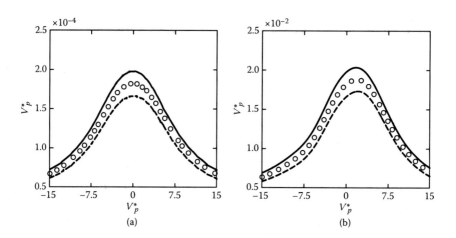

FIGURE 7.10 The y component translational velocity as a function of the particle's location y^*_p under $E^* = 7.7 \times 10^{-4}$ (a) and $E^* = 7.7 \times 10^{-2}$ (b). Circles, solid line, and dashed line represent, respectively, $\sigma^*_w = 0$, $\sigma^*_w = -0.1\sigma^*_p$, and $\sigma^*_w = 0.1\sigma^*_p$. $\theta^*_{p0} = 0$, $x^*_{p0} = 0$, and $\kappa a = 2.05$. (From Ai, Y., and S. Qian. 2011. Direct numerical simulation of electrokinetic translocation of a cylindrical particle through a nanopore using a Poisson–Boltzmann approach. *Electrophoresis* 32:996–1005 with permission from Wiley-VCH.)

When $\sigma_w^* = -0.1\sigma_p^*$, the nanopore carries a surface charge opposite the particle's surface charge, which in turn generates an EOF in the same direction as the particle's translocation. Therefore, the particle's translocation is enhanced approximately 10% compared to the reference. On the contrary, $\sigma_w^* = 0.1\sigma_p^*$ implies that the nanopore induces an EOF opposite to the particle's translocation, which indicates an approximate 10% retardation effect. As the magnitude of the nanopore's surface charge density increases, the retardation effect arising from the induced EOF could even prevent the particle's translocation through the nanopore. It is concluded that the enhancement ratio or retardation ratio of the particle's y component translational velocity is independent of the applied electric field and mainly depends on the ratio σ_w^*/σ_p^* when the EDLs of the particle and the nanopore are not overlapped. Although the nanopore's surface charge can significantly affect the particle's translocation, it has limited effect on the current deviation.

Considering the particle initially located at $x_{p0}^* = 5$ presenting an initial orientation $\theta_{p0}^* = 60°$ under a relatively low electric field, $E^* = 7.7 \times 10^{-4}$, Figure 7.11a shows the effect of the nanopore's surface charge density on particle trajectory through the nanopore. It is found that the particle trajectories through nanopores of $\sigma_w^* = 0$, $\sigma_w^* = -0.1\sigma_p^*$, and $\sigma_w^* = 0.1\sigma_p^*$ are nearly identical before the particle enters the nanopore; however, they deviate from each other after exiting the nanopore. Figure 7.11b shows the corresponding variations of the particle's orientation along the particle's location. The particle's orientations are also identical before entering the nanopore. When the particle exits the nanopore, the particle's angle for $\sigma_w^* = -0.1\sigma_p^*$ is larger than that for $\sigma_w^* = 0$, which is also larger than that for $\sigma_w^* = 0.1\sigma_p^*$. As mentioned, a larger particle's orientation leads to a more pronounced lateral movement, which is able to explain the three particle trajectories in Figure 7.11a. Figure 7.11c shows the corresponding particle's y component translational velocity as a function of the particle's location y_p^*.

The effect due to the EOF arising from the nanopore's surface charge on the particle velocity is similar to that in Figure 7.10a when the particle moves toward the nanopore. Due to the significant lateral movement when the particle exits the nanopore, the particle's y component translational velocities become nearly identical. Figure 7.11d shows the corresponding current deviations as a function of the particle's location y_p^*. Before the particle enters the nanopore, the current deviation is insensitive to the nanopore's surface charge. Because the particle exits the three nanopores with different orientations, the current deviation shows a dependence on the nanopore's surface charge. Because of the lateral movement, the difference in the three current deviations becomes increasingly smaller. When the applied electric field is relatively high, $E^* = 7.7 \times 10^{-2}$, the effect of the EOF on the particle's y component translational velocity is nearly identical to that in Figure 7.10b. As the particle is aligned to the applied electric field due to the DEP effect, the effect of the nanopore's surface charge shows limited influence on the particle's trajectory, orientation, and the resulting current deviation.

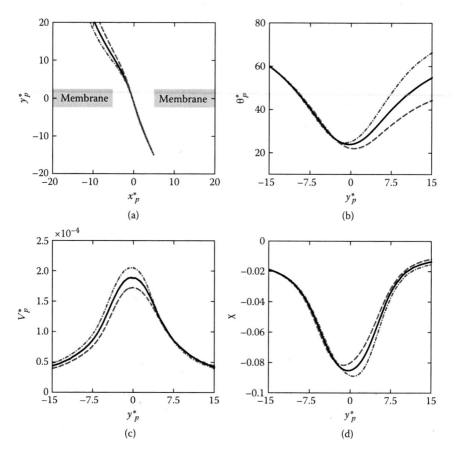

FIGURE 7.11   Trajectory (a), orientation (b), y component translational velocity of the particle (c), and current deviation (d), as a function of the particle's location y_p^* under $E^* = 7.7 \times 10^{-4}$. Solid, dashed, and dash-dotted lines represent, respectively, $\sigma_w^* = 0$, $\sigma_w^* = 0.1\sigma_p^*$, and $\sigma_w^* = -0.1\sigma_p^*$. $\theta_{p0}^* = 60°$, $x_{p0}^* = 5$, and $\kappa a = 2.05$. (From Ai, Y., and S. Qian. 2011. Direct numerical simulation of electrokinetic translocation of a cylindrical particle through a nanopore using a Poisson-Boltzmann approach. *Electrophoresis* 32:996–1005 with permission from Wiley-VCH.)

7.5   CONCLUDING REMARKS

A transient numerical model was developed to simultaneously solve the PB equation for the ionic concentrations and the electric field due to the surface charges, the Laplace equation for the externally applied electric field, the modified Stokes equations for the flow field, and the governing equations for the particle translation and rotation using the ALE method. Different from the thin EDL approximation in the previous chapters, the effect of finite EDL on electrokinetic particle translocation through a nanopore was taken into account. The developed continuum-based model was valid when the EDL of the particle was not disturbed by the external fields and nearby EDLs of the solid boundaries. When the axis of the particle initially coincided with the nanopore's centerline, the particle translocated along the nanopore's centerline in the absence of any rotational motion. Current blockade due to the translocating particle through the nanopore was predicted and was in qualitative agreement with existing experimental observations (Storm et al. 2005; Kim et al. 2007).

When the applied electric field was relatively low, the particle's initial orientation induced significant rotational motion during the particle's translocation process. The particle experienced lateral movement because of the nonzero initial orientation. In addition, the direction of the lateral movement highly depended on the particle's initial orientation. It was also found that the particle's initial orientation could significantly affect the current deviation. However, the particle's initial lateral offset from the nanopore's centerline had a minor effect on particle translocation and the resulting current deviation. When the applied electric field was relatively high, the dominant DEP effect aligned the particle parallel to the applied electric field quickly. As a result, the particle's initial orientation only affected the particle's y component translational velocity and the ionic current through the nanopore at the beginning of particle translocation.

The nanopore's surface charge can also affect particle translocation through the nanopore by the generated EOF. In addition, the enhancement ratio or retardation ratio of particle translocation was linearly proportional to the ratio of the nanopore's surface charge to the particle's surface charge when the particle mainly translocated along the nanopore's centerline and the conditions for the PB model were satisfied. However, the EOF had a limited effect on the current deviation.

REFERENCES

Ai, Y., and S. Qian. 2011. Direct numerical simulation of electrokinetic translocation of a cylindrical particle through a nanopore using a Poisson–Boltzmann approach. *Electrophoresis* 32:996–1005.

Ai, Y., M. Zhang, S. W. Joo, M. A. Cheney, and S. Qian. 2010. Effects of electroosmotic flow on ionic current rectification in conical nanopores. *Journal of Physical Chemistry C* 114 (9):3883–3890.

Aksimentiev, A., J. B. Heng, G. Timp, and K. Schulten. 2004. Microscopic kinetics of DNA translocation through synthetic nanopores. *Biophysical Journal* 87 (3):2086–2097.

Bayley, H. 2006. Sequencing single molecules of DNA. *Current Opinion in Chemical Biology* 10 (6):628–637.

Chang, H., F. Kosari, G. Andreadakis, M. A. Alam, G. Vasmatzis, and R. Bashir. 2004. DNA-mediated fluctuations in ionic current through silicon oxide nanopore channels. *Nano Letters* 4 (8):1551–1556.

Chen, L., and A. T. Conlisk. 2010. DNA nanowire translocation phenomena in nanopores. *Biomedical Microdevices* 12 (2):235–245.

Choi, Y., L. A. Baker, H. Hillebrenner, and C. R. Martin. 2006. Biosensing with conically shaped nanopores and nanotubes. *Physical Chemistry Chemical Physics* 8 (43):4976–4988.

Comer, J., V. Dimitrov, Q. Zhao, G. Timp, and A. Aksimentiev. 2009. Microscopic mechanics of hairpin DNA translocation through synthetic nanopores. *Biophysical Journal* 96 (2):593–608.

Corry, B., S. Kuyucak, and S. H. Chung. 2000. Invalidity of continuum theories of electrolytes in nanopores. *Chemical Physics Letters* 320 (1–2):35–41.

Ennis, J., and J. L. Anderson. 1997. Boundary effects on electrophoretic motion of spherical particles for thick double layers and low zeta potential. *Journal of Colloid and Interface Science* 185 (2):497–514.

Gu, L.-Q., and J. W. Shim. 2010. Single molecule sensing by nanopores and nanopore devices. *Analyst* 135 (3):441–451.

Henry, D. C. 1931. The cataphoresis of suspended particles. Part I. The equation of cataphoresis. *Proceedings of the Royal Society of London, Series A* 133:106–129.

Howorka, S., and Z. Siwy. 2009. Nanopore analytics: Sensing of single molecules. *Chemical Society Reviews* 38 (8):2360–2384.

Hsu, J. P., Z. S. Chen, D. J. Lee, S. Tseng, and A. Su. 2008. Effects of double-layer polarization and electroosmotic flow on the electrophoresis of a finite cylinder along the axis of a cylindrical pore. *Chemical Engineering Science* 63 (18):4561–4569.

Hsu, J. P., and C. Y. Kao. 2002. Electrophoresis of a finite cylinder along the axis of a cylindrical pore. *Journal of Physical Chemistry B* 106 (41):10605–10609.

Hsu, J. P., and C. C. Kuo. 2006. Electrophoresis of a finite cylinder positioned eccentrically along the axis of a long cylindrical pore. *Journal of Physical Chemistry B* 110 (35):17607–17615.

Hsu, J. P., C. C. Kuo, and M. H. Ku. 2006. Electrophoresis of a toroid along the axis of a cylindrical pore. *Electrophoresis* 27 (16):3155–3165.

Hsu, J. P., C. C. Kuo, and M. H. Ku. 2008. Electrophoresis of a charge-regulated toroid normal to a large disk. *Electrophoresis* 29 (2):348–357.

Kim, Y. R., J. Min, I. H. Lee, S. Kim, A. G. Kim, K. Kim, K. Namkoong, and C. Ko. 2007. Nanopore sensor for fast label-free detection of short double-stranded DNAs. *Biosensors and Bioelectronics* 22 (12):2926–2931.

Lathrop, D. K., E. N. Ervin, G. A. Barrall, M. G. Keehan, R. Kawano, M. A. Krupka, H. S. White, and A. H. Hibbs. 2010. Monitoring the escape of DNA from a nanopore using an alternating current signal. *Journal of the American Chemical Society* 132 (6):1878–1885.

Liu, H., H. H. Bau, and H. H. Hu. 2004. Electrophoresis of concentrically and eccentrically positioned cylindrical particles in a long tube. *Langmuir* 20 (7):2628–2639.

Liu, H., S. Qian, and H. H. Bau. 2007. The effect of translocating cylindrical particles on the ionic current through a nanopore. *Biophysical Journal* 92 (4):1164–1177.

Martin, C. R., and Z. S. Siwy. 2007. Learning nature's way: Biosensing with synthetic nanopores. *Science* 317 (5836):331–332.

Moy, G., B. Corry, S. Kuyucak, and S. H. Chung. 2000. Tests of continuum theories as models of ion channels. I. Poisson-Boltzmann theory versus Brownian dynamics. *Biophysical Journal* 78 (5):2349–2363.

Mukhopadhyay, R. 2009. DNA sequencers: the next generation. *Analytical Chemistry* 81 (5):1736–1740.

Ohshima, H. 1998. Surface charge density surface potential relationship for a cylindrical particle in an electrolyte solution. *Journal of Colloid and Interface Science* 200 (2):291–297.

Qian, S., and S. W. Joo. 2008. Analysis of self-electrophoretic motion of a spherical particle in a nanotube: Effect of nonuniform surface charge density. *Langmuir* 24 (9):4778–4784.

Qian, S., S. W. Joo, Y. Ai, M. A. Cheney, and W. Hou. 2009. Effect of linear surface-charge non-uniformities on the electrokinetic ionic-current rectification in conical nano-pores. *Journal of Colloid and Interface Science* 329 (2):376–383.

Qian, S., S. W. Joo, W. Hou, and X. Zhao. 2008. Electrophoretic motion of a spherical particle with a symmetric nonuniform surface charge distribution in a nanotube. *Langmuir* 24 (10):5332–5340.

Qian, S., A. H. Wang, and J. K. Afonien. 2006. Electrophoretic motion of a spherical par-ticle in a converging-diverging nanotube. *Journal of Colloid and Interface Science* 303 (2):579–592.

Rhee, M., and M. A. Burns. 2006. Nanopore sequencing technology: Research trends and applications. *Trends in Biotechnology* 24:580–586.

Sigalov, G., J. Comer, G. Timp, and A. Aksimentiev. 2007. Detection of DNA sequences using an alternating electric field in a nanopore capacitor. *Nano Letters* 8 (1):56–63.

Storm, A. J., C. Storm, J. H. Chen, H. Zandbergen, J. F. Joanny, and C. Dekker. 2005. Fast DNA translocation through a solid-state nanopore. *Nano Letters* 5 (7):1193–1197.

van Dorp, S., U. F. Keyser, N. H. Dekker, C. Dekker, and S. G. Lemay. 2009. Origin of the electrophoretic force on DNA in solid-state nanopores. *Nature Physics* 5 (5):347–351.

Zhao, X., C. M. Payne, and P. T. Cummings. 2008. Controlled translocation of DNA seg-ments through nanoelectrode gaps from molecular dynamics. *Journal of Physical Chemistry C* 112 (1):8–12.

Zhu, J., and X. Xuan. 2009. Particle electrophoresis and dielectrophoresis in curved micro-channels. *Journal of Colloid and Interface Science* 340 (2):285–290.

8 Electrokinetic Translocation of a Cylindrical Particle through a Nanopore *Poisson–Nernst–Planck Multi-Ion Model*

A small direct current (DC) voltage imposed across a nanopore merged in an aqueous electrolyte generates an ionic current flowing through the pore. The induced ionic current is sensitive to the size and shape of the nanopore. A nanoparticle, such as a DNA molecule, passing through a nanopore by electrophoresis will modulate the ionic current through the pore, yielding information on the translocating nanoparticle through the change of the recorded ionic current signal. Nanopore-based sensing has the potential to become a direct, fast, and inexpensive DNA sequencing technology.

In this chapter, dynamics of particle translocation and its corresponding ionic current through a nanopore are theoretically investigated using a transient continuum-based model, the Poisson–Nernst–Planck (PNP) multi-ion model. The model includes the Poisson equation for electrostatics, the Nernst–Planck equations for ionic mass transport, the Navier–Stokes equations for hydrodynamics, Newton's equation for particle translation, and the Euler equation for a particle's rotational motion. The model takes into account the full fluid-particle-electric field-ionic concentration field interactions with no assumption made concerning the thickness of the electrical double layer (EDL), the magnitudes of the surface charge densities on the particle and the nanopore, and the magnitude of the electric field imposed. The fully coupled system is numerically solved using the arbitrary Lagrangian–Eulerian (ALE) finite element method.

When the imposed electric field is relatively low, the translocating particle blocks the ionic current through the nanopore, resulting in "current blockade," and could even be trapped at the entrance of the nanopore if the EDL of the charged particle is relatively thick. When the electrical field imposed is relatively high, the charged particle can always pass through the nanopore by electrophoresis, and current enhancement is predicted if the EDL of the particle is relatively thick. The obtained diverse results, current blockade and current enhancement, qualitatively agree with the existing experimental observations. The particle's surface charge and initial orientation with respect to the centerline of the nanopore significantly affect both the dynamics of the particle translocation and the modulation of the ionic current through a nanopore. Due to the induced dielectrophoresis effect, a relatively high electric field tends to align the particle with its longest axis parallel to the local electric field quickly. The particle's initial lateral offset from the centerline of the nanopore has an insignificant effect on particle translocation and the ionic current through the nanopore.

8.1   INTRODUCTION

When individual nanoparticles such as DNA molecules, proteins, and organic polymers are electrophoretically driven through a single nanopore by an externally imposed electric field, the translocating nanoparticle modulates the ionic current flowing through the nanopore. The resulting distinct change in the ionic current enables detection and characterization of single nanoparticles for various bioanalytical applications (Choi et al. 2006; Martin and Siwy 2007; Howorka and Siwy 2009; Purnell and Schmidt 2009; Gu and Shim 2010). In particular, such nanopore-based technology can be used to detect and characterize DNA for DNA sequencing (Bayley 2006; Rhee and Burns 2006; Mukhopadhyay 2009). With the development of state-of-the-art nanofabrication technologies, the feasibility of the nanopore-based sensing technique has been experimentally demonstrated (Meller, Nivon, and Branton 2001; Chang et al. 2004; Heng et al. 2004; Storm, Chen, et al. 2005; Storm, Storm, et al. 2005; Kim et al. 2007; Lathrop et al. 2010). The obtained experimental results showed that the ionic current response to the translocating nanoparticle depended on many factors, such as the intensity of the electric field imposed (Meller, Nivon, and Branton 2001; Li et al. 2003; Aksimentiev et al. 2004; Heng et al. 2004; Storm, Storm, et al. 2005); the salt concentration (Chang et al. 2004; Fan et al. 2005); the nanopore's size and surface charge (Li et al. 2003; Aksimentiev et al. 2004; Chang et al. 2004; Heng et al. 2004); and the length and size of the nanoparticle (Heng et al. 2004; Meller, Nivon, and Branton 2001; Storm, Storm, et al. 2005). Depending on the experimental conditions, both current blockade and current enhancement have been experimentally observed during translocation of a single molecule.

To better understand these diverse results of the effect of the translocating nanoparticles on ionic currents, electrokinetic translocation of a nanoparticle through a nanopore has been theoretically investigated using molecular dynamics (MD) simulation (Aksimentiev et al. 2004; Sigalov et al. 2007; Zhao, Payne, and Cummings 2008; Comer et al. 2009). MD simulation provides unequivocal insights into the atomic-level dynamics of nanoparticle translocation through a nanopore and is even able to capture the conformational change of DNA molecules during the translocation. However, the time scale covered by MD simulations is currently limited to about 100 ns. To accelerate the translocation events that normally take microseconds, most of the MD simulations were performed at a much higher electric field than the actual one used in experiments. However, many experimental studies have demonstrated that the current response to the translocating particle depends on the electric field imposed (Meller, Nivon, and Branton 2001; Li et al. 2003; Aksimentiev et al. 2004; Heng et al. 2004; Storm, Storm, et al. 2005). Therefore, direct comparison between experiments and results obtained from MD simulations is not appropriate.

Due to the limitations of MD simulations, continuum-based models have been developed to simulate electrokinetic particle translocation through a nanopore (Hsu, Ku, and Kao 2004; Hsu, Yeh, and Ku 2006; Qian, Wang, and Afonien 2006; Liu, Qian, and Bau 2007; Qian and Joo 2008; Qian et al. 2008, 2009; Hsu, Chen, and Tseng 2009; Chen and Conlisk 2010). A quasi-static method, which assumes the electrostatics, the ionic mass transport, the momentum transport in fluid, and the particle motion are under steady state at any particle position, has been used to determine the particle's translational velocity along the axis of the nanopore based on the force balance on the particle. Despite the quasi-static approximation and neglect of particle rotation, a remarkable agreement between the numerical predictions and the existing experimental results is obtained. Thus, the continuum-based model is still able to capture the essential physics of the nanoparticle translocation process. However, the quasi-static method is unable to capture the dynamics of particle translocation through the nanopore, especially the particle's rotational dynamics, which may affect the current response.

In this chapter, a transient continuum-based model, consisting of the Nernst–Planck equations for ionic mass transport, the Poisson equation for electrostatics, the Navier–Stokes and continuity equations for the fluid flow field, Newton's equation for particle translation, and the Euler equation for particle rotation, is developed to capture the dynamics of the electrokinetic particle translocation through a nanopore. In contrast to the Poisson–Boltzmann (PB) model in Chapter 7, the model described in this chapter makes no assumption concerning the EDL thickness, the magnitude of the surface charge density on the charged surface, and the magnitude of the imposed electric field. The details of the mathematical model and its implementation in COMSOL are described, respectively, in Sections 8.2 and 8.3. Section 8.4 includes some representative simulation results on the dynamics of nanoparticle translocation and its effect on the ionic current through a nanopore under various conditions. Section 8.5 provides a conclusion.

8.2   MATHEMATICAL MODEL

To reduce computational effort, here we only consider a two-dimensional (2D) geometry. Figure 8.1 schematically depicts a slit nanochannel of length h and width $2b$ connecting two identical reservoirs of width $2W$ and height H filled

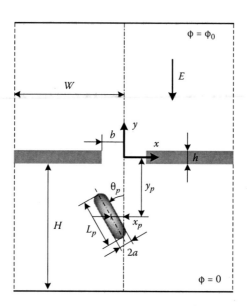

FIGURE 8.1 Schematics of nanoparticle translocation through a nanopore. (From Ai, Y., and S. Qian. 2011. Electrokinetic particle translocation through a nanopore. *Physical Chemistry Chemical Physics* 13:4060–4071 with permission from RCS.)

with a binary KCl aqueous solution of density ρ, dynamic viscosity μ, and relative permittivity ε_f. The two reservoirs are large enough to maintain bulk ionic concentration C_0 at places far away from the nanopore. A Cartesian coordinate system (x, y) with the origin fixed at the center of the nanopore is used to describe all the variables. A negatively charged, rigid, nanoparticle of length L_p, capped with two half circles of radius a, is initially located at (x_{p0}, y_{p0}) and has an angle θ_{p0} with respect to the centerline of the nanopore. $\theta_{p0} > 0$ represents the angle counterclockwise with respect to the centerline of the nanopore. A potential difference ϕ_0 is applied across two electrodes positioned inside the two reservoirs, inducing a negative axial electric field $\mathbf{E}$ across the nanopore to generate an ionic current through the nanopore and drive the negatively charged particle translocating through the nanopore by electrophoresis. We assume that the nanopore and the nanoparticle have uniform surface charge densities of σ_w and σ_p, respectively.

8.2.1 Dimensional Form of Mathematical Model

The electrostatics and ionic mass transport within the electrolyte solution are governed by the verified PNP equations (White and Bund 2008; Qian, Das, and Luo 2007; Qian et al. 2009; Ai et al. 2010):

$$-\varepsilon_0\varepsilon_f\nabla^2\phi = F(c_1z_1 + c_2z_2), \qquad (8.1)$$

$$\frac{\partial c_i}{\partial t} + \nabla \bullet \mathbf{N}_i = 0, \; i = 1 \text{ and } 2,$$ (8.2)

where ε_0 is the absolute permittivity of the vacuum; ϕ is the electric potential within the fluid; F is the Faraday constant; c_1 and c_2 are, respectively, the molar concentrations of the cations (K^+) and anions (Cl^-) in the electrolyte solution; z_1 and z_2 are, respectively, the valences of cations ($z_1 = 1$ for K^+) and anions ($z_2 = -1$ for Cl^-). $\mathbf{N}_i = \mathbf{u}c_i - D_i \nabla c_i - z_i \frac{D_i}{RT} F c_i \nabla \phi$ is the ionic flux density of the ith ionic species, in which $\mathbf{u}$ is the fluid velocity; D_i is the diffusivity of the ith ionic species; R is the universal gas constant; and T is the absolute temperature of the electrolyte solution.

Appropriate boundary conditions are required to solve these PNP equations. The electric potentials applied at the ends of the two reservoirs are, respectively, $\phi(x, -(H+h/2)) = 0$ and $\phi(x, (H+h/2)) = \phi_0$. Surface charge density boundary conditions on the particle surface and the nanopore are, respectively, $-\varepsilon_0 \varepsilon_f \mathbf{n} \bullet \nabla \phi = \sigma_p$ and $-\varepsilon_0 \varepsilon_f \mathbf{n} \bullet \nabla \phi = \sigma_w$, where $\mathbf{n}$ is the unit normal vector directed from the boundary into the fluid. All the other boundaries are assumed to be uncharged. The boundary conditions for the ionic concentrations at the ends of the two reservoirs are $c_i(x, \pm(H+h/2)) = C_0$, $i = 1$ and 2. The normal ionic flux on the particle surface only includes the convective flux, $\mathbf{n} \bullet \mathbf{N}_i = \mathbf{n} \bullet (\mathbf{u}c_i)$, $i = 1$ and 2. The normal ionic fluxes on all the other boundaries are set as zero.

The Reynolds number of electrokinetic flow in a nanopore is extremely small; it is thus appropriate to model the flow field using the modified Stokes equations by neglecting the inertial terms in the Navier–Stokes equations,

$$\rho \frac{\partial \mathbf{u}}{\partial t} = -\nabla p + \mu \nabla^2 \mathbf{u} - F(z_1 c_1 + z_2 c_2)\nabla \phi,$$ (8.3)

and the continuity equation for the incompressible fluid,

$$\nabla \bullet \mathbf{u} = 0.$$ (8.4)

In these equations, p is the hydrostatic pressure. The first, second, and third terms on the right-hand side of Equation (8.3) denote, respectively, the pressure, viscous, and electrostatic forces.

A nonslip boundary condition is imposed on the surfaces of the nanopore and the membrane. Normal flow with $p = 0$ is imposed at the ends of the two reservoirs. A slip boundary condition is applied at the side boundaries of the two reservoirs, which are far away from the nanopore. As the particle simultaneously translates and rotates, the flow boundary condition on the particle surface is given as

$$\mathbf{u} = \mathbf{U}_p + \boldsymbol{\omega}_p \times (\mathbf{x}_s - \mathbf{x}_p),$$ (8.5)

where $\mathbf{U}_p$ is the translational velocity; $\boldsymbol{\omega}_p$ is the rotational velocity; and $\mathbf{x}_s$ and $\mathbf{x}_p$ are, respectively, the position vectors of the surface and center of mass of the particle. The particle's translational velocity is governed by Newton's second law,

$$m_p \frac{d\mathbf{U}_p}{dt} = \mathbf{F}_H + \mathbf{F}_E, \tag{8.6}$$

where m_p is the mass of the particle. $\mathbf{F}_H$ and $\mathbf{F}_E$ are, respectively, the hydrodynamic force and the electrostatic force acting on the particle; they are obtained, respectively, by integrating the hydrodynamic stress tensor $\mathbf{T}^H$ and the Maxwell stress tensor $\mathbf{T}^E$ over the particle surface:

$$\mathbf{F}_H = \int \mathbf{T}^H \bullet \mathbf{n} d\Gamma = \int \left[-p\mathbf{I} + \mu \left(\nabla \mathbf{u} + \nabla \mathbf{u}^T \right) \right] \bullet \mathbf{n} d\Gamma, \tag{8.7}$$

$$\mathbf{F}_E = \int \mathbf{T}^E \bullet \mathbf{n} d\Gamma = \int \varepsilon_0 \varepsilon_f \left[\mathbf{E}\mathbf{E} - \frac{1}{2} (\mathbf{E} \bullet \mathbf{E}) \mathbf{I} \right] \bullet \mathbf{n} d\Gamma. \tag{8.8}$$

In these equations, $\mathbf{E} = -\nabla \phi$ is the electric field related to the electric potential, and Γ is the particle's surface.

The rotational velocity of the particle is determined by

$$I_p \frac{d\boldsymbol{\omega}_p}{dt} = \int (\mathbf{x}_s - \mathbf{x}_p) \times \left(\mathbf{T}^H \bullet \mathbf{n} + \mathbf{T}^E \bullet \mathbf{n} \right) d\Gamma, \tag{8.9}$$

where I_p is the particle's moment of inertia. The right-hand side of Equation (8.9) is the total torque acting on the particle. The center of mass $\mathbf{x}_p$ and the orientation θ_p of the particle are expressed, respectively, by

$$\mathbf{x}_p = \mathbf{x}_{p0} + \int_0^t \mathbf{U}_p dt, \tag{8.10}$$

$$\boldsymbol{\theta}_p = \boldsymbol{\theta}_{p0} + \int_0^t \boldsymbol{\omega}_p dt, \tag{8.11}$$

where $\mathbf{x}_{p0}$ and θ_{p0} denote, respectively, the initial position and orientation of the particle.

The induced ionic current through the nanopore is given as

$$I = \int F\left(z_1 \mathbf{N}_1 + z_2 \mathbf{N}_2 \right) \bullet \mathbf{n} dS, \tag{8.12}$$

where S denotes the opening of either reservoir due to the current conservation.

8.2.2 DIMENSIONLESS FORM OF MATHEMATICAL MODEL

The bulk concentration C_0 as the ionic concentration scale, RT/F as the potential scale, the particle radius a as the length scale, $U_0 = \varepsilon_0 \varepsilon_f R^2 T^2 / (\mu a F^2)$ as the velocity scale, and $\mu U_0 / a$ as the pressure scale are chosen to normalize the governing equations given in Section 8.2.1:

$$-\nabla^{*2}\phi^* = \frac{(\kappa a)^2}{2}(c_1^* z_1 + c_2^* z_2), \tag{8.13}$$

$$\frac{\partial c_i^*}{\partial t^*} + \nabla^* \bullet (\mathbf{u}^* c_i^* - D_i^* \nabla^* c_i^* - z_i D_i^* c_i^* \nabla^* \phi^*) = 0, \ i = 1 \text{ and } 2, \tag{8.14}$$

$$\mathrm{Re}\frac{\partial \mathbf{u}^*}{\partial t^*} = -\nabla^* p^* + \nabla^{*2}\mathbf{u}^* - \frac{(\kappa a)^2}{2}(z_1 c_1^* + z_2 c_2^*)\nabla^* \phi^*, \tag{8.15}$$

$$\nabla^* \bullet \mathbf{u}^* = 0. \tag{8.16}$$

In these equations, $\kappa^{-1} = \lambda_D = \sqrt{\varepsilon_0 \varepsilon_f RT / (2F^2 C_0)}$ is the Debye length, and $\mathrm{Re} = \rho U_0 a / \mu$. The corresponding dimensionless boundary conditions are

$$\phi^*(x^*, \ -(H^* + h^*/2)) = \phi^*(x^*, \ (H^* + h^*/2)) - \phi_0^* = 0, \tag{8.17}$$

$$-\mathbf{n} \bullet \nabla^* \phi^* = \sigma_p^*, \tag{8.18}$$

$$-\mathbf{n} \bullet \nabla^* \phi^* = \sigma_w^*. \tag{8.19}$$

$$c_i^*(x^*, \ \pm(H^* + h^*/2)) = 1, \ i = 1 \text{ and } 2, \tag{8.20}$$

$$\mathbf{n} \bullet \mathbf{N}_i^* = \mathbf{n} \bullet (\mathbf{u}^* c_i^*), \ i = 1 \text{ and } 2, \tag{8.21}$$

$$\mathbf{u}^* = \mathbf{U}_p^* + \omega_p^* \times (\mathbf{x}_s^* - \mathbf{x}_p^*), \tag{8.22}$$

where the surface charge density scale is $\varepsilon_0 \varepsilon_f RT / (Fa)$. The force and torque are, respectively, normalized by μU_0 and $\mu U_0 a$, yielding the dimensionless equations for particle motion:

$$m_p^* \frac{d\mathbf{U}_p^*}{dt^*} = \mathbf{F}_H^* + \mathbf{F}_E^*, \tag{8.23}$$

$$I_p^* \frac{d\omega_p^*}{dt^*} = \int \left(\mathbf{x}_s^* - \mathbf{x}_p^* \right) \times \left(\mathbf{T}^{H*} \bullet \mathbf{n} + \mathbf{T}^{E*} \bullet \mathbf{n} \right) d\Gamma^*, \qquad (8.24)$$

where the mass and the moment of inertia are, respectively, normalized by $a\mu/U_0$ and $a^3\mu/U_0$ in 2D:

$$\mathbf{T}^{H*} = -p^*\mathbf{I} + \left(\nabla^*\mathbf{u}^* + \nabla^*\mathbf{u}^{*T} \right)$$

and

$$\mathbf{T}^{E*} = \mathbf{E}^*\mathbf{E}^* - \frac{1}{2}\left(\mathbf{E}^* \bullet \mathbf{E}^* \right)\mathbf{I}.$$

The dimensionless ionic current flowing through the nanopore normalized by $FU_0C_0a^2$ is

$$I^* = \int (z_1\mathbf{N}_1^* + z_2\mathbf{N}_2^*) \bullet \mathbf{n}\, dS^*. \qquad (8.25)$$

8.3 NUMERICAL IMPLEMENTATION IN COMSOL AND CODE VALIDATION

The strongly coupled governing equations are numerically solved by COMSOL Multiphysics® 3.5a using the predefined ALE method. The dimensions used in the present study are $b = 5$ nm, $h = 5$ nm, $W = 25$ nm, $H = 40$ nm, $a = 1$ nm, and $L_p = 10$ nm. The physical parameters used in the simulation are $\varepsilon_f = 80$, $\rho = 1 \times 10^3$ kg m^3, $\mu = 1 \times 10^{-3}$ Pa·s, D_1 (diffusivity of K$^+$) = 1.95×10^{-9} m^2/s, D_2 (diffusivity of Cl$^-$) = 2.03×10^{-9} m^2/s, $\sigma_p = -0.01$ C/m^2, and $T = 300$ K. Table 8.1 lists all the constants that need to be defined in COMSOL Multiphysics. Table 8.2 shows detailed instructions for setting up the dimensionless model in the graphical user interface (GUI) of COMSOL Multiphysics. The GUI model can be saved as a MATLAB M-file, which could be further modified to achieve an automated remeshing for the tracking of long-term particle motion. The full COMSOL MATLAB script M-file is as follows:

```
%%%%%%%%%%%%%%%%%%%%%%%%%%PNP model.m%%%%%%%%%%%%%%%%%%%%%%%%%%
%%%%%%%%%%%%%%%%C0=100mM
%%%%%%%%%%%%%%%%ang=60
%%%%%%%%%%%%%%%%xp=2.5
%%%%%%%%%%%%%%%%yp=-15
%%%%%%%%%%%%%%%%E=2000KV/m
%%%%%%%%%%%%%%%%cw=0
```

(Continued on Page 278)

TABLE 8.1
Constant Table

Variable	Value or Expression	Description
D1	1.95e-9 [m^2/s]	Diffusivity of K^+
D2	2.03e-9 [m^2/s]	Diffusivity of Cl^-
z1	1	Valence of K^+
z2	−1	Valence of Cl^-
F	96485.3415 [C/mol]	Faraday constant
R0	8.314472 [J/K/mol]	Gas constant
T	300 [K]	Temperature
eps0	8.854187817e-12 [F/m]	Permittivity of vacuum
eps_r	80	Relative permittivity of fluid
eta	1.0e-3 [pa*s]	Fluid viscosity
rho	1e3 [kg/m^3]	Density of the fluid and the particle
V0	2000000*85e-9[V]	Applied voltage for electric field 2000 kV/m
cp	−0.01 [C/m^2]	Surface charge density of the particle
cw	0 [C/m^2]	Surface charge density of the nanopore
c0	100 mM	Bulk concentration
a	1e-9 [m]	Particle radius
lamda	sqrt(eps_r*eps0*R0*T/2/F^2/c0)	Debye length
Lp	10*a	Length of the particle
Mass1	rho*pi*a^2	Mass of the two semicircles
Mass2	rho*2*a*(Lp-2*a)	Mass of the middle region of the particle
Massd	(Mass1+Mass2)*U0/eta/a	Dimensionless particle mass
Ird	(Mass1*(0.5*a^2+(Lp-2*a)^2/4+(Lp-2*a)*4*a/(3*pi))+Mass2*1/12*((Lp-2*a)^2+4*a^2))*U0/eta/a^3	Dimensionless moment of inertia
xpin	0	Initial x location of the particle
ypin	−15	Initial y location of the particle
angin	60*pi/180	Initial orientation of the particle
ai	a/lamda	κa
U0	eps0*eps_r*R0^2*T^2/eta/a/F^2	Velocity scale
charges	F*a/R0/T/eps_r/eps0	Surface charge density scale
Re	rho*U0*a/eta	Reynolds number
cpd	cp*charges	Dimensionless particle's surface charge density
cwd	cw*charges	Dimensionless nanopore's surface charge density
D1d	D1*eta*F^2/(eps0*eps_r*R0^2*T^2)	Dimensionless diffusivity of K^+
D2d	D2*eta*F^2/(eps0*eps_r*R0^2*T^2)	Dimensionless diffusivity of Cl^-
V0d	V0*F/R0/T	Dimensionless applied voltage

TABLE 8.2
Model Setup in the GUI of COMSOL Multiphysics 3.5a

Model Navigator	Select **2D** in space dimension and click **Multiphysics** button. Select **COMSOL Multiphysics|Deformed Mesh|Moving Mesh (ALE)|Transient analysis**. Click **Add** button. Select **Chemical Engineering Module|Mass Transport|Nernst– Planck without Electroneutrality|Transient analysis**. Remove the predefined variables in the **Dependent variables** and enter c1 and c2. Click **Add** button. Select **COMSOL Multiphysics|Electromagnetics|Electrostatics**. Click **Add** button. Select **COMSOL Multiphysics|Fluid Dynamics|Incompressible Navier–Stokes|Transient analysis**. Click **Add** button.
Option Menu|**Constants**	Define variables in Table 8.1.
Physics Menu| **Subdomain Setting**	ale model
	Subdomain 1
	Free displacement
	chekf mode
	Subdomain 1
	Tab c1
	$\delta ts = 1$; $D = D1d$; $R = 0$; $um = D1d/F$; $z = z1$; $u = u$; $v = v$; $V = V$
	Tab c2
	$\delta ts = 1$; $D = D2d$; $R = 0$; $um = D2d/F$; $z = z2$; $u = u$; $v = v$; $V = V$
	Tab Init
	$c1(t_0) = 1$; $c2(t_0) = 1$
	es mode
	Subdomain 1
	$d = 1$; $\rho = 0.5*(ai)^2*(z1*c1+z2*c2)$; $\varepsilon r = 1/eps0$
	ns mode
	Subdomain 1
	Tab Physics
	$\rho = Re$; $\eta = 1$; $Fx = -0.5*ai^2*(z1*c1+z2*c2)*Vx$; $Fy = -0.5*ai^2*(z1*c1+z2*c2)*Vy$
	Tab Stabilization
	Deactivate Streamline diffusion and Crosswind diffusion
Physics Menu|**Boundary** **Setting**	ale mode
	Particle surface: Mesh velocity $vx = up-omega*(y-Yp)$
	$vy = vp+omega*(x-Xp)$
	Other boundaries: Mesh displacement $dx = 0$; $dy = 0$
	chekf mode
	Two ends of the nanopore: Concentration $c1 = 1$ and $c2 = 1$
	Particle surface: Convective flux for c1 and c2
	Other boundaries: Insulation/Symmetry for c1 and c2

TABLE 8.2 (CONTINUED)
Model Setup in the GUI of COMSOL Multiphysics 3.5a

	es mode Lower end of the nanopore: $V_0 = 0$ Upper end of the nanopore: $V_0 = V0d$ Particle surface: Surface charge $\rho_s = cpd$ Nanopore wall: Surface charge $\rho_s = cwd$ Others: Zero charge/Symmetry
	ns mode Two ends of the nanopore: Outlet|Pressure, no viscous stress $P_0 = 0$ Particle surface: Inlet|Velocity $u_0 = up - omega*(y - Yp)$ $\qquad\qquad\qquad v_0 = vp + omega*(x - Xp)$ Nanopore wall: Wall|No slip Reservoir wall: Wall|Symmetry boundary
Other Settings	Properties of ale model (**Physics** Menu|**Properties**) Smoothing method: Winslow Allow remeshing: On
	Properties of ns model (**Physics** Menu|**Properties**) Weak constraints: On Constraint type: Non-ideal
	Deactivate the unit system (**Physics** Menu|**Model Setting**) Base unit system: None

Options Menu|
Integration Coupling
Variables|Boundary
Variables
	Boundary selection: 8, 9, 17,18, 19, 20 (Particle surface)
	Name: Fh_x Expression: -lm6; Integration order: 4; Frame:
	  Frame(mesh)
	Name: Fh_y Expression: -lm7; Integration order: 4; Frame:
	  Frame(mesh)
	Name: Fe_x Expression: -(VTy+cpd*ny)*(VTx+cpd*nx)*ny
	−0.5*(VTx+cpd*nx)^2*nx+0.5*(VTy+cpd*ny)^2*nx; Integration
	  order: 4; Frame: Frame(mesh)
	Name: Fe_y Expression: -(VTy+cpd*ny)*(VTx+cpd*nx)*nx
	+0.5*(VTx+cpd*nx)^2*ny-0.5*(VTy+cpd*ny)^2*ny; Integration
	  order: 4; Frame: Frame(mesh)
	Name: Tr Expression: (x-Xp)*(-(VTy+cpd*ny)*(VTx+cpd*nx)*nx
	+0.5*(VTx+cpd*nx)^2*ny-0.5*(VTy+cpd*ny)^2*ny-lm7)
	-(y-Yp)*(-(VTy+cpd*ny)*(VTx+cpd*nx)*ny-0.5*(VTx+cpd*nx)^2*nx
	+0.5*(VTy+cpd*ny)^2*nx-lm6); Integration order: 4; Frame:
	  Frame(mesh)

Physics Menu|**Global** **Equations**	Name: up; Expression: Massd*upt-(Fh_x+Fe_x); Init (u): 0; Init (ut): 0 Name: Xp; Expression: Xpt-up; Init (u): xpin; Init (ut): 0 Name: vp; Expression: Massd*vpt-(Fh_y+Fe_y); Init (u): 0; Init (ut): 0 Name: Yp; Expression: Ypt-vp; Init (u): ypin; Init (ut): 0 Name: omega; Expression: Ird*omegat-Tr; Init (u): 0; Init (ut): 0 Name: ang; Expression: angt-omega; Init (u): angin; Init (ut): 0

continued

TABLE 8.2 (CONTINUED)
Model Setup in the GUI of COMSOL Multiphysics 3.5a

Mesh|Free Mesh Parameters	Tab subdomain
	Subdomain 1|Maximum element size: 0.7
	Tab boundary
	Particle surface: 8, 9, 17, 18, 19, 20|**Maximum element size**: 0.2
	Nanopore: 3, 5, 7, 10, 11, 12, 15, 16, 21, 22|**Maximum element size**: 0.2
Solve Menu|	**Solver Parameters**
	Tab General
	Times: Range(0, 2, 10000)
	Relative tolerance: 1e−3
	Absolute tolerance: 1e−5
	Tab Timing Stepping
	Activate Use stop condition
	minqual1_ale−0.5
	Click = **Solve Problem**
Postprocessing Menu|	Result check: up, Xp, vp, Yp, omega, ang
	Global Variables Plot|Select all the predefined quantities, click >>
	Click Apply
	Result check: Ionic current through the nanopore
	Boundary selection: 2 or 6 (Either end of the reservoirs)
	Boundary Integration|Expression:
	(ntflux_c1_chekf-ntflux_c2_chekf)*ny
	Click **Apply**

```
flclear fem

% COMSOL version
clear vrsn
vrsn.name = 'COMSOL 3.5';
vrsn.ext = '';
vrsn.major = 0;
vrsn.build = 494;
vrsn.rcs = '$Name: $';
vrsn.date = '$Date: 2008/09/19 16:09:48 $';
fem.version = vrsn;

% NEW:Define particle's initial x, y location and orientation
xp0=2.5;
yp0=-15;
angle0=60*pi/180;

% NEW: Length of the particle
Length_p=10;
```

```
% Constants
fem.const = {'D1','1.95e-9[m^2/s]', ...
             'D2','2.03e-9[m^2/s]', ...
             'F','96485.3415[C/mol]', ...
             'R0','8.314472[J/K/mol]', ...
             'T','300[K]', ...
             'z1','1', ...
             'z2','-1', ...
             'eps_r','80', ...
             'eta','1.0e-3[Pa*s]', ...
             'v0','2000000*85e-9[V]', ...
             'rho','1e3[kg/m^3]', ...
             'cp','-0.01[C/m^2]', ...
             'cw','0[C/m^2]', ...
             'c0','100[mol/m^3]', ...
             'a','1e-9[m]', ...
             'Mass1','rho*pi*a^2', ...
             'Mass2','rho*2*a*(Lp-2*a)', ...
             'Massd','(Mass1+Mass2)*U0/eta/a', ...
             'Lp',num2str(Length_p*1e-9), ...

'Ir','(Mass1*(0.5*a^2+(Lp-2*a)^2/4+(Lp-2*a)*4*a/
(3*pi))+Mass2*1/12*((Lp-2*a)^2+4*a^2))', ...
             'Ird','Ir*U0/eta/a^3', ...
             'U0','eps0*eps_r*R0^2*T^2/eta/a/F^2', ...
             'lamda','sqrt(eps_r*eps0*R0*T/2/F^2/c0)', ...
             'ai','a/lamda', ...
             'v0d','v0*F/R0/T', ...
             'charges','F*a/R0/T/eps_r/eps0', ...
             'cpd','cp*charges', ...
             'cwd','cw*charges', ...
             'Re','rho*U0*a/eta', ...
             'eps0','8.854187817e-12[F/m]', ...
             'D1_dless','D1*eta*F^2/(eps0*eps_r*R0^2*T^2)',
             ...
             'D2_dless','D2*eta*F^2/(eps0*eps_r*R0^2*T^2)',
             ...
             'xpin',num2str(xp0), ...
             'ypin',num2str(yp0), ...
             'angin',num2str(angle0)};

% Geometry
% NEW: Create the nanopore system
g1=rect2('10','5','base','center','pos',{'0','0'},
'rot','0');
g2=rect2('50','40','base','center','pos',{'0','22.5'},
'rot','0');
g3=rect2('50','40','base','center','pos',{'0',
'-22.5'},'rot','0');
```

```
g4=geomcomp({g1,g2,g3},'ns',{'g1','g2','g3'},'sf',
'g1+g2+g3','edge','none');
g5=geomdel(g4);
g6=fillet(g5,'radii',1.5,'point',[5,6,7,8]);

% NEW: Create the nanoparticle
g7=rect2('2','8','base','center','pos',{num2str(xp0),num2str
(yp0)},'rot','0');
g8=circ2('1','base','center','pos',{num2str(xp0),num2str
(yp0+4)},'rot','0');
g9=circ2('1','base','center','pos',{num2str(xp0),num2str
(yp0-4)},'rot','0');
g10=geomcomp({g7,g8,g9},'ns',{'g7','g8','g9'},'sf',
'g7+g8+g9','edge','none');
g11=geomdel(g10);
g11=rotate(g11,angle0,[xp0,yp0]);
g12=geomcomp({g6,g11},'ns',{'g6','g11'},'sf',
'g6-g11','edge','none');

% Geometry

% Analyze geometry
clear s
s.objs={g12};
s.name={'CO2'};
s.tags={'g12'};

fem.draw=struct('s',s);
fem.geom=geomcsg(fem);

% NEW: Display geometry with the sequence of the boundary
geomplot(fem, 'Labelcolor','r','Edgelabels','on','submode',
'off');
% (Default values are not included)

% NEW: Solve moving mesh
% Application mode 1
clear appl
appl.mode.class = 'MovingMesh';
appl.sdim = {'Xm','Ym','Zm'};
appl.shape = {'shlag(2,''lm1'')','shlag(2,''lm2'')',
'shlag(2,''x'')','shlag(2,''y'')'};
appl.gporder = {30,4};
appl.cporder = 2;
appl.assignsuffix = '_ale';
clear prop
prop.smoothing='winslow';
prop.analysis='transient';
prop.allowremesh='on';
prop.origrefframe='ref';
appl.prop = prop;
```

```
clear bnd
bnd.defflag = {{1;1}};
bnd.veldeform = {{0;0},{'up-omega*(y-
Yp)';'vp+omega*(x-Xp)'}};
bnd.wcshape = [1;2];
bnd.name = {'Fix','Particle'};
bnd.type = {'def','vel'};
bnd.veldefflag = {{1;1}};
bnd.ind = [1,1,1,1,1,1,1,2,2,1,1,1,1,1,1,1,2,2,1,1,2,2];
appl.bnd = bnd;
clear equ
equ.gporder = 2;
equ.shape = [3;4];
equ.ind = [1];
appl.equ = equ;
fem.appl{1} = appl;

% NEW: Solve ionic concentration using Nernst-Planck
equations
% Application mode 2
clear appl
appl.mode.class = 'ElectroKF_NernstPl';
appl.dim = {'c1','c2'};
appl.module = 'CHEM';
appl.gporder = 4;
appl.cporder = 2;
appl.assignsuffix = '_chekf';
clear prop
prop.equform='cons';
clear weakconstr
weakconstr.value = 'off';
weakconstr.dim = {'lm3','lm4'};
prop.weakconstr = weakconstr;
appl.prop = prop;
clear bnd
bnd.c0 = {0,{1;1},0};
bnd.name = {'Insulator','Bulk','Particle'};
bnd.type = {{'N0';'N0'},{'C';'C'},{'Nc';'Nc'}};
bnd.ind = [1,2,1,1,1,2,1,3,3,1,1,1,1,1,1,1,3,3,1,1,3,3];
appl.bnd = bnd;
clear equ
equ.D = {{'D1_dless';'D2_dless'}};
equ.V = {{'V';'V'}};
equ.init = 1;
equ.um = {{'D1_dless/F';'D2_dless/F'}};
equ.v = {{'v';'v'}};
equ.u = {{'u';'u'}};
equ.z = {{'z1';'z2'}};
equ.ind = [1];
appl.equ = equ;
```

```
fem.appl{2} = appl;

% NEW: Solve electrostatics using Poisson equation
% Application mode 3
clear appl
appl.mode.class = 'Electrostatics';
appl.sshape = 2;
appl.assignsuffix = '_es';
clear prop
clear weakconstr
weakconstr.value = 'off';
weakconstr.dim = {'lm5'};
prop.weakconstr = weakconstr;
appl.prop = prop;
clear bnd
bnd.V0 = {'v0d',0,0,0,0};
bnd.name = {'Up','Particle','wall','insulator','down'};
bnd.rhos = {0,'cpd','cwd',0,0};
bnd.type = {'V','r','r','nD0','V'};
bnd.ind = [4,5,3,4,3,1,3,2,2,3,3,3,4,4,3,3,2,2,3,3,2,2];
appl.bnd = bnd;
clear equ
equ.epsilonr = '1/eps0';
equ.rho = '0.5*(ai)^2*(z1*c1+z2*c2)';
equ.ind = [1];
appl.equ = equ;
fem.appl{3} = appl;

% NEW: Solve flow field using modified Navier-Stokes equa-
tions
% Application mode 4
clear appl
appl.mode.class = 'FlNavierStokes';
appl.shape = {'shlag(2,''lm6'')','shlag(2,''lm7'')',
'shlag(1,''lm8'')','shlag(2,''u'')','shlag(2,''v'')',
'shlag(1,''p'')'};
appl.gporder = {30,4,2};
appl.cporder = {2,1};
appl.sshape = 2;
appl.assignsuffix = '_ns';
clear prop
clear weakconstr
weakconstr.value = 'on';
weakconstr.dim = {'lm6','lm7','lm8'};
prop.weakconstr = weakconstr;
prop.constrtype='non-ideal';
appl.prop = prop;
clear bnd
bnd.v0 = {0,0,0,'vp+omega*(x-Xp)'};
bnd.type = {'sym','outlet','walltype','inlet'};
```

```
bnd.name = {'far','outlet','wall','particle'};
bnd.u0 = {0,0,0,'up-omega*(y-Yp)'};
bnd.wcshape = [1;2;3];
bnd.velType = {'U0in','U0in','U0in','u0'};
bnd.ind = [1,2,3,1,3,2,3,4,4,3,3,3,1,1,3,3,4,4,3,3,4,4];
appl.bnd = bnd;
clear equ
equ.gporder = {{2;2;3}};
equ.F_y = '-0.5*ai^2*(z1*c1+z2*c2)*Vy';
equ.rho = 0;
equ.cporder = {{1;1;2}};
equ.F_x = '-0.5*ai^2*(z1*c1+z2*c2)*Vx';
equ.shape = [4;5;6];
equ.cdon = 0;
equ.sdon = 0;
equ.ind = [1];
appl.equ = equ;
fem.appl{4} = appl;
fem.sdim = {{'Xm','Ym'},{'X','Y'},{'x','y'}};
fem.frame = {'mesh','ref','ale'};
fem.border = 1;
clear units;
units.basesystem = 'SI';
fem.units = units;


% NEW: Boundary integration to calculate the hydrodynamic
force
% NEW: and electrical force on the particle
% Coupling variable elements
clear elemcpl
% Integration coupling variables
clear elem
elem.elem = 'elcplscalar';
elem.g = {'1'};
src = cell(1,1);
clear bnd
bnd.expr = {{{},'-lm6'},{{},'-lm7'},{{}, ...
  '-(VTy+cpd*ny)*(VTx+cpd*nx)*ny-0.5*(VTx+cpd*nx)^2*nx+0.5*(
VTy+cpd*ny)^2*nx'},{{}, ...
  '-(VTy+cpd*ny)*(VTx+cpd*nx)*nx+0.5*(VTx+cpd*nx)^2*ny-
0.5*(VTy+cpd*ny)^2*ny'},{{}, ...

  '(x-Xp)*(-(VTy+cpd*ny)*(VTx+cpd*nx)*nx+0.5*(VTx+cpd*nx)^2
*ny-0.5*(VTy+cpd*ny)^2*ny-lm7)-(y-Yp)*(-
(VTy+cpd*ny)*(VTx+cpd*nx)*ny-0.5*(VTx+cpd*nx)^2*nx+0.5*(VTy+
cpd*ny)^2*nx-lm6)'}};
bnd.ipoints = {{{},'4'},{{},'4'},{{},'4'},{{},'4'},{{},'4'}};
bnd.frame = {{{},'mesh'},{{},'mesh'},{{},'mesh'},
{{},'mesh'},{{},'mesh'}};
```

```
bnd.ind = {{'1','2','3','4','5','6','7','10','11','12','13',
'14','15', ...
  '16','19','20'},{'8','9','17','18','21','22'}};
src{1} = {{},bnd,{}};
elem.src = src;
geomdim = cell(1,1);
geomdim{1} = {};
elem.geomdim = geomdim;
elem.var = {'Fh_x','Fh_y','Fe_x','Fe_y','Tr'};
elem.global = {'1','2','3','4','5'};
elemcpl{1} = elem;
fem.elemcpl = elemcpl;


% Descriptions
clear descr
descr.const= {'eps_r','Permittivity of medium','eta','Viscoc
ity','v0','Electrical potential','F','Faraday
constant','z1','valence of K','a','Particle
radius','z2','valence of Cl','R0','Gas
constant','cw','Charge density of wall','T','Temperature','c
p','Charge density of particle','c0','Concentration','D2','D
iffusion coefficient of Cl','D1','Diffusion coefficient of
K','rho','Density'};
fem.descr = descr;


% NEW: Solve six ODEs to determine the particle's transla-
tional velocity, rotational
% NEW: velocity, position and orientation
% ODE Settings
clear ode
ode.dim={'up','Xp','vp','Yp','omega','ang'};
ode.f={'Massd*upt-(Fh_x+Fe_x)','Xpt-up','Massd*vpt-(Fh_
y+Fe_y)','Ypt-vp','Ird*omegat-Tr','angt-omega'};
ode.init={'0','xpin','0','ypin','0','angin'};
ode.dinit={'0','0','0','0','0','0'};
clear units;
units.basesystem = 'SI';
ode.units = units;
fem.ode=ode;


% Multiphysics
fem=multiphysics(fem);


% NEW: Loop for continuous particle tracking
% NEW: Remeshing index
j=1;


% NEW: Define matrix for data storage
particle=zeros(10,7);
```

```
% NEW: Define data storage index
num=1;

% NEW: Define initial time
time_e=0;

% NEW: Define time_step
time_step=0.6;

% NEW: Update particle's position and orientation
xp=xp0;
yp=yp0;
angle=angle0;

% NEW: Define mesh size on particle surface and nanopore
mesh_p=0.2;
mesh_w=0.2;

% NEW: Define end condition of the computation (Loop con-
trol)
while yp<20

% Initialize mesh
fem.mesh=meshinit(fem, ...
                  'hauto',5, ...

    'hmaxedg', [3,mesh_w,5,mesh_w,7,mesh_w,8,mesh_p,9,mesh_p,10
    ,mesh_w,11,mesh_w,12,mesh_w,15,mesh_w,16,mesh_w,17,mesh_p,18
    ,mesh_p,19,mesh_p,20,mesh_p,21,mesh_w,22,mesh_w], ...
                  'hmaxsub', [1,1]);

% NEW: Refine mesh near the particle after each remeshing
fem.mesh=meshrefine(fem, ...
                    'mcase',0, ...
                    'boxcoord', [xp-2-Length_p*sin(angle)/2
xp+2+Length_p*sin(angle)/2 yp-2-Length_p*cos(angle)/2
yp+2+Length_p*cos(angle)/2], ...
    'rmethod','regular');

% Extend mesh
fem.xmesh=meshextend(fem);

% NEW: Find out the mesh quality
quality=meshqual(fem.mesh);
minimum=min(quality)
used=minimum-0.1;

% NEW: Computation before first remeshing
```

```
if j==1

% Solve problem
fem.sol=femtime(fem, ...

'solcomp',{'ang','up','vp','omega','lm2','lm1','lm8','lm7','
lm6','v','u','Yp','V','c1','c2','p','y','Xp','x'}, ...

'outcomp',{'ang','up','vp','omega','lm2','lm1','lm8','lm7',
'lm6','v','u','Yp','V','c1','c2','p','Y','X','y','Xp','x'},
...
                    'tlist',[time_e:time_step:1e6], ...
                    'tout','tsteps', ...
                    'tsteps','intermediate', ...
                    'rtol',1e-3, ...
                    'atol',1e-5, ...
                    'stopcond',strcat('minquall_ale-
                    ',num2str(used))));
else

% NEW: Map the solution in the deformed mesh into the new
geometry with undeformed mesh
init = asseminit(fem,'init',fem0.sol,'xmesh',fem0.xmesh,'fra
mesrc','ale','domwise','on');

% NEW: Computation after first remeshing
% Solve problem
fem.sol=femtime(fem, ...
                    'init',init, ...

'solcomp',{'ang','up','vp','omega','lm2','lm1','lm8','lm7','
lm6','v','u','Yp','V','c1','c2','p','y','Xp','x'}, ...

'outcomp',{'ang','up','vp','omega','lm2','lm1','lm8','lm7',
'lm6','v','u','Yp','V','c1','c2','p','Y','X','y','Xp','x'},
...
                    'tlist',[time_e:time_step:1e6], ...
                    'tout','tsteps', ...
                    'tsteps','intermediate', ...
                    'rtol',1e-3, ...
                    'atol',1e-5, ...
                    'stopcond',strcat('minquall_ale-
                    ',num2str(used))));
end

% NEW: Save current fem structure for restart purposes
fem0=fem;

% NEW: Get the last time from solution
time_e = fem.sol.tlist(end);
```

```
% NEW: Get the length of the time steps
Leng =length(fem.sol.tlist);

% NEW: Adaptive time step to ensure 100 outputs in one
remeshing
time_step=Leng*time_step/100;

% Global variables plot
data=postglobalplot(fem,{'Xp','Yp','ang','up','vp','omega'},
...
                    'linlegend','on', ...
                    'title','Particle', ...
                    'Outtype','postdata', ...
                    'axislabel',{'Time','Parameters'});

% NEW: Store the interested data
for n=15:Leng
    particle(num,1)=data.p(1,n);
    particle(num,2)=data.p(2,n);
    particle(num,3)=data.p(2,n+Leng);
    particle(num,4)=data.p(2,n+2*Leng);
    particle(num,5)=data.p(2,n+3*Leng);
    particle(num,6)=data.p(2,n+4*Leng);
    particle(num,7)=data.p(2,n+5*Leng);

% NEW: Current at anode
% NEW: Boundary integrate
I1=postint(fem,'(ntflux_c1_chekf-ntflux_c2_chekf)*ny', ...
                    'unit','mol/(m*s)', ...
                    'dl',6, ...
                    'edim',1, ...
                    'solnum',n);
particle(num,8)=I1;

% NEW: Current at cathode
% NEW: Boundary integrate
I2=postint(fem,'(ntflux_c1_chekf-ntflux_c2_chekf)*ny', ...
                    'unit','mol/(m*s)', ...
                    'dl',2, ...
                    'edim',1, ...
                    'solnum',n);

particle(num,9)=I2;

% NEW: current deviation
particle(num,10)=(abs(I1)-abs(I2))/abs(I1)*100;

    num=num+1;
end
```

```
% NEW: update the particle's position and orientation
% NEW: Output them in the MATLAB
xp=data.p(2,Leng)
yp=data.p(2,2*Leng)
angle=data.p(2,3*Leng)


% NEW: Output the COMSOL Multiphysics GUI .mph file if
necessary
flsave(strcat('100mM_ang_60_xp_2p5_yp_-15_cw_0_E_2000KV_
yp_',num2str(yp),'_angle_',num2str(angle),'.mph'),fem);

% NEW: Write the data into a file
dlmwrite(strcat('100mM_ang_60_xp_2p5_yp_-15_
cw_0_E_2000KV','.dat'),particle,',');


j=j+1;

% Generate geom from mesh
fem = mesh2geom(fem, ...
                'frame','ale', ...
                'srcdata','deformed', ...
                'destfield',{'geom','mesh'}, ...
                'srcfem',1, ...
                'destfem',1);
end
% NEW: End of this script file
```

The developed model is validated by simulating the electrokinetic translation of a sphere along the axis of a cylindrical nanotube. The approximation solution of the particle's steady velocity is available when the EDL is not distorted by the external electric field, flow field, nearby solid boundaries, and EDLs, and the zeta potential of the particle is relatively low ($\zeta/(RT/F) < 1$) (Ennis and Anderson 1997). Figure 8.2 shows the effect of the ratio of the particle radius to the pore radius, a/b on the particle's steady axial velocity normalized by $\varepsilon_0\varepsilon_f\zeta E/\mu$, when $a = 1$ nm, $\kappa a = 2.05$, $\zeta = 1$ mV, and $E = 1{,}000$ kV/m. When the boundary effect is negligible, the numerical predictions are in good agreement with the approximation solution. Once the boundary effect becomes more significant, the numerical predictions begin to deviate from the approximation solution.

8.4 RESULTS AND DISCUSSION

First, the results obtained from the PB model in Chapter 7 are compared to those of the PNP model described in the current chapter. The electrokinetic translocation of a nanoparticle through a nanopore is investigated as functions of the following five parameters: the ratio of the particle radius to the Debye length κa; the applied electric field E^*; the initial angle θ_{p0}^*; the initial lateral offset from the centerline of the nanopore x_{p0}^*; and the surface charge density of the nanopore σ_w^*.

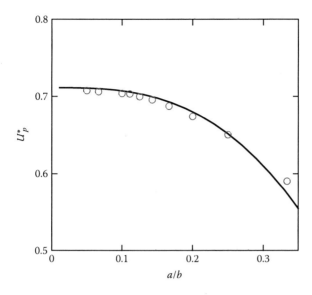

FIGURE 8.2 Axial electrophoretic velocity of a sphere of radius a translating along the axis of an uncharged cylindrical nanotube of radius b as a function of the ratio a/b. The conditions were $a = 1$ nm, $\kappa a = 2.05$, the zeta potential of the particle $\zeta = 1$ mV, and the axial electric field imposed $E = 1,000$ KV/m. Solid line and circles represent, respectively, the approximation solution and our numerical results. (From Ai, Y., and S. Qian. 2011. Electrokinetic particle translocation through a nanopore. *Physical Chemistry Chemical Physics* 13:4060–4071 with permission from RCS.)

8.4.1 COMPARISON BETWEEN PB MODEL AND PNP MODEL

Figure 8.3 compares the particle's y component velocity obtained from the PB model and the PNP model under two different applied electric fields, $E^* = 7.7 \times 10^{-4}$ ($E = 20$ kV/m; Figure 8.3a) and $E^* = 7.7 \times 10^{-2}$ ($E = 2,000$ kV/m; Figure 8.3b). The nanopore is assumed to be uncharged, $\sigma_w^* = 0$. The initial lateral offset from the centerline of the nanopore and the initial orientation of the particle are, respectively, $x_{p0}^* = 0$ and $\theta_{p0}^* = 0$. Therefore, the particle only translates along the centerline of the nanopore without any rotation and lateral movement. When the EDL of the particle is relatively thin ($\kappa a > 1$), the results obtained from the PB model agree with those obtained by the PNP model. As the EDL of the particle becomes increasingly thick ($\kappa a < 1$), the results obtained by the PB model gradually deviate from those obtained by the PNP model, especially under a relatively high electric field, as shown in Figure 8.3b.

Current deviations, $\chi = (I^* - I_0^*)/I_0^*$, corresponding to the cases in Figure 8.3 obtained by the PB model and the PNP model are compared in Figure 8.4. In this equation, I_0^* refers to the base current when the particle is far away from the nanopore and is numerically obtained based on Equation (8.25) without including the particle in the simulation. When the EDL of the particle is relatively thin

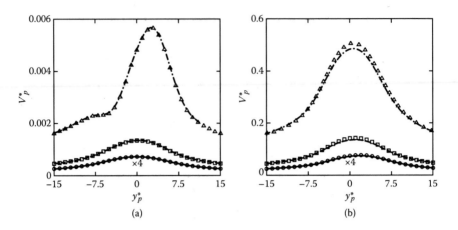

FIGURE 8.3 Translational velocity of the particle as a function of the particle's axial location y_p^* under two different electric fields: $E^* = 7.7 \times 10^{-4}$ ($E = 20$ kV/m; a), and $E^* = 7.7 \times 10^{-2}$ ($E = 2,000$ kV/m; b). $x_{p0}^* = 0$, $\theta_{p0}^* = 0$. Lines and symbols represent, respectively, the results obtained by the PB model and the PNP model. Solid line (circles), dashed line (squares), and dash-dotted line (triangles), represent, respectively, $\kappa a = 2.05$, 1.03, and 0.65. A scale of 4 was applied to the solid line and triangles for a clear visualization.

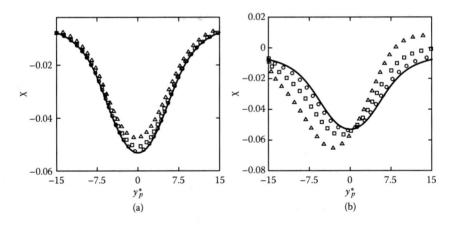

FIGURE 8.4 Current deviation as a function of the particle's location y_p^* under two different electric fields: $E^* = 7.7 \times 10^{-4}$ (a), and $E^* = 7.7 \times 10^{-2}$ (b). $x_{p0}^* = 0$, $\theta_{p0}^* = 0$. Lines and symbols represent, respectively, the results obtained by the PB model and the PNP model. Solid line (circles), dashed line (squares), and dash-dotted line (triangles), represent, respectively, $\kappa a = 2.05$, 1.03, and 0.65.

($\kappa a > 1$), the results obtained by the PB model almost match those obtained by the PNP model. The PB model shows that the current deviation is independent of κa. However, the results from the PNP model depict that the current deviation strongly depends on κa, which also was confirmed by Liu, Qian, and Bau (2007). When the EDL of the particle becomes increasingly thick ($\kappa a < 1$), the current deviation predicted by the PB model significantly deviates from that obtained by the PNP model.

The results from the PNP model show that the current deviation also depends on the applied electric field intensity, owing to the double-layer polarization, discussed further in this chapter. However, the results from the PB model show that the current deviation is independent of the applied electric field intensity. The PB model only predicts current blockade during the translocation process, while the PNP model successfully predicts both current blockade and current enhancement, shown in Figure 8.4b. Therefore, it is obvious to conclude that the PB model is only valid when the EDL of the particle is relatively thin and is not significantly distorted by the external field. The results discussed next were obtained from the PNP model.

8.4.2 Effect of Ratio of Particle Radius to Debye Length

We further investigate the effect of κa on the y component particle velocity under two different applied electric fields, $E^* = 7.7 \times 10^{-4}$ (Figure 8.5a) and $E^* = 7.7 \times 10^{-2}$ (Figure 8.5b). The particle only translates along the centerline of the uncharged nanopore when $x_{p0}^* = 0$ and $\theta_{p0}^* = 0$. When the applied electric field is relatively low, $E^* = 7.7 \times 10^{-4}$, the particle velocity shows symmetry with respect to $y_p^* = 0$ when κa is relatively large ($\kappa a = 2.05$ and $\kappa a = 1.03$ in Figure 8.5a). As the EDL of the particle increases, the particle velocity becomes asymmetric with respect to $y_p^* = 0$ ($\kappa a = 0.65$ in Figure 8.5a). As the EDL of the particle further increases,

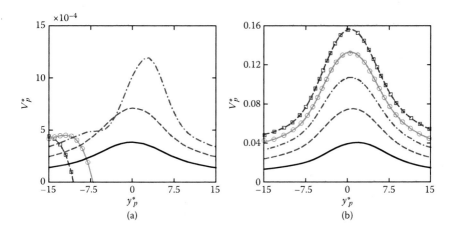

FIGURE 8.5 The y component translational velocity of the particle as a function of the particle's location y_p^* under $E^* = 7.7 \times 10^{-4}$ ($E = 20$ kV/m; a) and $E^* = 7.7 \times 10^{-2}$ ($E = 2,000$ kV/m; b). $x_{p0}^* = 0$, $\theta_{p0}^* = 0$. Solid line, dashed line, dash-dotted line, solid line with circles, and dashed line with squares represent, respectively, $\kappa a = 2.05$, 1.03, 0.65, 0.46, and 0.32. (From Ai, Y., and S. Qian. 2011. Electrokinetic particle translocation through a nanopore. *Physical Chemistry Chemical Physics* 13:4060–4071 with permission from RCS.)

the particle cannot enter the nanopore anymore and is trapped near the entrance of the nanopore ($\kappa a = 0.46$ and $\kappa a = 0.32$ in Figure 8.5a). If the applied electric field is increased 100 times, the particle velocity is accordingly increased 100 times when κa is relatively large ($\kappa a = 2.05$ and $\kappa a = 1.03$ in Figure 8.5b). However, the particle velocity is always symmetric with respect to $y_p^* = 0$ even when $\kappa a < 1$. In addition, the particle trapping near the entrance of the nanopore is not predicted under a relatively high electric field. The particle velocity increases as κa decreases, which is attributed to the increase in the zeta potential of a particle with a fixed surface charge density (Ohshima 1998). This kind of phenomenon has also been predicted by the quasi-static model (Liu, Qian, and Bau 2007).

To explain the two distinct particle behaviors under different electric fields, the ionic concentration difference $c_1^* - c_2^*$ and the flow field near the particle for $x_p^* = 0$, $y_p^* = -7$, $\theta_p^* = 0$, $\kappa a = 0.46$, and $\sigma_w^* = 0$ are shown in Figure 8.6. As the particle is negatively charged, more positive ions are accumulated near the negatively charged particle, as shown in Figures 8.6a and 8.6b. When the particle is about to enter the nanopore, the EDL of the particle has already reached the inside of the nanopore, leading to an enrichment of positive ions within the nanopore.

When the applied electric field is relatively low, $E^* = 7.7 \times 10^{-4}$, an electroosmotic flow (EOF) opposite to the particle translocation is induced inside the nanopore, as shown in Figure 8.6c, which accordingly slows the particle translocation. As the particle translocates further toward the nanopore, more positive ions are presented inside the nanopore, resulting in a more significant EOF. Once the opposite EOF overpowers the electrical driving force, the particle is trapped near the entrance of the nanopore. The particle-trapping phenomenon has been experimentally observed and utilized to concentrate nanoparticles (Plecis, Schoch, and Renaud 2005; Wang, Stevens, and Han 2005). When the applied electric field is relatively high, $E^* = 7.7 \times 10^{-2}$, positive ions could be considerably repelled out of the nanopore, as shown in Figure 8.6b. As a result, the opposite EOF is remarkably reduced and becomes insufficient to trap particles at the entrance of the nanopore. Therefore, a relatively high electric field is usually used in nanopore-based sensing, in which nanoparticles must translocate through a nanopore to induce a current change for detection. The redistribution of ions near a charged surface due to the presence of an external electric field is called double-layer polarization.

As stated, nanopore-based sensing is based on the current change arising from the translocating particle inside the nanopore. Figure 8.7 shows the current deviations corresponding to the cases shown in Figure 8.5. When the applied electric field is relatively low, $E^* = 7.7 \times 10^{-4}$, current blockade is predicted as shown in Figure 8.7a, which is in qualitative agreement with the existing experimental observations (Meller, Nivon, and Branton 2001; Li et al. 2003; Storm, Storm, et al. 2005; Storm, Chen, et al. 2005). Because the particle cannot translocate through the nanopore when $\kappa a = 0.46$ and 0.32, the corresponding current deviations are not shown in Figure 8.7a. When the particle is inside the nanopore, a thicker EDL implies a higher ionic concentration inside the nanopore, which thus reduces the

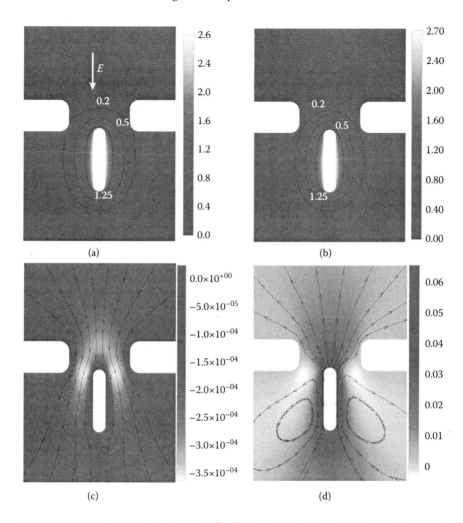

FIGURE 8.6   Spatial distributions of $(c_1^* - c_2^*)$ (a and b) and flow field (c and d) around the particle under $E^* = 7.7 \times 10^{-4}$ (a and c) and $E^* = 7.7 \times 10^{-2}$ (b and d). $x_{p0}^* = 0$, $y_p^* = -7$, $\theta_{p0}^* = 0$, and $\kappa a = 0.46$. The graded bars in (c) and (d) represent the y component fluid velocity, and the lines with arrows denote the streamlines of the flow field. (From Ai, Y., and S. Qian. 2011. Electrokinetic particle translocation through a nanopore. *Physical Chemistry Chemical Physics* 13:4060–4071 with permission from RCS.)

magnitude of current deviation. When the applied electric field is relatively high, $E^* = 7.7 \times 10^{-2}$, current blockade is also observed under a relatively high κa, as shown in Figure 8.7b. As κa decreases, the current deviation becomes asymmetric with respect to $y_p^* = 0$. As κa further decreases, current blockade becomes even more pronounced when the particle enters the nanopore; current enhancement is predicted however when the particle exits the nanopore. When the particle is about to enter the nanopore, the nanopore is predominantly occupied with more

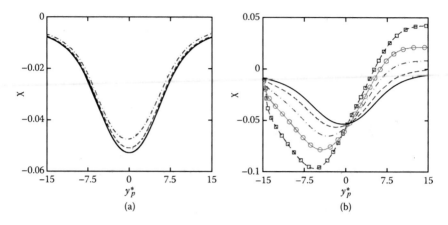

FIGURE 8.7 Current deviation χ as a function of the particle's location y_p^* under $E^* = 7.7 \times 10^{-4}$ (a) and $E^* = 7.7 \times 10^{-2}$ (b). $x_{p0}^* = 0$ and $\theta_{p0}^* = 0$. Solid line, dashed line, dash-dotted line, solid line with circles, and dashed line with squares represent, respectively, $\kappa a = 2.05, 1.03, 0.65, 0.46$, and 0.32. (From Ai, Y., and S. Qian. 2011. Electrokinetic particle translocation through a nanopore. *Physical Chemistry Chemical Physics* 13:4060–4071 with permission from RCS.)

positive ions. However, the applied electric field could remarkably repel the positive ions out of the nanopore, which thus increases the current deviation. When the particle is leaving the nanopore, the positive ions within the EDL of the negatively charged particle are pushed inside the nanopore by the imposed electric field. As a result, current enhancement is expected, which has been predicted by a MD simulation (Aksimentiev et al. 2004) and experimentally observed (Heng et al. 2004).

8.4.3 Effect of Particle's Initial Orientation

Next, we investigate the effect of the particle's initial orientation on its translocation process and the corresponding current deviation under two different electric fields, $E^* = 7.7 \times 10^{-4}$ and $E^* = 7.7 \times 10^{-2}$, when $x_{p0}^* = 0$, $\theta_{p0}^* = 60°$, $\kappa a = 1.03$, and $\sigma_w^* = 0$. Figure 8.8 shows the particle trajectories under the two electric fields. Apparently, the particle rotation plays an important role in its translocation process. When the applied electric field is relatively low, $E^* = 7.7 \times 10^{-4}$, the particle rotates clockwise as it moves toward the nanopore. The hydrodynamic interaction between the particle and the solid boundary of the nanopore further enhances particle rotation. When the particle exits the nanopore, the particle slightly rotates counterclockwise. However, it cannot recover its initial orientation.

Interestingly, we also observe a lateral movement arising from the nonzero initial orientation. When the particle's initial orientation is positive (negative), the lateral movement is toward the negative (positive) x direction. If the applied electric field is increased 100 times, the duration of the particle translocation is also shortened 100 times. In addition, the particle rotates clockwise quickly at the beginning and

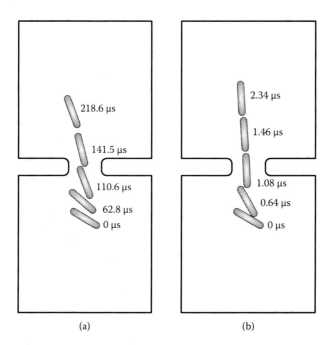

(a)   (b)

FIGURE 8.8   Trajectories of the particle under $E^* = 7.7 \times 10^{-4}$ (a) and $E^* = 7.7 \times 10^{-2}$ (b). $x^*_{p0} = 0$, $\theta^*_{p0} = 60°$ and $\kappa a = 1.03$. (From Ai, Y., and S. Qian. 2011. Electrokinetic particle translocation through a nanopore. *Physical Chemistry Chemical Physics* 13:4060–4071 with permission from RCS.)

stays aligned with its longest axis parallel to the electric field thereafter. The alignment behavior is attributed to the dominant DEP effect, which has been demonstrated in Chapter 4. When the applied electric field is relatively low, the uniform electric field due to the surface charge of the particle becomes dominant. Accordingly, the DEP effect is negligible when the applied electric field is relatively low. Thus, quick particle alignment to the applied electric field is not observed in Figure 8.8a.

Figure 8.9a shows the y component particle velocity as a function of the particle's axial location y^*_p under two different electric fields, $E^* = 7.7 \times 10^{-4}$ and $E^* = 7.7 \times 10^{-2}$, when $x^*_{p0} = 0$, $\kappa a = 1.03$, and $\sigma^*_w = 0$. When $\theta^*_{p0} = 60°$, particle velocity is significantly reduced compared to the case of $\theta^*_{p0} = 0$ as the particle translocates toward the nanopore. Once the particle exits the nanopore, the particle velocity for $\theta^*_{p0} = 60°$ begins to match that for $\theta^*_{p0} = 0$ because of a decrease in the particle's orientation.

Figure 8.9b shows the rotational velocity under the two electric fields when $\theta^*_{p0} = 60°$. The magnitude of the rotational velocity under $E^* = 7.7 \times 10^{-2}$ is much higher than that under $E^* = 7.7 \times 10^{-4}$. The magnitude of the rotational velocity increases as the particle translocates toward the nanopore and maximizes near the entrance of the nanopore. Subsequently, the magnitude of the rotational velocity begins to decrease. The particle experiences a counterclockwise rotation when it

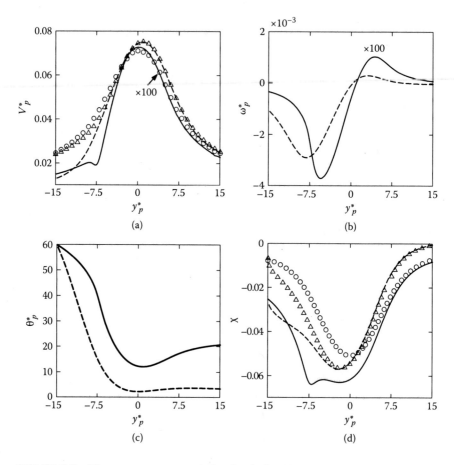

FIGURE 8.9 The y component translational velocity (a), rotational velocity (b), angle of the particle (c), and current deviation (d) as a function of the particle's location y_p^* under $E^* = 7.7 \times 10^{-4}$ (solid line and circles) and $E^* = 7.7 \times 10^{-2}$ (dashed line and triangles). Symbols and lines represent, respectively, $\theta_{p0}^* = 0$ and $60°$. $x_{p0}^* = 0$, and $\kappa a = 1.03$. A scale of 100 was applied to the solid line and circles in (a) and solid line in (b) for a clear visualization. (From Ai, Y., and S. Qian. 2011. Electrokinetic particle translocation through a nanopore. *Physical Chemistry Chemical Physics* 13:4060–4071 with permission from RCS.)

exits the nanopore under $E^* = 7.7 \times 10^{-4}$. However, the rotational velocity becomes almost zero after it exits the nanopore under $E^* = 7.7 \times 10^{-2}$.

The results mentioned are also confirmed by the evolution of the particle's orientation, as shown in Figure 8.9c. The quantitative analysis of the effect due to the particle's initial orientation on the current deviation is shown in Figure 8.9d. The nonzero initial orientation, $\theta_{p0}^* = 60°$, dramatically obstructs the ionic current through the nanopore, which in turn increases the magnitude of the current deviation compared to the cases of $\theta_{p0}^* = 0$. When the applied electric field is relatively low, $E^* = 7.7 \times 10^{-4}$, the current deviation of the particle with $\theta_{p0}^* = 60°$ begins to match the case of $\theta_{p0}^* = 0$ when it exits the nanopore due to the decrease in the particle's orientation and the lateral movement. When the applied electric field is relatively high, $E^* = 7.7 \times 10^{-2}$, the DEP effect could align the particle to the applied electric field quickly, after which the current deviation of the particle with $\theta_{p0}^* = 60°$ perfectly matches the case of $\theta_{p0}^* = 0$. It is clear that the particle's initial orientation plays an important role in its translocation process and the resulting current change.

8.4.4 EFFECT OF INITIAL LATERAL OFFSET FROM NANOPORE'S CENTERLINE

The particle is initially located away from the centerline of the nanopore to study the effect due to the initial lateral offset on particle translocation and the resulting ionic current. The trajectories of the particle under two different electric fields, $E^* = 7.7 \times 10^{-4}$ and $E^* = 7.7 \times 10^{-2}$, when $x_{p0}^* = 2.5$, $\theta_{p0}^* = 0$, $\kappa a = 1.03$, and $\sigma_w^* = 0$ are, respectively, shown in Figures 8.10a and 8.10b. Because of the initial lateral offset, the particle rotates counterclockwise and moves laterally toward the centerline of the nanopore when it starts to move. When the applied electric field is relatively low, $E^* = 7.7 \times 10^{-4}$, the particle exits the nanopore with a positive orientation, as shown in Figure 8.10a. When the applied electric field is relatively high, $E^* = 7.7 \times 10^{-2}$, the particle is aligned to the local external electric field, which however is not parallel to the centerline of the nanopore. Thus, the particle exits the nanopore with a negative orientation, as shown in Figure 8.10b. It is also found that the initial lateral offset exhibits a limited effect on the particle velocity and the current deviation. If the particle is also presenting a nonzero initial orientation, $\theta_{p0}^* = 60°$, the particle translocation process is similar to that shown in Figure 8.8 and thus is not discussed here.

Figure 8.11a shows the y component particle velocity as a function of the particle's axial location under $E^* = 7.7 \times 10^{-4}$ and $E^* = 7.7 \times 10^{-2}$ when $x_{p0}^* = 2.5$, $\theta_{p0}^* = 60°$, $\kappa a = 1.03$, and $\sigma_w^* = 0$. The asymmetry in the velocity profile is mainly due to the nonzero initial orientation, as explained in Section 8.4.3. The variations of the particle's rotational velocity and orientation, shown in Figures 8.11b and 8.11c, are also similar to those shown in Figures 8.9b and 8.9c, respectively. The slight difference between the current deviation shown in Figures 8.9d and 8.11d is due to the initial lateral offset. Overall, the initial lateral offset is a minor effect on particle translocation and ionic current through a nanopore.

8.4.5 EFFECT OF NANOPORE'S SURFACE CHARGE DENSITY

In all the studies discussed, the nanopore is assumed to be uncharged. However, most nanopores gain surface charges when they are immersed in ionic solutions. In this section, the effect of a nanopore's surface charge on particle translocation and the resulting ionic current is investigated. Figure 8.12 shows the y component particle velocity as a function of the particle's axial location under $E^* = 7.7 \times 10^{-4}$ and $E^* = 7.7 \times 10^{-2}$ when $x_{p0}^* = 0$, $\theta_{p0}^* = 0$, and $\kappa a = 1.03$. The case of $\sigma_w^* = 0$

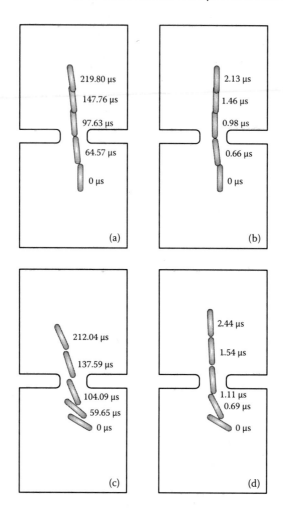

FIGURE 8.10   Trajectories of the particle under $E^* = 7.7 \times 10^{-4}$ (a and c) and $E^* = 7.7 \times 10^{-2}$ (b and d). $x^*_{p0} = 2.5$ and $\theta^*_{p0} = 0$ in (a) and (b); $x^*_{p0} = 2.5$ and $\theta^*_{p0} = 60°$ in (c) and (d). $\kappa a = 1.03$. (From Ai, Y., and S. Qian. 2011. Electrokinetic particle translocation through a nanopore. *Physical Chemistry Chemical Physics* 13:4060–4071 with permission from RCS.)

is regarded as a reference to highlight the effect of the nanopore's surface charge. Clearly, the reference velocity profile is symmetric with respect to $y^*_p = 0$.

When the applied electric field is relatively low, $E^* = 7.7 \times 10^{-4}$, a positively charged nanopore generates an EOF in the same direction as particle translocation. As a result, the nanopore with a surface charge density of $\sigma^*_w = -0.1\sigma^*_p$ leads to approximately a 10% enhancement in particle velocity in the reservoir region. When $\kappa a = 1.03$, the EDLs of the particle and the nanopore are partially overlapped, which leads to a particle-nanopore electrostatic interaction.

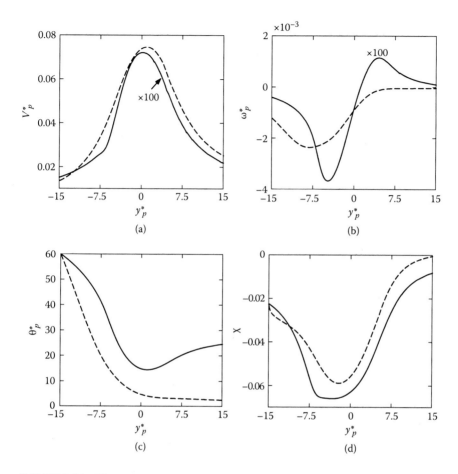

FIGURE 8.11 The y component translational velocity (a), rotational velocity (b), angle of the particle (c), and current deviation (d) as a function of the particle's location y_p^* under $E^* = 7.7 \times 10^{-4}$ (solid line) and $E^* = 7.7 \times 10^{-2}$ (dashed line). $x_{p0}^* = 2.5$, $\theta_{p0}^* = 60°$, and $\kappa a = 1.03$. A scale of 100 was applied to the solid line in (a) and (b) for a clear visualization. (From Ai, Y., and S. Qian. 2011. Electrokinetic particle translocation through a nanopore. *Physical Chemistry Chemical Physics* 13:4060–4071 with permission from RCS.)

When the particle and the nanopore carry surface charges with opposite polarity, the particle-nanopore electrostatic interaction exerts an attractive force on the particle, which enhances particle translocation when $y_p^* < 0$ and retards particle translocation when $y_p^* > 0$. If the nanopore carries a surface charge density of $\sigma_w^* = 0.1\sigma_p^*$, the induced EOF is opposite to the particle translocation, which accordingly slows particle translocation approximately 10% in the reservoir region compared to the reference case. In addition, the particle-nanopore electrostatic interaction becomes a repulsive force acting on the particle. Accordingly, the repulsive particle-nanopore electrostatic interaction retards

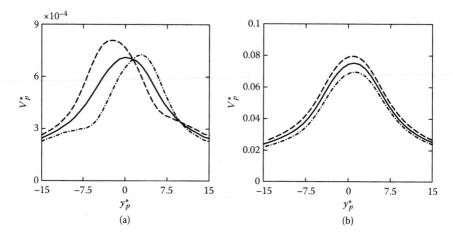

FIGURE 8.12 The y component translational velocity of the particle as a function of the particle's location y_p^* under $E^* = 7.7 \times 10^{-4}$ (a) and $E^* = 7.7 \times 10^{-2}$ (b). $x_{p0}^* = 0$, $\theta_{p0}^* = 0$ and $\kappa a = 1.03$. Solid line, dashed line, and dash-dotted line represent, respectively, $\sigma_w^* = 0$, $\sigma_w^* = -0.1\sigma_p^*$ and $\sigma_w^* = 0.1\sigma_p^*$. (From Ai, Y., and S. Qian. 2011. Electrokinetic particle translocation through a nanopore. *Physical Chemistry Chemical Physics* 13:4060–4071 with permission from RCS.)

particle translocation when $y_p^* < 0$ and enhances particle translocation when $y_p^* > 0$. When the applied electric field is relatively high, $E^* = 7.7 \times 10^{-2}$, the electrical driving force and EOF dominate over the particle-nanopore electrostatic interaction. Consequently, the nanopore carrying a surface charge of $\sigma_w^* = 0.1\sigma_p^*$ ($\sigma_w^* = -0.1\sigma_p^*$) retards (enhances) particle translocation approximately 10% compared to the reference case, as shown in Figure 8.12b. In addition, the velocity profile is always symmetric with respect to $y_p^* = 0$, which is independent of the nanopore's surface charge. Therefore, the particle-nanopore electrostatic interaction can only significantly affect particle translocation through a nanopore when the EDLs of the particle and the nanopore are overlapped and the applied electric field is relatively low.

Considering a nonzero initial orientation ($\theta_{p0}^* = 60°$) and a nonzero lateral offset ($x_{p0}^* = 2.5$), the effect due to the nanopore's surface charge on the particle transport and the ionic current through the nanopore is further examined. Figure 8.13a shows the trajectories of the particle through three nanopores with different surface charges when $E^* = 7.7 \times 10^{-4}$ and $\kappa a = 1.03$. The case with $\sigma_w^* = 0$ is also included as a reference. Before entering the nanopore, the particle trajectories of the three cases are almost identical, as shown in Figure 8.13a. Because of the positive initial orientation, the front end of the particle is close to the left boundary of the nanopore when it enters the nanopore. As a result, the nanopore with $\sigma_w^* = -0.1\sigma_p^*$ ($\sigma_w^* = 0.1\sigma_p^*$) attracts (pushes) the particle toward the negative (positive) x direction, clearly shown in Figure 8.13a. Figure 8.13b shows the corresponding variations of the particle's orientation.

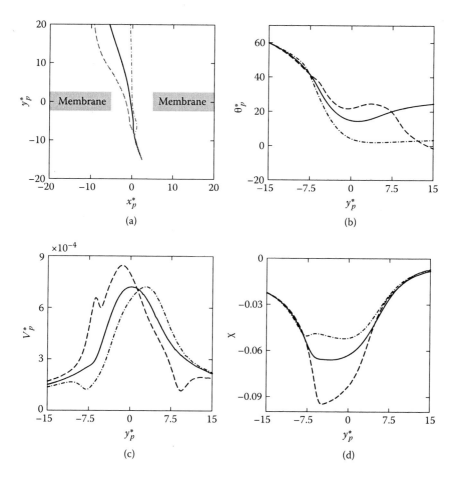

FIGURE 8.13 Trajectory (a), orientation (b), y component translational velocity of the particle (c), and current deviation (d) as a function of the particle's location y_p^* under $E^* = 7.7 \times 10^{-4}$, $\theta_{p0}^* = 60°$, $x_{p0}^* = 2.5$, and $\kappa a = 1.03$. Solid line, dashed line, and dash-dotted line represent, respectively, $\sigma_w^* = 0$, $\sigma_w^* = -0.1\sigma_p^*$, and $\sigma_w^* = 0.1\sigma_p^*$. (From Ai, Y., and S. Qian. 2011. Electrokinetic particle translocation through a nanopore. *Physical Chemistry Chemical Physics* 13:4060–4071 with permission from RCS.)

Similarly, the particle's orientations for the three cases coincide with each other before entering the nanopore. The attractive electrostatic interaction between the particle and the nanopore with $\sigma_w^* = -0.1\sigma_p^*$ causes the particle to rotate counterclockwise when $y_p^* < 0$ and clockwise when $y_p^* > 0$. The repulsive particle-nanopore electrostatic interaction when $\sigma_w^* = 0.1\sigma_p^*$ pushes the particle toward the centerline of the nanopore. As a result, the particle nearly translocates along the centerline with very small rotation when exiting the nanopore. Figure 8.13c shows the variation of the y component translational velocity, which is quite similar to Figure 8.12a. The sharp spine in the velocity profile for $\sigma_w^* = -0.1\sigma_p^*$ is due to the strong particle-nanopore electrostatic interaction. Figure 8.13d shows the corresponding current deviations as a function of the particle's axial location. It is found that the current deviation is not sensitive to the nanopore's surface charge when the particle is far away from the nanopore. However, the current deviations significantly differ from each other when the particles are inside the three different nanopores. Figure 8.13b shows that the particle's orientation for $\sigma_w^* = -0.1\sigma_p^*$ is larger than that for $\sigma_w^* = 0$, which is larger than that for $\sigma_w^* = 0.1\sigma_p^*$. A larger orientation implies a stronger blockade of ionic current through the nanopore. When the applied electric field is relatively high, $E^* = 7.7 \times 10^{-2}$, the particle-nanopore electrostatic interaction is limited, as mentioned. As a result, the y component particle velocity is mainly affected by the extra EOF, which is similar to Figure 8.12b. In addition, the particle is aligned to the local electric field quickly, and the effect of the nanopore's surface charge on the particle's trajectory, orientation, and current deviation are thus limited.

8.5   CONCLUDING REMARKS

A PNP multi-ion model has been developed to study electrokinetic particle translocation through a nanopore. There were no limitations concerning the EDL thickness, the magnitudes of the surface charges, and the magnitude of the imposed electric field in the present numerical model. The PB model developed in Chapter 7 began to deviate significantly from the PNP model when $\kappa a < 1$ and the EDL was significantly distorted by the externally imposed high electric field.

When the particle's initial orientation and lateral offset were both zero, $x_{p0}^* = 0$ and $\theta_{p0}^* = 0$, the particle only translated along the centerline of the nanopore without any rotational motion. If the applied electric field was relatively low, the particle velocity was symmetric when κa was relatively large. As κa decreased, the particle velocity became asymmetric with respect to the center of the nanopore. The particle could even be trapped near the entrance of the nanopore as κa further decreased under a relatively low imposed electric field. Current blockade was predicted under a relatively low electric field, which was in qualitative agreement with the existing experimental observations.

In addition, the current deviation depended on κa. When the applied electric field was relatively high, the particle velocity was symmetric with respect to the center of the nanopore, and particle trapping could not be expected. When κa was relatively large, current blockade was predicted as well. As κa decreased, current enhancement was predicted, which has been experimentally observed. It has also been found that the current deviation highly depends on the applied electric field strength.

When either the particle's initial orientation or lateral offset was nonzero, the particle's rotational motion came into play. The particle's initial orientation played an important role in the particle transport process and the resulting ionic current through the nanopore. When the applied electric field was relatively low, the particle velocity profile became asymmetric with respect to $y_p^* = 0$, which was attributed to the nonzero initial angle of the particle. The angle of the particle gradually decreased as the particle entered the nanopore and then slightly increased as it exited the nanopore. Due to the nonzero initial angle of the particle, the magnitude of the current deviation before the particle entered the nanopore was larger than that with a zero initial angle. When the applied electric field was relatively high, the particle was aligned with its longest axis parallel to the electric field very quickly owing to the DEP effect. The initial lateral offset from the centerline of the nanopore had a limited effect on the particle translocation process.

Furthermore, the nanopore's surface charge can affect the particle transport in terms of the induced EOF and particle-nanopore electrostatic interaction. However, the latter effect was only dominant when the applied electric field was relatively low. Under the condition mentioned, the particle's trajectory, orientation, velocity, and current deviation could be significantly altered due to the particle-nanopore electrostatic interaction. When the applied electric field was relatively high, the induced EOF dominated the particle-nanopore electrostatic interaction. As a result, the EOF linearly affected the particle velocity according to the surface charge ratio of the nanopore and the particle. The particle's trajectory, orientation, and current deviation seemed insensitive to the nanopore's surface charge under a relatively high electric field.

REFERENCES

Ai, Y., and S. Qian. 2011. Electrokinetic particle translocation through a nanopore. *Physical Chemistry Chemical Physics* 13:4060–4071.

Ai, Y., M. Zhang, S. W. Joo, M. A. Cheney, and S. Qian. 2010. Effects of electroosmotic flow on ionic current rectification in conical nanopores. *Journal of Physical Chemistry C* 114 (9):3883–3890.

Aksimentiev, A., J. B. Heng, G. Timp, and K. Schulten. 2004. Microscopic kinetics of DNA translocation through synthetic nanopores. *Biophysical Journal* 87 (3):2086–2097.

Bayley, H. 2006. Sequencing single molecules of DNA. *Current Opinion in Chemical Biology* 10 (6):628–637.

Chang, H., F. Kosari, G. Andreadakis, M. A. Alam, G. Vasmatzis, and R. Bashir. 2004. DNA-mediated fluctuations in ionic current through silicon oxide nanopore channels. *Nano Letters* 4 (8):1551–1556.

Chen, L., and A. T. Conlisk. 2010. DNA nanowire translocation phenomena in nanopores. *Biomedical Microdevices* 12 (2):235–245.

Choi, Y., L. A. Baker, H. Hillebrenner, and C. R. Martin. 2006. Biosensing with conically shaped nanopores and nanotubes. *Physical Chemistry Chemical Physics* 8 (43):4976–4988.

Comer, J., V. Dimitrov, Q. Zhao, G. Timp, and A. Aksimentiev. 2009. Microscopic mechanics of hairpin DNA translocation through synthetic nanopores. *Biophysical Journal* 96 (2):593–608.

Ennis, J., and J. L. Anderson. 1997. Boundary effects on electrophoretic motion of spherical particles for thick double layers and low zeta potential. *Journal of Colloid and Interface Science* 185 (2):497–514.

Fan, R., R. Karnik, M. Yue, D. Y. Li, A. Majumdar, and P. D. Yang. 2005. DNA translocation in inorganic nanotubes. *Nano Letters* 5 (9):1633–1637.

Gu, L.-Q., and J. W. Shim. 2010. Single molecule sensing by nanopores and nanopore devices. *Analyst* 135 (3):441–451.

Heng, J. B., C. Ho, T. Kim, R. Timp, A. Aksimentiev, Y. V. Grinkova, S. Sligar, K. Schulten, and G. Timp. 2004. Sizing DNA using a nanometer-diameter pore. *Biophysical Journal* 87 (4):2905–2911.

Howorka, S., and Z. Siwy. 2009. Nanopore analytics: Sensing of single molecules. *Chemical Society Reviews* 38 (8):2360–2384.

Hsu, J. P., Z. S. Chen, and S. Tseng. 2009. Effect of electroosmotic flow on the electrophoresis of a membrane-coated sphere along the axis of a cylindrical pore. *Journal of Physical Chemistry B* 113 (21):7701–7708.

Hsu, J. P., M. H. Ku, and C. Y. Kao. 2004. Electrophoresis of a spherical particle along the axis of a cylindrical pore: Effect of electroosmotic flow. *Journal of Colloid and Interface Science* 276 (1):248–254.

Hsu, J. P., L. H. Yeh, and M. H. Ku. 2006. Electrophoresis of a spherical particle along the axis of a cylindrical pore filled with a Carreau fluid. *Colloid and Polymer Science* 284 (8):886–892.

Kim, Y. R., J. Min, I. H. Lee, S. Kim, A. G. Kim, K. Kim, K. Namkoong, and C. Ko. 2007. Nanopore sensor for fast label-free detection of short double-stranded DNAs. *Biosensors and Bioelectronics* 22 (12):2926–2931.

Lathrop, D. K., E. N. Ervin, G. A. Barrall, M. G. Keehan, R. Kawano, M. A. Krupka, H. S. White, and A. H. Hibbs. 2010. Monitoring the escape of DNA from a nanopore using an alternating current signal. *Journal of the American Chemical Society* 132 (6):1878–1885.

Li, J. L., M. Gershow, D. Stein, E. Brandin, and J. A. Golovchenko. 2003. DNA molecules and configurations in a solid-state nanopore microscope. *Nature Materials* 2 (9):611–615.

Liu, H., S. Qian, and H. H. Bau. 2007. The effect of translocating cylindrical particles on the ionic current through a nanopore. *Biophysical Journal* 92 (4):1164–1177.

Martin, C. R., and Z. S. Siwy. 2007. Learning nature's way: Biosensing with synthetic nanopores. *Science* 317 (5836):331–332.

Meller, A., L. Nivon, and D. Branton. 2001. Voltage-driven DNA translocations through a nanopore. *Physical Review Letters* 86 (15):3435–3438.

Mukhopadhyay, R. 2009. DNA sequencers: The next generation. *Analytical Chemistry* 81 (5):1736–1740.

Ohshima, H. 1998. Surface charge density surface potential relationship for a cylindrical particle in an electrolyte solution. *Journal of Colloid and Interface Science* 200 (2):291–297.

Plecis, A., R. B. Schoch, and P. Renaud. 2005. Ionic transport phenomena in nanofluidics: Experimental and theoretical study of the exclusion-enrichment effect on a chip. *Nano Letters* 5 (6):1147–1155.

Purnell, R. F., and J. J. Schmidt. 2009. Discrimination of single base substitutions in a DNA strand immobilized in a biological nanopore. *ACS Nano* 3 (9):2533–2538.

Qian, S., B. Das, and X. B. Luo. 2007. Diffusioosmotic flows in slit nanochannels. *Journal of Colloid and Interface Science* 315 (2):721–730.

Qian, S., and S. W. Joo. 2008. Analysis of self-electrophoretic motion of a spherical particle in a nanotube: Effect of nonuniform surface charge density. *Langmuir* 24 (9):4778–4784.

Qian, S., S. W. Joo, Y. Ai, M. A. Cheney, and W. Hou. 2009. Effect of linear surface-charge non-uniformities on the electrokinetic ionic-current rectification in conical nanopores. *Journal of Colloid and Interface Science* 329 (2):376–383.

Qian, S., S. W. Joo, Y. Ai, M. A. Cheney, and W. S. Hou. 2009. Effect of linear surface-charge non-uniformities on the electrokinetic ionic-current rectification in conical nanopores. *Journal of Colloid and Interface Science* 329 (2):376–383.

Qian, S., S. W. Joo, W. Hou, and X. Zhao. 2008. Electrophoretic motion of a spherical particle with a symmetric nonuniform surface charge distribution in a nanotube. *Langmuir* 24 (10):5332–5340.

Qian, S., A. H. Wang, and J. K. Afonien. 2006. Electrophoretic motion of a spherical particle in a converging-diverging nanotube. *Journal of Colloid and Interface Science* 303 (2):579–592.

Rhee, M., and M. A. Burns. 2006. Nanopore sequencing technology: Research trends and applications. *Trends in Biotechnology* 24:580–586.

Sigalov, G., J. Comer, G. Timp, and A. Aksimentiev. 2007. Detection of DNA sequences using an alternating electric field in a nanopore capacitor. *Nano Letters* 8 (1):56–63.

Storm, A. J., J. H. Chen, H. W. Zandbergen, and C. Dekker. 2005. Translocation of double-strand DNA through a silicon oxide nanopore. *Physical Review E* 71 (5):051903.

Storm, A. J., C. Storm, J. H. Chen, H. Zandbergen, J. F. Joanny, and C. Dekker. 2005. Fast DNA translocation through a solid-state nanopore. *Nano Letters* 5 (7):1193–1197.

Wang, Y. C., A. L. Stevens, and J. Y. Han. 2005. Million-fold preconcentration of proteins and peptides by nanofluidic filter. *Analytical Chemistry* 77 (14):4293–4299.

White, H. S., and A. Bund. 2008. Ion current rectification at nanopores in glass membranes. *Langmuir* 24 (5):2212–2218.

Zhao, X., C. M. Payne, and P. T. Cummings. 2008. Controlled translocation of DNA segments through nanoelectrode gaps from molecular dynamics. *Journal of Physical Chemistry C* 112 (1):8–12.

9 Field Effect Control of DNA Translocation through a Nanopore

In this chapter, regulation of DNA nanoparticle translocation through a nanopore by a nanofluidic field effect transistor (FET) is numerically investigated using a continuum-based mathematical model, which includes the Poisson equation for electrostatics, the Nernst–Planck equations for ionic mass transport, and the Navier–Stokes and continuity equations for fluid flow. The field effect modulates the surface potential of the nanopore's inner wall and accordingly the electroosmotic flow (EOF) inside the nanopore as well as the particle-nanopore electrostatic interaction. The EOF enhances (retards) the nanoparticle translocation process when the direction of the induced EOF is the same as (opposite to) that of the DNA electrophoretic motion.

The effect of the particle-nanopore electrostatic interaction highly depends on the ratio of the particle size to the electrical double layer (EDL) thickness. When the EDLs of the DNA nanoparticle and the nanopore wall are overlapped and the electric field imposed is relatively low, the effect of the particle-nanopore electrostatic interaction dominates over the EOF effect, which might lead to DNA nanoparticle trapping inside the nanopore. However, the particle-nanopore electrostatic interaction is negligible if the EDLs are not overlapped. The nanofluidic FET offers a flexible and electrically compatible approach to control DNA nanoparticle translocation actively through a nanopore for DNA sequencing.

9.1 INTRODUCTION

DNA sequencing is the determination of the precise sequence of nucleotides adenine (A), guanine (G), cytosine (C), and thymine (T) in a DNA molecule, which is important in basic biological research and other applied fields, such as the famous Human Genome Project launched by the U.S. National Institutes of Health (NIH) in the late 1980s (Luria 1989). Since 1970, researchers have been striving to develop a high-throughput and affordable DNA sequencing technique (Mukhopadhyay 2009). Among various DNA sequencing techniques, the nanopore-based DNA sequencing technique has emerged as one of the most promising approaches to achieve the goal (Meller, Nivon, and Branton 2001; Saleh and Sohn 2003; Storm, Storm, et al. 2005; Rhee and Burns 2006; Healy, Schiedt, and Morrison 2007; Dekker 2007; Griffiths 2008; Howorka and Siwy 2009).

In the nanopore-based DNA sequencing technique, a low voltage is imposed across a nanopore immersed in an aqueous electrolyte, creating an ionic current flowing through the nanopore and simultaneously driving the DNA nanoparticle through the nanopore by electrophoresis. The ionic current through the nanopore is sensitive to the size and shape of the nanopore. If single bases or strands of DNA pass (or part of the DNA molecule passes) through the nanopore by electrophoresis, this can create a change in the magnitude of the ionic current through the nanopore. The ionic current is blocked when a DNA molecule enters the nanopore, and returns to the baseline current after it exits the nanopore. Since the A, C, G, and T nucleotides on the DNA molecule carry different surface charges, each nucleotide may obstruct the nanopore at a different characteristic degree, resulting in different magnitudes of ionic current. The amount of ionic current flowing through the nanopore at any given moment therefore varies depending on whether the nanopore is blocked by an A, a C, a G, or a T nucleotide base.

The distinctive change in the ionic current through the nanopore arising from the DNA translocation through the nanopore represents a direct reading of the DNA sequence. Therefore, it is hypothesized that the sequence of nucleotide bases in a single DNA nanoparticle can be recorded by monitoring the current modulations (Meller, Nivon, and Branton 2001; Chang et al. 2004; Heng et al. 2004). This method examines the electronic signals or relies on physical properties, in contrast to existing paradigms based on chemical techniques. Thus, nanopore-based DNA sequencing involves no sample amplification, leading to a label-free and single-molecule approach. Furthermore, the sequencing time of nucleic acids is within microseconds. In addition, the estimated cost of nanopore-based sequencing of a human genome is within the range of $1,000, which meets the goals set by the NIH in 2004. This cost is believed to be sufficiently low to revolutionize genomic medicine (Branton et al. 2008; Clarke et al. 2009; Bayley 2006).

All these encouraging benefits of the nanopore-based DNA sequencing stimulate a fast-growing research area related to nanopore-based analysis. However, one of the major challenges of the nanopore-based technique is that DNA nanoparticles translocate through the nanopore too fast for detection. As a result, an extremely high temporal resolution is indispensable for precise detection of each nucleotide base, which requires an extremely high bandwidth for the sensing system (Bayley 2006). Therefore, the key to nanopore-based DNA sequencing is the ability to regulate the translocation process with a nanometer-scale spatial accuracy.

Although one can reduce the voltage across the nanopore to slow the DNA translocation, it also reduces the ionic current and accordingly decreases the signal-to-noise ratio. In addition, the DNA translocation through the nanopore per unit time would also significantly decrease as the applied voltage decreases. To achieve high throughput, a relatively high electric field imposed across the nanopore is thus typically used. To date, several other methods have been proposed to slow the DNA translocation through the nanopore to obtain a detectable current signal. For example, an extra mechanical force could be exerted on DNA nanoparticles using optical tweezers to slow DNA translocation at the expense of a highly focused laser (Trepagnier et al. 2007). Chemical functionalization could increase the energy barrier of the nanopore to slow DNA translocation (Kim et al. 2007).

It is also found that the electrophoretic velocity of the DNA translocation highly depends on the ionic concentration, which could be utilized to control DNA translocation (Ghosal 2007). Regulation of DNA translocation through a nanopore is also achieved by adjusting the viscosity of the aqueous solution and accordingly the viscous force acting on the DNA nanoparticle (Kawano et al. 2009). The temperature of the electrolyte solution also affects the viscosity and thus the DNA translocation (Fologea et al. 2005). One order of magnitude decrease in the DNA translocation has been achieved by simultaneously controlling the electrolyte temperature, the electrolyte viscosity, the ionic concentration, and the applied voltage (Fologea et al. 2005). One can also slow the DNA nanoparticle translocation by adding organic salts in the electrolyte solution (de Zoysa et al. 2009).

Similar to the metal oxide semiconductor field effect transistors (MOSFETs) except that the medium is made of electrolyte ions instead of electrons or holes, a dielectric nanopore with an electrically addressable gate electrode, a so-called nanofluidic FET, has been fabricated using state-of-the-art nanofabrication technologies (Karnik et al. 2005; Kalman et al. 2009; Nam et al. 2009; Taniguchi et al. 2009; Joshi et al. 2010). The gate potential applied to the gate electrode can effectively control the surface potential at the nanopore-fluid interface (Schasfoort et al. 1999). The nanopore's inner wall becomes positively (negatively) charged when a positive (negative) gate potential is imposed. Since the EOF inside the nanopore highly depends on the surface potential, one can control both direction and strength of the EOF by regulation of the gate potential. The field effect has been utilized to regulate EOF in microfluidic devices (Schasfoort et al. 1999; Vajandar et al. 2009), ionic mass transport, and ionic conductance in nanofluidic devices (Karnik et al. 2005; Kalman et al. 2009; Nam et al. 2009; Joshi et al. 2010; Daiguji 2010). The field effect regulation of the electrokinetic transport of charged dye nanoparticles in a nanopore has been experimentally demonstrated (Oh et al. 2008, 2009). Nanofluidic FET offers a more flexible and electrically compatible approach for the control of the surface potential than the chemical functionalization method. This chapter provides a profound theoretical analysis of the field effect regulation of DNA nanoparticle translocation through a nanopore.

In a previous study, three different models were developed to study the electrokinetic translocation of a cylindrical nanoparticle through a nanopore without the field effect control (Liu, Qian, and Bau 2007). It was found that the multi-ion model (MIM), including the Poisson equation for the electrostatics, the Nernst–Planck equations for the transport of ions, and the Navier–Stokes and continuity equations for the hydrodynamic field, successfully captures the essential physics of the DNA translocation process for an arbitrary EDL thickness. In contrast, the simplified models based on the Poisson–Boltzmann model (PBM) and the Smoluchowski's slip velocity model (SVM) are not appropriate under thick EDL. The continuum-based MIM model has also been used to study the ionic current rectification phenomenon in a nanopore, and its predictions qualitatively agreed with the experimental data obtained from the literature (Ai, Zhang, et al. 2010; White and Bund 2008). The validity of the continuum model has also been confirmed when the pore's radius was close to the Debye length (Corry, Kuyucak, and Chung 2000; Stein, Kruithof, and Dekker 2004; Schoch, van Lintel, and Renaud 2005; Pennathur and Santiago 2005).

Therefore, the continuum MIM model is adopted to explore the field effect regulation of DNA translocation through a nanopore. To elucidate the mechanism of field effect regulation of nanoparticle translocation through a nanopore, DNA nanoparticle translocation through a nanopore is investigated as a function of four main factors: the electric field imposed across the nanopore, the gate potential applied on the gate electrode, the ratio of the particle radius to the Debye length, and the permittivity of the dielectric nanopore.

9.2 MATHEMATICAL MODEL

Figure 9.1a schematically depicts a charged nanoparticle located along the axis of a nanopore of length L_c and radius b filled with a binary KCl aqueous solution

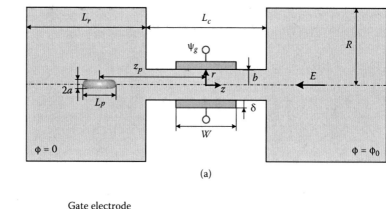

(a)

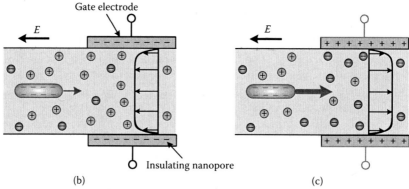

(b)　　　　　　　　　　　　　　　　　　　　(c)

FIGURE 9.1　Schematics of DNA translocation through a nanofluidic FET (a). The EOF retards the negatively charged DNA translocation when the gate potential is negative (b) and enhances the DNA translocation when the gate potential is positive (c). (From Ai, Y., J. Liu, B. Zhang, and S. Qian. 2010. Field effect regulation of DNA translocation through a nanopore. *Analytical Chemistry* 82, 8217–8225 with permission from the American Chemical Society.)

of density ρ, dynamic viscosity μ, and relative permittivity ε_f. The cylindrical nanopore connects two identical reservoirs on either side. For simplification of the numerical simulation, an axisymmetric model is used in the present study, and all variables are defined in a cylindrical coordinate system (r, z) with the origin fixed at the center of the nanopore. The r and z coordinates are, respectively, perpendicular and parallel to the axis of the nanopore. The axial length L_r and radius R of the reservoirs are sufficiently large to maintain a bulk ionic concentration C_0 at places far away from the nanopore.

Usually, a very long DNA strand is coiled randomly inside the reservoir. However, since the nanopore is very small (the pore diameter is less than 10 nm), the DNA nanoparticle is elongated or stretched to translocate through the nanopore (Storm, Chen, et al. 2005; Storm, Storm, et al. 2005). The uncoiled DNA nanoparticle is very similar to a nanometer-size cylinder and therefore is approximated as a cylindrical nanoparticle of length L_p and radius a, having two hemispherical caps of radius a at both ends. The nanoparticle's axis is assumed to coincide with the nanopore's axis. Therefore, the nanoparticle only translates along the axis of the nanopore without rotational motion.

We further assume that the DNA nanoparticle bears a uniform surface charge density σ_p, resulting in an EDL formed in the vicinity of the charged nanoparticle's surface. In addition, an EDL is formed next to the nanopore's inner wall, bearing a surface charge density of σ_w. A negative axial electric field $\mathbf{E}$ is imposed across the nanopore to drive the negatively charged DNA translocation along the axis of the nanopore and simultaneously generate an ionic current through the nanopore. In the middle region of the nanopore, a gate electrode of length W is coated on the outer surface of the dielectric nanopore of thickness δ. A gate potential ψ_g is applied to the gate electrode to modulate the surface potential of the nanopore's inner surface next to the gate electrode, which in turn regulates the EOF and accordingly the DNA translocation through the nanopore.

When a negative gate potential is applied to the gate electrode, more cations are accumulated in the vicinity of the nanopore's inner surface next to the gate electrode, as shown in Figure 9.1b. An EOF opposite to the particle electrophoretic motion is induced, thus retarding the DNA translocation. In contrast, when a positive gate potential is applied to the gate electrode, anions are predominantly occupied in the EDL region of the nanopore where the gate electrode is located. Figure 9.1c shows that the generated EOF, in the same direction as the DNA translocation, enhances DNA translocation through the nanopore.

Assuming quasi-steady state and no chemical reactions, the electrostatics and ion transport within the electrolyte solution are governed by the verified Poisson–Nernst–Planck (PNP) equations (White and Bund 2008; Qian, Das, and Luo 2007; Qian et al. 2009; Ai, Zhang, et al. 2010):

$$-\varepsilon_0 \varepsilon_f \nabla^2 \phi = F(c_1 z_1 + c_2 z_2), \tag{9.1}$$

$$\nabla \bullet \mathbf{N}_i = \nabla \bullet (\mathbf{u} c_i - D_i \nabla c_i - z_i \frac{D_i}{RT} F c_i \nabla \phi) = 0, \ i = 1 \text{ and } 2, \tag{9.2}$$

where ε_0 is the absolute permittivity of the vacuum; ϕ is the electric potential within the fluid; F is the Faraday constant; c_1 and c_2 are, respectively, the molar concentrations of the cations (K^+) and anions (Cl^-) in the electrolyte solution; z_1 and z_2 are, respectively, the valences of cations ($z_1 = 1$ for K^+) and anions ($z_2 = -1$ for Cl^-); N_i is the ionic flux density of the ith ionic species; u is the fluid velocity; D_i is the diffusivity of the ith ionic species; R is the universal gas constant; and T is the absolute temperature of the electrolyte solution. Hereafter, **bold** letters denote vectors or tensors. The thickness of the EDL is characterized by the Debye length, $\lambda_D = \kappa^{-1} = \sqrt{\varepsilon_0 \varepsilon_f RT / (2F^2 C_0)}$, based on the bulk ionic concentration.

Axial symmetric boundary conditions for all the physical fields are applied on the axis of the nanopore. The boundary condition for the ionic concentrations at the ends of the two reservoirs is specified as $c_i(r, \pm(L_r + L_c / 2)) = C_0$, $i = 1$ and 2. The normal ionic flux on the particle surface only includes the convective flux, $\mathbf{n} \bullet \mathbf{N}_i = \mathbf{n} \bullet (\mathbf{u}c_i)$, $i = 1$ and 2, where $\mathbf{n}$ is the unit normal vector directed from the particle surface into the fluid. The normal ionic fluxes on all the other boundaries are set as zero.

The boundary conditions for the electric potentials at the ends of the two reservoirs are, respectively, $\phi(r, -(L_r + L_c / 2)) = 0$ and $\phi(r, (L_r + L_c / 2)) = \phi_0$. The surface charge boundary conditions on the nanoparticle's surface and the nanopore's inner wall expect that next to the gate electrode are, respectively, $-\varepsilon_0\varepsilon_f\mathbf{n} \bullet \nabla\phi = \sigma_p$ and $-\varepsilon_0\varepsilon_f\mathbf{n} \bullet \nabla\phi = \sigma_w$. The following boundary condition is imposed at the nanopore's inner wall right next to the gate electrode:

$$-\varepsilon_0\varepsilon_f\mathbf{n} \bullet \nabla\phi + \varepsilon_0\varepsilon_d\mathbf{n} \bullet \nabla\psi = \sigma_w, \tag{9.3}$$

and

$$\phi = \psi, \tag{9.4}$$

where ε_d is the permittivity of the dielectric nanopore material. The insulating boundary condition, $\mathbf{n} \bullet \nabla\phi = 0$, is applied in the other boundaries in contact with the fluid. Here, ψ is the electric potential inside the dielectric nanopore wall sandwiched between the gate electrode and the fluid, governed by the Laplace equation:

$$\nabla^2\psi = 0. \tag{9.5}$$

The gate potential $\psi = \psi_g$ is applied on the gate electrode. The other boundaries of the insulator are set to be the zero normal electric field (i.e., $\mathbf{n} \bullet \nabla\psi = 0$).

Since typically the Reynolds number associated with the electrokinetic flow inside a nanopore is extremely small, it is appropriate to model the flow field using the modified Stokes equations by neglecting the inertial terms in the Navier–Stokes equations,

$$-\nabla p + \mu\nabla^2\mathbf{u} - F(z_1c_1 + z_2c_2)\nabla\phi = 0, \tag{9.6}$$

and the continuity equation for the incompressible fluid,

$$\nabla \bullet \mathbf{u} = 0. \tag{9.7}$$

In these equations, p is the pressure. The first, second, and third terms in Equation (9.6) represent, respectively, the pressure, viscous, and electrostatic forces.

A no-slip boundary condition is applied on the inner surface of the nanopore and the reservoir walls. A normal flow with $p = 0$ is applied at the ends of the two reservoirs. A slip boundary condition is applied at the side boundaries of the two reservoirs, which are far away from the nanopore. As the DNA nanoparticle only translates along the axis of the nanopore, the fluid velocity on the surface of the particle is $\mathbf{u}(r,z) = U_p \mathbf{e}_z$, where U_p is the axial velocity of the particle, and $\mathbf{e}_z$ is the axial unit vector. The axial velocity of the particle is determined based on the balance of the z component force acting on the particle using a quasi-static method (Qian, Wang, and Afonien 2006; Qian and Joo 2008; Qian et al. 2008; Hsu, Chen, and Tseng 2009; Hsu, Hsu, and Liu 2010),

$$F_E + F_H = 0. \tag{9.8}$$

In this equation,

$$F_E = \int \varepsilon_0 \varepsilon_f \left[\frac{\partial \phi}{\partial z} \frac{\partial \phi}{\partial r} n_r + \frac{1}{2}(\frac{\partial \phi}{\partial z})^2 n_z - \frac{1}{2}(\frac{\partial \phi}{\partial r})^2 n_z \right] d\Gamma \tag{9.9}$$

is the axial electrical force based on the integration of the Maxwell stress tensor over the particle surface (Ai, Beskok, et al. 2009; Ai, Joo, et al. 2009; Ai, Park, et al. 2010; Ai, Qian, et al. 2010; Ai and Qian 2010), and

$$F_H = \int \left[-pn_z + 2\mu \frac{\partial u_z}{\partial z} n_z + \mu(\frac{\partial u_r}{\partial z} + \frac{\partial u_z}{\partial r})n_r \right] d\Gamma \tag{9.10}$$

is the z component hydrodynamic force acting on the particle. Respectively, u_r and u_z are the r and z components of the fluid velocity $\mathbf{u}$; n_r and n_z are, respectively, the r and z components of the unit vector $\mathbf{n}$; Γ denotes the surface of the DNA nanoparticle.

The induced ionic current through the nanopore is

$$I = \int F(z_1 \mathbf{N}_1 + z_2 \mathbf{N}_2) \bullet \mathbf{n}\, dS, \tag{9.11}$$

where S denotes the opening of either reservoir due to current conservation.

The bulk concentration C_0 as the ionic concentration scale, the particle radius a as the length scale, RT/F as the potential scale, $U_0 = \varepsilon_0 \varepsilon_f R^2 T^2/(\mu a F^2)$ as the velocity scale, and $\mu U_0/a$ as the pressure scale are chosen to normalize all the governing equations mentioned:

$$-\nabla^{*2}\phi^* = \frac{(\kappa a)^2}{2}(c_1^* z_1 + c_2^* z_2), \tag{9.12}$$

$$\nabla^* \bullet \mathbf{N}_i^* = \nabla^* \bullet (\mathbf{u}^* c_i^* - D_i^* \nabla^* c_i^* - z_i D_i^* c_i^* \nabla^* \phi^*) = 0, \ i = 1 \text{ and } 2, \qquad (9.13)$$

$$\nabla^{*2} \psi^* = 0, \qquad (9.14)$$

$$-\nabla^* p^* + \nabla^{*2} \mathbf{u}^* - \frac{(\kappa a)^2}{2}(z_1 c_1^* + z_2 c_2^*)\nabla^* \phi^* = 0, \qquad (9.15)$$

$$\nabla^* \bullet \mathbf{u}^* = 0. \qquad (9.16)$$

The corresponding boundary conditions are normalized as

$$c_i^*(r^*, \ \pm(L_r^* + L_c^* / 2)) = 1, \ i = 1 \text{ and } 2, \qquad (9.17)$$

$$\mathbf{n} \bullet \mathbf{N}_i^* = \mathbf{n} \bullet (\mathbf{u}^* c_i^*), \ i = 1 \text{ and } 2, \qquad (9.18)$$

$$\phi^*(r^*, \ -(L_r^* + L_c^* / 2)) = \phi^*(r^*, \ (L_r^* + L_c^* / 2)) - \phi_0^* = 0, \qquad (9.19)$$

$$-\mathbf{n} \bullet \nabla^* \phi^* = \sigma_p^*, \qquad (9.20)$$

$$-\mathbf{n} \bullet \nabla^* \phi^* = \sigma_w^*. \qquad (9.21)$$

$$-\mathbf{n} \bullet \nabla^* \phi^* + \frac{\varepsilon_d}{\varepsilon_f} \mathbf{n} \bullet \nabla^* \psi^* = \sigma_w^*, \qquad (9.22)$$

where the characteristic surface charge density is $\varepsilon_0 \varepsilon_f RT / (Fa)$. The normalized z component electrical and hydrodynamic forces are, respectively,

$$F_E^* = \int \left[\frac{\partial^* \phi^*}{\partial^* z^*} \frac{\partial^* \phi^*}{\partial^* r^*} n_r + \frac{1}{2}(\frac{\partial^* \phi^*}{\partial^* z^*})^2 n_z - \frac{1}{2}(\frac{\partial^* \phi^*}{\partial^* r^*})^2 n_z \right] d\Gamma^*, \qquad (9.23)$$

$$F_H^* = \int \left[-p^* n_z + 2\frac{\partial^* u_z^*}{\partial^* z^*} n_z + (\frac{\partial^* u_r^*}{\partial^* z^*} + \frac{\partial^* u_z^*}{\partial^* r^*})n_r \right] d\Gamma^*. \qquad (9.24)$$

The dimensionless ionic current through the nanopore normalized by $FU_0 C_0 a^2$ is

$$I^* = \int (z_1 \mathbf{N}_1^* + z_2 \mathbf{N}_2^*) \bullet \mathbf{n} \, dS^*. \qquad (9.25)$$

9.3 IMPLEMENTATION IN COMSOL MULTIPHYSICS AND CODE VALIDATION

The dimensions of the geometry are δ = 5 nm, W = 20 nm, $L_r = L_c$ = 40 nm, b = 4 nm, and R = 40 nm. The gate electrode is located at -10 nm $\leq z \leq 10$ nm. The radius of a single DNA nanoparticle is around a = 1 nm, and the length of one nucleotide unit is around 0.33 nm (Mandelkern et al. 1981). Here, we assume the total length of the DNA nanoparticle is L_p = 10 nm. Figure 9.2 shows the dimensionless geometry of a DNA nanoparticle within a nanofluidic FET drawn in COMSOL Multiphysics®. Table 9.1 lists all the variables used in the modeling, which will be defined in COMSOL Multiphysics. Table 9.2 provides the detailed instructions for setting up the dimensionless model in the graphical user interface (GUI) of COMSOL 3.5a. However, the model established in the COMSOL Multiphysics GUI can only simulate a single case, in which the particle's location is fixed. As the variation of the particle velocity along the axis of the nanopore and the corresponding ionic current response are of great interest in this study, one needs to simulate many cases with different particle locations along the axis of the nanopore. The COMSOL MATLAB script can easily carry out the geometry automation, which is difficult in the COMSOL Multiphysics GUI.

To implement the geometry automation in a COMSOL MATLAB script, it is not necessary to write the entire script. One can set up the model in the COMSOL Multiphysics GUI first and then save the file as a model M-file (.m file). All the operations and settings in the COMSOL Multiphysics GUI will be translated into COMSOL MATLAB script commands in the generated .m file. Apparently, it is more efficient to establish one's own .m file by modifying the generated one. Because all the operations in the COMSOL Multiphysics GUI are recorded in the .m file, one may need to remove some useless script commands manually if some unnecessary operations have been made during the setup in the COSMOL

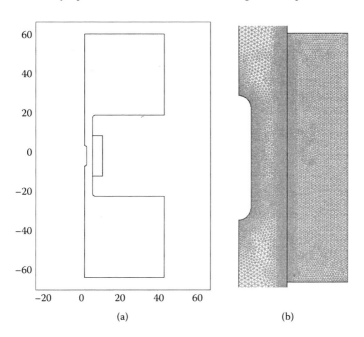

(a)   (b)

FIGURE 9.2   (a) Axially symmetric geometry of a DNA nanoparticle within a nanofluidic FET in COMSOL Multiphysics. (b) Mesh around the DNA and inside the dielectric nanopore.

TABLE 9.1
Constant Table

Variable	Value or Expression	Description
D1	1.95e-9 [m^2/s]	Diffusivity of K^+
D2	2.03e-9 [m^2/s]	Diffusivity of Cl^-
z1	1	Valence of K^+
z2	−1	Valence of Cl^-
F	96485.3415 [C/mol]	Faraday constant
R0	8.314472 [J/K/mol]	Gas constant
T	300 [K]	Temperature
eps0	8.854187817e-12 [F/m]	Permittivity of vacuum
eps_r	80	Relative permittivity of fluid
eps_d	3.9	Relative permittivity of nanopore
eta	1.0e-3 [pa*s]	Fluid viscosity
rho	1e3 [kg/m^3]	Fluid density
V0	0.0012 [V]	Applied voltage
Vg	0.517041 [V]	Applied gate potential
cp	−0.01 [C/m^2]	Surface charge density of the particle
cw	0 [C/m^2]	Surface charge density of the nanopore
c0	100 [mM]	Bulk concentration
a	1e-9 [m]	Particle radius
lamda	sqrt(eps_r*eps0*R0*T/2/F^2/c0)	Debye length
ai	a/lamda	κa
U0	eps0*eps_r*R0^2*T^2/eta/a/F^2	Velocity scale
charges	F*a/R0/T/eps_r/eps0	Surface charge density scale
Re	rho*U0*a/eta	Reynolds number
cpd	cp*charges	Dimensionless particle's surface charge density
cwd	cw*charges	Dimensionless nanopore's surface charge density
D1d	D1*eta*F^2/(eps0*eps_r*R0^2*T^2)	Dimensionless diffusivity of K^+
D2d	D2*eta*F^2/(eps0*eps_r*R0^2*T^2)	Dimensionless diffusivity of Cl^-
V0d	V0*F/R0/T	Dimensionless applied voltage
Vgd	Vg*F/R0/T	Dimensionless applied gate potential

Multiphysics GUI. It is highly recommended to save as the .m file once the model is correctly set up in COMSOL Multiphysics GUI. Subsequently, one can modify the .m file based on one's own interest.

Here, we are interested in the geometry automation. Therefore, we define a vector z_p in the .m file to store all the particle's axial locations in the simulation. Accordingly, a loop (FOR loop) is added in the .m file to go through all the particle's locations for geometry automation. Furthermore, one can add another loop outside the loop for geometry automation to investigate another parameter in the simulation. For example, we can add another loop to go through three different

TABLE 9.2
Model Setup in COMSOL Multiphysics 3.5a

Model Navigator	Select **Axial symmetry (2D)** and click **Multiphysics** button.
	Select **Chemical Engineering Module|Mass Transport|Nernst–Planck without Electroneutrality**. Remove the predefined variables in the **Dependent variables** and enter c1 and c2. Click **Add** button.
	Select **COMSOL Multiphysics|Electromagnetics|Electrostatics**. Click **Add** button.
	Select **COMSOL Multiphysics|Fluid Dynamics|Incompressible Navier–Stokes**. Click **Add** button.
Option Menu|	Define variables in Table 9.1.
Constants	
Physics Menu|	chekf mode
Subdomain Setting	Subdomain 1
	Tab c1
	$D = D1\_dless; R = 0; um = D1\_dless/F; z = z1; u = u; v = v; V = V$
	Tab c2
	$D = D2\_dless; R = 0; um = D2\_dless/F; z = z2; u = u; v = v; V = V$
	Deactivate subdomain 2
	es mode
	Subdomain 1
	$\rho = 0.5*(ai)^2*(z1*c1+z2*c2); \varepsilon r = 1/eps0$
	Subdomain 2
	$\rho = 0; \varepsilon r = 1/eps0*eps\_d/eps\_r$
	ns mode
	Subdomain 1
	Tab **Physics**
	$\rho = 0; \eta = 1; Fr = -0.5*ai^2*(z1*c1+z2*c2)*Vr;$
	$Fz = -0.5*ai^2*(z1*c1+z2*c2)*Vz$
	Tab **Stabilization**
	Deactivate **Streamline diffusion** and **Crosswind diffusion**
	Deactivate subdomain 2
Physics Menu|	chekf mode
Boundary Setting	Axis: Axial symmetry
	Two ends of the nanopore: Concentration c1 = 1 and c2 = 1
	Particle surface: Convective flux for c1 and c2
	Walls of the nanopore and the reservoir: Insulation/Symmetry for c1 and c2
	Keep deactivated boundaries unchanged
	es mode
	Axis: Axial symmetry
	Lower end of the nanopore: $V_0 = 0$
	Upper end of the nanopore: $V_0 = V_{0\_}dless$
	Particle surface: Surface charge $\rho_s = cp\_dless$

continued

TABLE 9.2 (CONTINUED)
Model Setup in COMSOL Multiphysics 3.5a

Nanopore wall without next to the gate electrode:
Surface charge ρ_s = cw_dless
Nanopore wall next to the gate electrode:
Surface charge ρ_s = cw_dless
Gate electrode: Electric potential V_0 = Vg_dless
Others: Zero charge/Symmetry

ns mode
Axis: Axial symmetry
Two ends of the nanopore: Outlet|Pressure, no viscous stress P_0 = 0
Particle surface: Inlet|Velocity u_0 = 0; v_0 = vp where vp is the particle
    velocity
Nanopore wall: Wall|No slip
Reservoir wall: Wall|Slip
Keep deactivated boundaries unchanged

Options|Integration Coupling Variables|Boundary Variables	Boundary selection: 5, 16, 17 (Particle surface) Name: Fh_z; Expression: -T_z_ns*2*pi*r; Integration order: 4 Name: Fe_z; Expression: (-Vz*Vr*nr-0.5*Vz^2*nz+0.5*Vr^2*nz)*2*pi*r; Integration order: 4
Physics Menu| Global Equations	Name: vp; Expression: (Fh_z+Fe_z)*1e20; Init (u): 0; Init (ut): 0
Mesh|Free Mesh Parameters ...	Tab **subdomain** Subdomain 1|**Maximum element size**: 0.7 Subdomain 2|**Maximum element size**: 0.4
	Tab **boundary** Boundary 5, 6, 7, 9, 11, 12, 16, 17, 18, 19|**Maximum element size**: 0.1
Solve Menu| Postprocessing	Click **solve problem (=)** Result check: vp **Plot Parameters|Surface|Expression**: vp Click Apply Result check: Ionic current through the nanopore **Plot Parameters|Boundary Integration|Expression**: 2*pi*r*(ntflux_c1_chekf-ntflux_c2_chekf)*nz Check **Compute surface integral (for axisymmetric modes)** Click **Apply**

bulk concentrations to study the effects of the bulk concentration on the DNA translocation through a nanopore. The full COMSOL MATLAB script .m file, with the authors' commentary beginning with "% NEW", is as follows:

```
%%%%%%%%%%%%%%%%%%%%%%%%%%%%%Geometry automatio
n.m%%%%%%%%%%%%%%%%%%%%%%%%%

flclear fem
```

```
% NEW: Define three different bulk concentrations to
% NEW: investigate the effect
% NEW: of the bulk concentration on the DNA translocation
% NEW: through a nanopore
conc=[100,400,1000];

% NEW: FOR loop to go through all the bulk concentrations
for j=1:length(conc)

% NEW: Define all the particle's locations in a vector
zp=-35:1:35;

% NEW: Data storage for output
data=zeros(length(zp),7);

% NEW: FOR loop to go through all the particle's locations
for i=1:length(zp)

% NEW: Get a particle's axial location in one single
% NEW: simulation
zp0=zp(i);

% NEW: Particle length
Length_p=10;

% Constants
% NEW: Function "num2str" is used to transfer the variables
% NEW: defined in .m file
% NEW: to COMSOL constants
fem.const = {'D1','1.95e-9[m^2/s]', ...
             'D2','2.03e-9[m^2/s]', ...
             'z1','1', ...
             'z2','-1', ...
             'F','96485.3415[C/mol]', ...
             'R0','8.314472[J/K/mol]', ...
             'T','300[K]', ...
             'eps0','8.854187817e-12[F/m]', ...
             'eps_r','80', ...
             'eps_d','3.9', ...
             'eta','1.0e-3[Pa*s]', ...
             'rho','1e3[kg/m^3]', ...
             'V0','10000*120e-9[V]', ...
             'Vg','0.517041[V]', ...
             'cp','-0.01[C/m^2]', ...
             'cw','0[C/m^2]', ...
             'c0',num2str(conc(j)), ...
             'a','1e-9[m]', ...
             'lamda','sqrt(eps_r*eps0*R0*T/2/F^2/c0)', ...
             'ai','a/lamda', ...
             'U0','eps0*eps_r*R0^2*T^2/eta/a/F^2', ...
             'charges','F*a/R0/T/eps_r/eps0', ...
```

```
                'Re','rho*U0*a/eta', ...
                'cp_dless','cp*charges', ...
                'cw_dless','cw*charges', ...
                'D1_dless','D1*eta*F^2/(eps0*eps_r*R0^2*T^2)',
                ...
                'D2_dless','D2*eta*F^2/(eps0*eps_r*R0^2*T^2)',
                ...
                'V0_dless','V0*F/R0/T', ...
                'Vg_dless','Vg*F/R0/T'};

% Geometry
carr={curve2([0,40],[-60,-60]), ...
      curve2([40,40],[-60,-20]), ...
      curve2([40,4],[-20,-20]), ...
      curve2([4,4],[-20,20]), ...
      curve2([4,40],[20,20]), ...
      curve2([40,40],[20,60]), ...
      curve2([40,0],[60,60]), ...
      curve2([0,0],[60,-60])};

g1=geomcoerce('curve',carr);
g2=geomcoerce('solid',{g1});
g3=fillet(g2,'radii',1.5,'point',[3,4]);

% NEW: Particle's axial location zp0 is used for geometry
% NEW: automation
g7=rect2('2','8','base','center','pos',{'0',num2str(zp0)},
  'rot','0');
g8=circ2('1','base','center','pos',{'0',num2str(zp0+4)},'
  rot','0');
g9=circ2('1','base','center','pos',{'0',num2str(zp0-4)},
  'rot','0');

g10=geomcomp({g7,g8,g9},'ns',{'g7','g8','g9'},'sf',
  'g7+g8+g9','edge','none');
g11=geomdel(g10);

g12=geomcomp({g3,g11},'ns',{'g3','g11'},'sf',
  'g3-g11','edge','none');

g3=rect2('5','20','base','center','pos',{'6.5','0'},
  'rot','0');

% Analyzed geometry
clear s
s.objs={g12,g3};
s.name={'CO1','R1'};
s.tags={'g12','g3'};

fem.draw=struct('s',s);
fem.geom=geomcsg(fem);
```

```
% NEW: Define maximum mesh sizes for different boundaries
mesh_p=0.2;
mesh_w=0.1;
mesh_o=0.4;

% Initialize mesh
fem.mesh=meshinit(fem, ...
                  'hauto',5, ...
'hmaxedg', [5,mesh_p,16,mesh_p,17,mesh_p,6,mesh_w,9,mesh_w,11
  ,mesh_w,12,mesh_w,18,mesh_w,19,mesh_w,7,mesh_w,2,mesh_o,4,
  mesh_o], ...
                  'hmaxsub', [1,0.5,2,0.4]);

 % Refine mesh
fem.mesh=meshrefine(fem, ...
                    'mcase',0, ...
                    'boxcoord',[0 2 zp0-7 zp0+7], ...
                    'rmethod','regular');

% NEW: Display geometry with the sequence of the boundary
geomplot(fem, 'Labelcolor','r','Edgelabels','on','submode',
  'off');

% NEW: Solve ionic concentration using Nernst-Planck
% NEW: equations
% Application mode 1
clear appl
appl.mode.class = 'ElectroKF_NernstPl';
appl.mode.type = 'axi';
appl.dim = {'c1','c2'};
appl.module = 'CHEM';
appl.gporder = 4;
appl.cporder = 2;
appl.assignsuffix = '_chekf';
clear prop
prop.analysis='static';
prop.equform='cons';
appl.prop = prop;
clear bnd
bnd.c0 = {0,{1;1},0,0,0};
bnd.name = {'axis','bulk','particle','Insulator','dielect
  ric'};
bnd.type = {{'ax';'ax'},{'C';'C'},{'Nc';'Nc'},{'N0';'N0'},
  {'cont';'cont'}};
bnd.ind = [1,2,1,2,3,4,4,5,4,5,4,4,5,4,4,3,3,4,4];
appl.bnd = bnd;
clear equ
equ.D = {{'D1_dless';'D2_dless'},{1;1}};
equ.V = {{'V';'V'},0};
equ.init = {1,0};
```

```
equ.um = {{'D1_dless/F';'D2_dless/F'},{1;1}};
equ.v = {{'v';'v'},0};
equ.u = {{'u';'u'},0};
equ.usage = {1,0};
equ.z = {{'z1';'z2'},{1;1}};
equ.ind = [1,2];
appl.equ = equ;
fem.appl{1} = appl;

% NEW: Solve electrostatics using Poisson equation
% Application mode 2
clear appl
appl.mode.class = 'Electrostatics';
appl.mode.type = 'axi';
appl.border = 'on';
appl.assignsuffix = '_es';
clear prop
clear weakconstr
weakconstr.value = 'off';
weakconstr.dim = {'lm3'};
prop.weakconstr = weakconstr;
appl.prop = prop;
clear bnd
bnd.V0 = {0,0,'V0_dless',0,0,0,0,'Vg_dless',0};
bnd.name = {'axis','down','up','particle','membrane','zero',
   'insulator','gate', ...
      'interface'};
bnd.rhos = {0,0,0,'cp_dless','cw_dless',0,0,0,'cw_dless'};
bnd.type = {'ax','V','V','r','r','nD0','nD0','V','r'};
bnd.ind = [1,2,1,3,4,5,9,6,5,6,5,5,8,7,7,4,4,5,5];
appl.bnd = bnd;
clear equ
equ.epsilonr = {'1/eps0','1/eps0*eps_d/eps_r'};
equ.rho = {'0.5*(ai)^2*(z1*c1+z2*c2)',0};
equ.ind = [1,2];
appl.equ = equ;
fem.appl{2} = appl;

% NEW: Solve flow field using modified Navier-Stokes
% NEW: equations
% Application mode 3
clear appl
appl.mode.class = 'FlNavierStokes';
appl.mode.type = 'axi';
appl.gporder = {4,2};
appl.cporder = {2,1};
appl.assignsuffix = '_ns';
clear prop
prop.analysis='static';
clear weakconstr
```

```
weakconstr.value = 'off';
weakconstr.dim = {'lm4','lm5','lm6'};
prop.weakconstr = weakconstr;
appl.prop = prop;
clear bnd
bnd.v0 = {0,0,'vp',0,0,0};
bnd.symtype = {'ax','sym','sym','sym','sym','sym'};
bnd.type = {'sym','outlet','inlet','walltype','int','wallt
  ype'};
bnd.velType = {'U0in','U0in','u0','U0in','U0in','U0in'};
bnd.walltype = {'noslip','noslip','noslip','noslip',
  'noslip','slip'};
bnd.name = {'axis','outlet','particle','membrane',
  'dielectric','far'};
bnd.ind = [1,2,1,2,3,4,4,5,4,5,4,4,5,6,6,3,3,4,4];
appl.bnd = bnd;
clear equ
equ.gporder = {{1;1;2}};
equ.F_z = {'-0.5*ai^2*(z1*c1+z2*c2)*Vz',0};
equ.rho = {0,1};
equ.cporder = {{1;1;2}};
equ.F_r = {'-0.5*ai^2*(z1*c1+z2*c2)*Vr',0};
equ.cdon = {0,1};
equ.sdon = {0,1};
equ.usage = {1,0};
equ.ind = [1,2];
appl.equ = equ;
fem.appl{3} = appl;
fem.sdim = {'r','z'};
fem.frame = {'ref'};
fem.border = 1;

% NEW: Boundary integration to calculate the hydrodynamic
% NEW: force
% NEW: and electrical force on the particle
% Coupling variable elements
clear elemcpl
% Integration coupling variables
clear elem
elem.elem = 'elcplscalar';
elem.g = {'1'};
src = cell(1,1);
clear bnd
bnd.expr = {{{},'-T_z_ns*2*pi*r'},{{}, ...
  '2*pi*r*(-Vz*Vr*nr-0.5*Vz^2*nz+0.5*Vr^2*nz)'}};
bnd.ipoints = {{{},'4'},{{},'4'}};
bnd.frame = {{'ref','ref'},{'ref','ref'}};
bnd.ind = {{'1','2','3','4','6','7','8','9','10','11','12',
            '13','14', ...
            '15','18','19'},{'5','16','17'}};
```

```
src{1} = {{},bnd,{}};
elem.src = src;
geomdim = cell(1,1);
geomdim{1} = {};
elem.geomdim = geomdim;
elem.var = {'Fh_z','Fe_z'};
elem.global = {'1','2'};
elem.maxvars = {};
elemcpl{1} = elem;
fem.elemcpl = elemcpl;


% NEW: Determine the particle velocity by balancing the
% NEW: hydrodynamic force and electrical force on the
% NEW: particle
% NEW: Multiplier 1e20 is to make sure (Fh_z+Fe_z) is
% NEW: extremely close to zero
% ODE Settings
clear ode
ode.dim={'vp'};
ode.f={'(Fh_z+Fe_z)*1e20'};
ode.init={'0'};
ode.dinit={'0'};
fem.ode=ode;


% Multiphysics
fem=multiphysics(fem);


% Extend mesh
fem.xmesh=meshextend(fem);


% Solve problem
fem.sol=femstatic(fem, ...
'solcomp',{'v','u','V','c1','c2','p','vp'}, ...
'outcomp',{'v','u','V','c1','c2','p','vp'}, ...
'blocksize','auto');


% Save current fem structure for restart purposes
fem0=fem;


% NEW: Get velocities and displacments
data_comsol = postglobaleval(fem);


% NEW: Store the interested data
data(i,1)=zp0;
data(i,2)=data_comsol.y;


% NEW: force
% Integrate
I1=postint(fem,'T_z_ns*2*pi*r', ...
```

```
                'unit','', ...
                'recover','off', ...
                'dl',[5,16,17], ...
                'edim',1);

I2=postint(fem,'2*pi*r*(-Vz*Vr*nr-0.5*Vz^2*nz+0.5*Vr^2*nz)',
                ...
                'unit','', ...
                'recover','off', ...
                'dl',[5,16,17], ...
                'edim',1);
data(i,3)=I1;
data(i,4)=I2;


% NEW: Current
% NEW: Anode
% Integrate
I3=postint(fem,'2*pi*r*(ntflux_c1_chekf-ntflux_c2_
                chekf)*nz', ...
                'unit','mol/s', ...
                'recover','off', ...
                'dl',4, ...
                'edim',1);

% NEW: Cathode
I4=postint(fem,'2*pi*r*(ntflux_c1_chekf-ntflux_c2_
                chekf)*nz', ...
                'unit','mol/s', ...
                'recover','off', ...
                'dl',2, ...
                'edim',1);
data(i,5)=I3;
data(i,6)=I4;


% NEW: Current deviation
data(i,7)=(abs(I3)-abs(I4))/abs(I3)*100;

% NEW: Write the data into a file
dlmwrite(strcat('ka_E_10KV_cw_0_gate_20_perm_3p9_concentrati
    on_',num2str(conc(j)),'.dat'),data,',');

% NEW: Output the COMSOL Multiphysics GUI .mph file if
% NEW: necessary
flsave(strcat('ka_E_10KV_cw_0_gate_20_perm_3p9_concentration
    _',num2str(conc(j)),'_zp_',num2str(zp0),'.mph'),fem);

end
end
% NEW: End of this script file
```

A validation of the numerical results against existing approximate results can be obtained for the special case of a sphere translating along the axis of an uncharged cylindrical pore when the EDL of the particle is not affected by the solid boundary, and the zeta potential of the particle ζ is relatively small ($\zeta/(RT/F) \ll 1$) (Ennis and Anderson 1997). The implementation of the code validation in the COMSOL Multiphysics GUI is very similar to Table 9.2 except that the geometry is different, and the electric potential inside the dielectric nanopore wall is not considered. Figure 9.3 depicts particle velocity, normalized by $\varepsilon_0\varepsilon_f\zeta E/\mu$, as a function of the ratio of the particle's radius to the pore's radius a/b when $a = 1$ nm, bulk concentration $C_0 = 400$ mM ($\kappa a = 2.05$), $\zeta = 1$ mV, and $E = 10$ kV/m. The numerical results (circles) are in good agreement with the approximation solution (solid line) when the pore size is much larger than the particle size. However, the approximation solution underestimates the particle velocity as a/b increases since the Poisson–Boltzmann model used to derive the approximation solution becomes inappropriate.

9.4 RESULTS AND DISCUSSION

The obtained results are presented in a dimensional form to indicate the practical DNA translational velocity through a nanopore. To highlight the field effect on the DNA translocation, the nanopore is assumed to be originally uncharged. Four

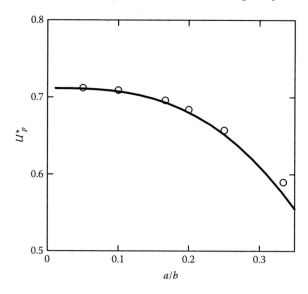

FIGURE 9.3 Electrophoretic velocity of a sphere of radius a translating along the axis of an uncharged cylindrical tube of radius b as a function of the ratio a/b. The conditions were $a = 1$ nm, $\kappa a = 2.05$, the zeta potential of the particle $\zeta = 1$ mV, and the axial electric field imposed $E = 10$ KV/m. The solid line and circles represent, respectively, the approximation solution and our numerical results. (From Ai, Y., J. Liu, B. Zhang, and S. Qian. 2010. Field effect regulation of DNA translocation through a nanopore. *Analytical Chemistry* 82, 8217–8225 with permission from the American Chemical Society.)

important factors are investigated in this section: the applied electric field, the applied gate potential, the ratio of the particle radius to the Debye length κa, and the relative permittivity of the dielectric nanopore.

9.4.1 EFFECT OF GATE POTENTIAL

Figure 9.4 depicts the z component particle velocity as a function of the particle's axial location along the axis of the nanopore under two different electric fields, $E = 10$ kV/m (a) and $E = 1,000$ kV/m (b) when $C_0 = 100$ mM ($\kappa a = 1.03$) and $\varepsilon_d = 3.9$ (the corresponding dielectric nanopore material is silicon dioxide). Under $E = 10$ kV/m, when the gate electrode is ineffective, referring to a floating gate electrode, the particle's translational velocity is symmetric with respect to $z_p = 0$ (circles in Figure 9.4a). As the cross section of the nanopore is much smaller than that of the reservoir, the electric field is enhanced inside the nanopore, leading to a particle velocity inside the nanopore much higher than that inside the reservoirs. However, DNA translocation still takes a long time due to the relatively low electric field imposed.

To further enhance the DNA translocation, a positive gate potential is applied on the gate electrode to generate an EOF in the same direction of the DNA translocation. Figure 9.4a depicts that a positive gate potential ($\psi_g = 0.52$ V, solid line; $\psi_g = 1.03$ V, dashed line) can attract the DNA into the nanopore quickly compared to the case with a floating gate electrode. Obviously, a higher gate potential leads to a more significant EOF. As a result, faster attraction of the DNA nanoparticle into the nanopore is observed when a higher gate potential is applied. The particle velocity attains a maximum at nearly $z_p = -10$ nm; however, this decreases as the DNA translocates further. It is predicted that the DNA is trapped right after passing the center of the nanopore. The negative particle velocity means that the DNA is not able to pass through the nanopore. When the applied electric field is increased 100 times to $E = 1,000$ kV/m, the particle velocity is also increased about 100 times when the gate electrode is floating, as shown in Figure 9.4b. In addition, the particle velocity profile keeps symmetric with respect to $z_p = 0$. When a positive gate potential is applied, the particle velocity is enhanced along the entire nanopore, with the maximum velocity occurring at $z_p < 0$ (solid line), compared to the case with a floating gate electrode. However, the DNA trapping at $z_p > 0$ is not predicted under a relatively high electric field. In contrast, a negative gate potential generates an EOF opposite to the DNA translocation. Accordingly, the DNA translocation is retarded (dashed line) compared to the case with a floating gate electrode. Also, the particle velocity profile becomes asymmetric, with the maximum velocity occurring at $z_p > 0$.

To further understand the field effect on the DNA translocation through the nanopore, the distributions of ionic concentration $c_1 - c_2$, the z component fluid velocity, and the electric potential in the fluid medium confined in the nanopore are shown in Figure 9.5 when $E = 10$ kV/m, $C_0 = 100$ mM ($\kappa a = 1.03$), $\psi_g = 0.52$ V, $\varepsilon_d = 3.9$, $z_p = -10$ nm (a) and 10 nm (b). When the particle is located at $z_p = -10$ nm, more cations are predominantly occupied within the EDL of the negatively charged DNA, as shown in Figure 9.5a (I). Oppositely, the EDL of the nanopore right next to the gate electrode is mainly occupied by anions due to the positive gate potential. As a result,

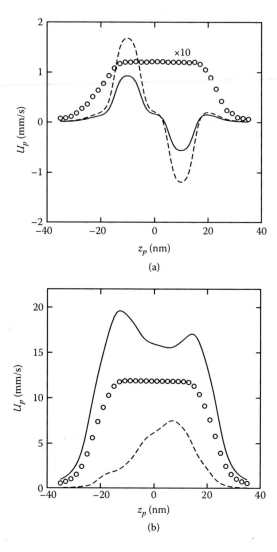

FIGURE 9.4 Variation of the particle velocity along the axis of the nanopore under $E = 10$ KV/m (a) and $E = 1,000$ KV/m (b). The circles and solid lines represent, respectively, ψ_g = floating and 0.52 V. The dashed lines represent, respectively, ψ_g = 1.03 V (a) and -0.52 V (b). $C_0 = 100$ mM ($\kappa a = 1.03$), and $\varepsilon_d = 3.9$. A scale of 10 was applied to the circles in (a) for a clear visualization. (From Ai, Y., J. Liu, B. Zhang, and S. Qian. 2010. Field effect regulation of DNA translocation through a nanopore. *Analytical Chemistry* 82, 8217–8225 with permission from the American Chemical Society.)

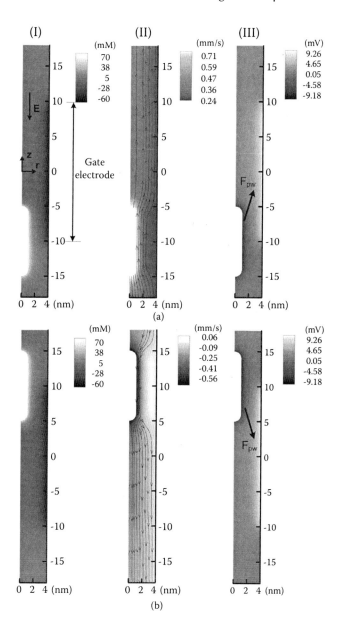

FIGURE 9.5 Distributions of $c_1 - c_2$ (I), the z component fluid velocity (II), and the electric potential (III) in the fluid confined in the nanopore when the particle's location $z_p = -10$ nm (a) and 10 nm (b). Lines with arrows in (II) represent the flow streamlines. $E = 10$ KV/m, $C_0 = 100$ mM ($\kappa a = 1.03$), $\psi_g = 0.52$ V, and $\varepsilon_d = 3.9$. (From Ai, Y., J. Liu, B. Zhang, and S. Qian. 2010. Field effect regulation of DNA translocation through a nanopore. *Analytical Chemistry* 82, 8217–8225 with permission from the American Chemical Society.)

the net EOF generated inside the nanopore is in the same direction as the DNA translocation, which in turn enhances DNA translocation, as shown in Figure 9.5a (II).

When the externally applied electric field is relatively low, the overall electric field inside the fluid medium confined in the nanopore is dominated by the electric field generated by the surface charges of the DNA nanoparticle and the nanopore next to the gate electrode. As the DNA nanoparticle and the nanopore next to the gate electrode carry charges of opposite signs, an attractive particle-nanopore electrostatic interaction force F_{pw} is induced, which pulls the DNA into the nanopore when $z_p < 0$, as shown in Figure 9.5a (III). The particle-nanopore electrostatic interaction force is maximized at $z_p = -10$ nm, leading to a maximum particle velocity at $z_p = -10$ nm, shown in Figure 9.4a. When the particle is located at $z_p = 10$ nm, Figure 9.5b (I) shows that the ionic distribution around the DNA and the nanopore is identical to the case of $z_p = -10$ nm. However, the attractive particle-nanopore electrostatic interaction retards the DNA translocation through the nanopore when $z_p > 0$, as shown in Figure 9.5b (III). In addition, the attractive particle-nanopore electrostatic interaction force dominates the electrical driving force and the hydrodynamic force arising from the EOF, which eventually traps the DNA inside the nanopore at $z_p = 3$ nm. When the particle is located at $z_p = 10$ nm, the attractive particle-nanopore electrostatic interaction peaks again, which reverses the DNA translocation. Therefore, the fluid velocity becomes negative due to the reversed particle motion, as shown in Figure 9.5b (II). It has also been experimentally confirmed that both the EOF and the particle-nanopore electrostatic interaction can significantly affect the nanoparticle translocation confined in nanopores (Oh et al. 2009). However, the field effect regulation of the DNA translocation mainly depends on the particle-nanopore electrostatic interaction for a relatively low electric field imposed.

Figure 9.6 illustrates the distributions of ionic concentration $c_1 - c_2$, the z component fluid velocity, and the electric potential in the fluid medium confined in the nanopore when $E = 1,000$ kV/m, $C_0 = 100$ mM ($\kappa a = 1.03$), $z_p = 10$ nm, $\varepsilon_d = 3.9$, $\psi_g = -0.52$ V (a) and 0.52 V (b). When the applied gate potential is $\psi_g = -0.52$ V, more cations are predominantly occupied within the EDLs of the DNA and the nanopore next to the gate electrode. In addition, $c_1 - c_2 > 0$ occurs in the gap region between the DNA and the nanopore, as shown in Figure 9.6a (I), implying that the two EDLs are overlapped. Because of the negative gate potential, the induced net EOF inside the nanopore is opposite to the DNA translocation, resulting in a retardation effect on the DNA translocation, as shown in Figure 9.6a (II). Because the DNA and the nanopore are both negatively charged, the particle-nanopore electrostatic interaction induces a repulsive force. Therefore, the particle-nanopore electrostatic interaction retards the DNA translocation when $z_p < 0$ and enhances the DNA translocation when $z_p > 0$. This explains the asymmetric velocity profile with the maximum velocity occurring in the region of $z_p > 0$ in Figure 9.4b (dashed line). As the applied electric field dominates the electric field arising from the surface charges of the DNA and the nanopore, the electrical driving force and the hydrodynamic force from the EOF overpower the particle-nanopore electrostatic interaction force. As a result, DNA trapping is not predicted when the applied electric field is relatively high.

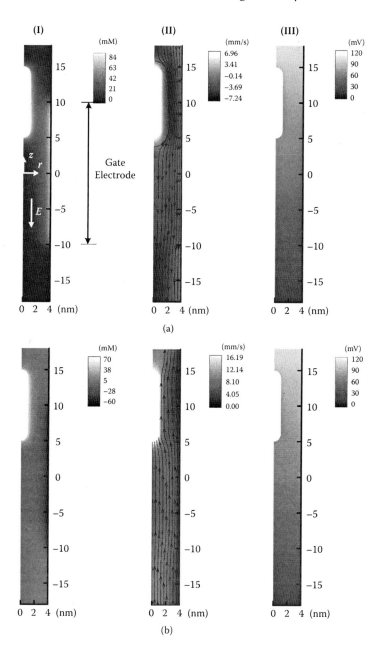

FIGURE 9.6   Distributions of $c_1 - c_2$ (I), the z component fluid velocity (II), and the electric potential (III) in the fluid confined in the nanopore with $\psi_g = -0.52$ V (a) and 0.52 V (b). Lines with arrows in (II) represent the flow streamlines. $E = 10$ KV/m, $C_0 = 100$ mM ($\kappa a = 1.03$), $z_p = 10$ nm, and $\varepsilon_d = 3.9$. (From Ai, Y., J. Liu, B. Zhang, and S. Qian. 2010. Field effect regulation of DNA translocation through a nanopore. *Analytical Chemistry* 82, 8217–8225 with permission from the American Chemical Society.)

When the applied gate potential is ψ_g = 0.52 V, cations and anions are, respectively, dominant within the EDLs of the DNA and the nanopore next to the gate electrode, as illustrated in Figure 9.6b (I). Owing to the positive gate potential, the generated EOF is in the same direction of the DNA translocation and thus facilitates the DNA translocation through the nanopore, as shown in Figure 9.6b (II). Here, the particle-nanopore electrostatic force acts as an attractive force. Therefore, it enhances DNA translocation when $z_p < 0$ and retards DNA translocation when $z_p > 0$, which also explains the asymmetric velocity profile with a maximum velocity occurring in the region of $z_p < 0$ in Figure 9.4b (solid line).

9.4.2 Effect of Ratio of Particle Radius to Debye Length

Figure 9.7 depicts the particle velocity as a function of the particle's axial location along the axis of the nanopore under E = 10 kV/m (a) and E = 1,000 kV/m (b) when ε_d = 3.9. When the gate electrode is floating, the particle velocity inside the nanopore remains a constant. In addition, the particle velocity for C_0 = 100 mM (κa = 1.03, circles) is larger than that for C_0 = 1,000 mM (κa = 3.26, squares). Note that the zeta potential of a particle with a fixed surface charge density decreases as κa increases (Ohshima 1998). As a result, the particle velocity decreases as κa increases, which has also been predicted in the previous study (Liu, Qian, and Bau 2007). When the applied electric field is relatively low (i.e., E = 10 kV/m), the particle-nanopore electrostatic interaction force dominates the electrical driving force and the hydrodynamic force arising from the induced EOF, as discussed previously. Therefore, a positive gate potential ψ_g = 0.52 V tends to enhance the

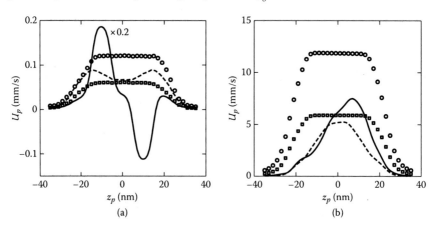

FIGURE 9.7 Variation of the particle velocity along the axis of the nanopore under E = 10 KV/m (a) and E = 1,000 KV/m (b). Symbols and lines are, respectively, ψ_g = floating, 0.52 V (a), and −0.52 V (b). Solid lines and circles represent C_0 = 100 mM (κa = 1.03); dashed lines and squares represent C_0 = 1,000 mM (κa = 3.26). ε_d = 3.9. A scale of 0.2 was applied to the solid line in (a) for clear visualization. (From Ai, Y., J. Liu, B. Zhang, and S. Qian. 2010. Field effect regulation of DNA translocation through a nanopore. *Analytical Chemistry* 82, 8217–8225 with permission from the American Chemical Society.)

DNA translocation when $z_p < 0$; however, it traps the DNA at $z_p = 3$ nm when $\kappa a = 1.03$ (solid line). When the bulk concentration is increased to $C_0 = 1,000$ mM ($\kappa a = 3.26$), the particle velocity becomes nearly symmetric with respect to $z_p = 0$.

As mentioned, the asymmetric velocity profile is attributed to the particle-nanopore electrostatic interaction. Accordingly, the symmetric velocity profile implies that the particle-nanopore electrostatic interaction is negligible when the bulk concentration is relatively high. The electric field generated by the surface charges decays quickly within the EDL and reaches zero in the bulk. A higher bulk concentration leads to a thinner EDL, which in turn decreases the degree of EDL overlapping. Once the EDLs of the DNA and the nanopore are not overlapped, they are not able to feel the existence of each other, resulting in a negligible particle–nanopore electrostatic interaction. Therefore, the enhancement of DNA translocation is mainly due to the induced EOF inside the nanopore. When the applied electric field is relatively high, $E = 1,000$ kV/m, a negative gate potential $\psi_g = -0.52$ V slows the DNA translocation as the direction of the EOF is opposite to that of the DNA translocation. The velocity profile, asymmetric when $\kappa a = 1.03$ (solid line), tends to be symmetric as the bulk concentration increases to obtain $\kappa a = 3.26$ (dashed line). A previous experimental study also confirmed that the particle–nanopore electrostatic interaction highly depends on the degree of EDL overlapping (Wanunu et al. 2008).

9.4.3 Effect of Nanopore's Permittivity

Figure 9.8 depicts the effect of the dielectric nanopore's relative permittivity on the z component particle velocity along the axis of the nanopore under

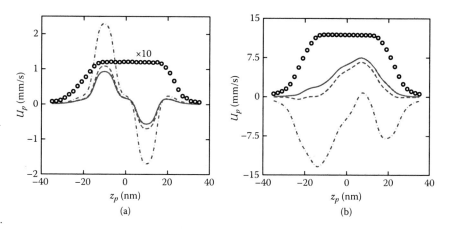

FIGURE 9.8 Variation of the particle velocity along the axis of the nanopore under $E = 10$ KV/m (a) and $E = 1,000$ KV/m (b). Circles and lines represent, respectively, a floating and gate potential $\psi_g = 0.52$ V (a) and -0.52 V (b). Solid, dashed, and dash-dotted lines represent, respectively, $\varepsilon_d = 3.9$, 4.7, and 11.7. $C_0 = 100$ mM ($\kappa a = 1.03$). A scale of 10 was applied to the circles in (a) for clear visualization. (From Ai, Y., J. Liu, B. Zhang, and S. Qian. 2010. Field effect regulation of DNA translocation through a nanopore. *Analytical Chemistry* 82, 8217–8225 with permission from the American Chemical Society.)

E = 10 kV/m (a) and E = 1,000 kV/m (b) when C_0 = 100 mM (κa = 1.03). Three different materials: silicon dioxide (ε_d = 3.9), silicon (ε_d = 4.7), and Pyrex glass (ε_d = 11.7), are considered to fabricate the dielectric nanopore. A previous study found that a nanopore with a higher relative permittivity has a stronger capacitive coupling, which accordingly gains a higher surface potential on the nanopore next to the gate electrode (Karnik et al. 2005). Under a relatively low electric field, E = 10 kV/m, a positive gate potential ψ_g = 0.52 V is imposed to attract DNA into the nanopore. It is obvious from Figure 9.8a that a higher relative permittivity of the nanopore induces a larger particle-nanopore electrostatic interaction and a stronger EOF and accordingly a larger particle velocity at z_p < 0. However, the DNA confined in nanopores with different permittivities is trapped at the same location, z_p = 3 nm. As stated previously, the change in the ionic current arising from DNA translocation is the basic principle of nanopore-based DNA sequencing.

The ionic current deviation, defined as $\chi = (I - I_0)/I_0 \times 100\%$, is used to quantify the change in the ionic current. In this equation, I_0 is the base current when the DNA is far away from the nanopore. Figure 9.9a shows the current deviation as a function of the particle's location along the axis of the nanopore when ε_d = 11.7. The current blockade due to the presence of DNA inside the nanopore is predicted, which is in qualitative agreement with the existing experimental results (Meller, Nivon, and Branton 2001; Li et al. 2003; Storm, Storm, et al. 2005; Storm, Chen, et al. 2005). Once the DNA is trapped inside the nanopore, the gate electrode is changed to be floating to allow the DNA to pass through the nanopore. Although the current deviation when

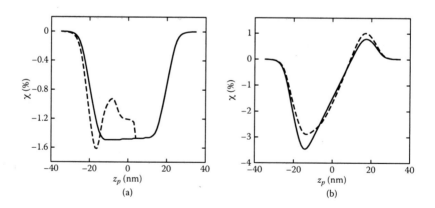

(a)   (b)

FIGURE 9.9 Current deviation χ as a function of the particle's location under E = 10 KV/m (a) and E = 1,000 KV/m (b). Solid lines represent the gate electrode is floating. Dashed line in (a) represents ψ_g = 0.52 V when z_p < 3 nm and ψ_g = floating when z_p > 3 nm; the dashed line in (b) represents ψ_g = −0.52 V. C_0 = 100 mM (κa = 1.03), and ε_d = 11.7 (a) and 4.7 (b). (From Ai, Y., J. Liu, B. Zhang, and S. Qian. 2010. Field effect regulation of DNA translocation through a nanopore. *Analytical Chemistry* 82, 8217–8225 with permission from the American Chemical Society.)

$\psi_g = 0.52$ V (dashed line) fluctuates compared to the case with a floating gate electrode (solid line), its maximum current deviation is slightly larger than that with a floating gate electrode.

Under a relatively high electric field, $E = 1,000$ kV/m, a negative gate potential $\psi_g = -0.52$ V is applied to slow the DNA translocation. A higher relative permittivity of the nanopore leads to a larger particle-nanopore electrostatic interaction and a stronger EOF. When the relative permittivity of the nanopore is high enough, for example, $\varepsilon_d = 11.7$, the DNA cannot even enter the nanopore, as shown in Figure 9.8b (dash-dotted line). To allow DNA to pass through the nanopore, one needs to reduce the gate potential. Figure 9.9b shows the current deviation as a function of the particle's location along the axis of the nanopore when $\varepsilon_d = 4.7$. Current blockade is observed in the region -30 nm $< z_p < 10$ nm; however, current enhancement is predicted in the region 10 nm $< z_p < 30$ nm.

Current enhancement has also been experimentally observed when the EDLs of the DNA and the nanopore are overlapped and the applied electric field is relatively high (Heng et al. 2004). It is attributed to the enhanced diffusive ionic current due to the EDL overlapping under a relatively high electric field (Liu, Qian, and Bau 2007). A nanopore with the same charge polarity as the particle tends to increase the current enhancement. Hence, the negative gate potential slightly decreases the current blockade and meanwhile increases the current enhancement, compared to the case with a floating electrode (solid line).

9.5   CONCLUDING REMARKS

In this chapter, a nanofluidic FET was proposed to regulate DNA translocation through the nanopore. A continuum-based mathematical model, composed of the coupled PNP equations and Navier–Stokes equations, was developed to predict DNA translocation and the ionic current through the nanopore. The field effect regulation of DNA translocation was attributed to two main factors: the EOF and the particle-nanopore electrostatic interaction. The former effect influences DNA translocation in a consistent direction, referring to a global effect. In contrast, the latter effect on DNA translocation highly depends on the DNA location, referring to a local effect. When the externally applied electric field is relatively low and the EDLs of the DNA and the nanopore are overlapped, the particle-nanopore electrostatic interaction force could dominate the electrical driving force and the hydrodynamic force arising from the EOF. As a result, DNA trapping inside the nanopore could be expected. When the applied electric field is relatively high and the EDLs are overlapped, the particle velocity becomes asymmetric due to the inversion of the particle-nanopore electrostatic interaction. However, DNA trapping inside the nanopore cannot be observed as the electrical driving force overpowers the particle-nanopore electrostatic interaction. When the EDLs are not overlapped, the particle-nanopore electrostatic interaction is negligible, indicating a symmetric velocity profile. A higher relative permittivity of the dielectric nanopore could enhance the field effect regulation of the DNA translocation.

REFERENCES

Ai, Y., A. Beskok, D. T. Gauthier, S. W. Joo, and S. Qian. 2009. DC electrokinetic transport of cylindrical cells in straight microchannels. *Biomicrofluidics* 3 (4):044110.

Ai, Y., S. W. Joo, Y. Jiang, X. Xuan, and S. Qian. 2009. Transient electrophoretic motion of a charged particle through a converging-diverging microchannel: Effect of direct current-dielectrophoretic force. *Electrophoresis* 30 (14):2499–2506.

Ai, Y., J. Liu, B. Zhang, and S. Qian. 2010. Field effect regulation of DNA translocation through a nanopore. *Analytical Chemistry* 82, 8217–8225.

Ai, Y., S. Park, J. Zhu, X. Xuan, A. Beskok, and S. Qian. 2010. DC electrokinetic particle transport in an L-shaped microchannel. *Langmuir* 26 (4):2937–2944.

Ai, Y., and S. Qian. 2010. DC dielectrophoretic particle-particle interactions and their relative motions. *Journal of Colloid and Interface Science* 346 (2):448–454.

Ai, Y., S. Qian, S. Liu, and S. W. Joo. 2010. Dielectrophoretic choking phenomenon in a converging-diverging microchannel. *Biomicrofluidics* 4 (1):013201.

Ai, Y., M. Zhang, S. W. Joo, M. A. Cheney, and S. Qian. 2010. Effects of electroosmotic flow on ionic current rectification in conical nanopores. *Journal of Physical Chemistry C* 114 (9):3883–3890.

Bayley, H. 2006. Sequencing single molecules of DNA. *Current Opinion in Chemical Biology* 10 (6):628–637.

Branton, D., D. W. Deamer, A. Marziali, H. Bayley, S. A. Benner, T. Butler, M. Di Ventra, S. Garaj, A. Hibbs, X. H. Huang, S. B. Jovanovich, P. S. Krstic, S. Lindsay, X. S. S. Ling, C. H. Mastrangelo, A. Meller, J. S. Oliver, Y. V. Pershin, J. M. Ramsey, R. Riehn, G. V. Soni, V. Tabard-Cossa, M. Wanunu, M. Wiggin, and J. A. Schloss. 2008. The potential and challenges of nanopore sequencing. *Nature Biotechnology* 26 (10):1146–1153.

Chang, H., F. Kosari, G. Andreadakis, M. A. Alam, G. Vasmatzis, and R. Bashir. 2004. DNA-mediated fluctuations in ionic current through silicon oxide nanopore channels. *Nano Letters* 4 (8):1551–1556.

Clarke, J., H. C. Wu, L. Jayasinghe, A. Patel, S. Reid, and H. Bayley. 2009. Continuous base identification for single-molecule nanopore DNA sequencing. *Nature Nanotechnology* 4 (4):265–270.

Corry, B., S. Kuyucak, and S. H. Chung. 2000. Tests of continuum theories as models of ion channels. II. Poisson-Nernst-Planck theory versus Brownian dynamics. *Biophysical Journal* 78 (5):2364–2381.

Daiguji, H. 2010. Ion transport in nanofluidic channels. *Chemical Society Reviews* 39 (3):901–911.

Dekker, C. 2007. Solid-state nanopores. *Nature Nanotechnology* 2 (4):209–215.

de Zoysa, R. S. S., D. A. Jayawardhana, Q. T. Zhao, D. Q. Wang, D. W. Armstrong, and X. Y. Guan. 2009. Slowing DNA translocation through nanopores using a solution containing organic salts. *Journal of Physical Chemistry B* 113 (40):13332–13336.

Ennis, J., and J. L. Anderson. 1997. Boundary effects on electrophoretic motion of spherical particles for thick double layers and low zeta potential. *Journal of Colloid and Interface Science* 185 (2):497–514.

Fologea, D., J. Uplinger, B. Thomas, D. S. McNabb, and J. L. Li. 2005. Slowing DNA translocation in a solid-state nanopore. *Nano Letters* 5 (9):1734–1737.

Ghosal, S. 2007. Effect of salt concentration on the electrophoretic speed of a polyelectrolyte through a nanopore. *Physical Review Letters* 98 (23):238104.

Griffiths, J. 2008. The realm of the nanopore. *Analytical Chemistry* 80 (1):23–27.

Healy, K., B. Schiedt, and A. P. Morrison. 2007. Solid-state nanopore technologies for nanopore-based DNA analysis. *Nanomedicine* 2 (6):875–897.

Heng, J. B., C. Ho, T. Kim, R. Timp, A. Aksimentiev, Y. V. Grinkova, S. Sligar, K. Schulten, and G. Timp. 2004. Sizing DNA using a nanometer-diameter pore. *Biophysical Journal* 87 (4):2905–2911.

Howorka, S., and Z. Siwy. 2009. Nanopore analytics: Sensing of single molecules. *Chemical Society Reviews* 38 (8):2360–2384.

Hsu, J. P., Z. S. Chen, and S. Tseng. 2009. Effect of electroosmotic flow on the electrophoresis of a membrane-coated sphere along the axis of a cylindrical pore. *Journal of Physical Chemistry B* 113 (21):7701–7708.

Hsu, J. P., W. L. Hsu, and K. L. Liu. 2010. Diffusiophoresis of a charge-regulated sphere along the axis of an uncharged cylindrical pore. *Langmuir* 26 (11):8648–8658.

Joshi, P., A. Smolyanitsky, L. Petrossian, M. Goryll, M. Saraniti, and T. J. Thornton. 2010. Field effect modulation of ionic conductance of cylindrical silicon-on-insulator nanopore array. *Journal of Applied Physics* 107 (5):054701.

Kalman, E. B., O. Sudre, I. Vlassiouk, and Z. S. Siwy. 2009. Control of ionic transport through gated single conical nanopores. *Analytical and Bioanalytical Chemistry* 394 (2):413–419.

Karnik, R., R. Fan, M. Yue, D. Y. Li, P. D. Yang, and A. Majumdar. 2005. Electrostatic control of ions and molecules in nanofluidic transistors. *Nano Letters* 5 (5):943–948.

Kawano, R., A. E. P. Schibel, C. Cauley, and H. S. White. 2009. Controlling the translocation of single-stranded DNA through alpha-hemolysin ion channels using viscosity. *Langmuir* 25 (2):1233–1237.

Kim, Y. R., J. Min, I. H. Lee, S. Kim, A. G. Kim, K. Kim, K. Namkoong, and C. Ko. 2007. Nanopore sensor for fast label-free detection of short double-stranded DNAs. *Biosensors and Bioelectronics* 22 (12):2926–2931.

Li, J. L., M. Gershow, D. Stein, E. Brandin, and J. A. Golovchenko. 2003. DNA molecules and configurations in a solid-state nanopore microscope. *Nature Materials* 2 (9):611–615.

Liu, H., S. Qian, and H. H. Bau. 2007. The effect of translocating cylindrical particles on the ionic current through a nanopore. *Biophysical Journal* 92 (4):1164–1177.

Luria, S. E. 1989. Human genome program. *Science* 246 (4932):873–874.

Mandelkern, M., J. G. Elias, D. Eden, and D. M. Crothers. 1981. The dimensions of DNA in solution. *Journal of Molecular Biology* 152 (1):153–161.

Meller, A., L. Nivon, and D. Branton. 2001. Voltage-driven DNA translocations through a nanopore. *Physical Review Letters* 86 (15):3435–3438.

Mukhopadhyay, R. 2009. DNA sequencers: The next generation. *Analytical Chemistry* 81 (5):1736–1740.

Nam, S. W., M. J. Rooks, K. B. Kim, and S. M. Rossnagel. 2009. Ionic field effect transistors with sub-10 nm multiple nanopores. *Nano Letters* 9 (5):2044–2048.

Oh, Y. J., T. C. Gamble, D. Leonhardt, C. H. Chung, S. R. J. Brueck, C. F. Ivory, G. P. Lopez, D. N. Petsev, and S. M. Han. 2008. Monitoring FET flow control and wall adsorption of charged fluorescent dye molecules in nanochannels integrated into a multiple internal reflection infrared waveguide. *Lab on a Chip* 8 (2):251–258.

Oh, Y. J., A. L. Garcia, D. N. Petsev, G. P. Lopez, S. R. J. Brueck, C. F. Ivory, and S. M. Han. 2009. Effect of wall-molecule interactions on electrokinetic transport of charged molecules in nanofluidic channels during FET flow control. *Lab on a Chip* 9 (11):1601–1608.

Ohshima, H. 1998. Surface charge density surface potential relationship for a cylindrical particle in an electrolyte solution. *Journal of Colloid and Interface Science* 200 (2):291–297.

Pennathur, S., and J. G. Santiago. 2005. Electrokinetic transport in nanochannels. 2. Experiments. *Analytical Chemistry* 77 (21):6782–6789.

Qian, S., B. Das, and X. B. Luo. 2007. Diffusioosmotic flows in slit nanochannels. *Journal of Colloid and Interface Science* 315 (2):721–730.

Qian, S., and S. W. Joo. 2008. Analysis of self-electrophoretic motion of a spherical particle in a nanotube: Effect of nonuniform surface charge density. *Langmuir* 24 (9):4778–4784.

Qian, S., S. W. Joo, Y. Ai, M. A. Cheney, and W. S. Hou. 2009. Effect of linear surface-charge non-uniformities on the electrokinetic ionic-current rectification in conical nanopores. *Journal of Colloid and Interface Science* 329 (2):376–383.

Qian, S., S. W. Joo, W. Hou, and X. Zhao. 2008. Electrophoretic motion of a spherical particle with a symmetric nonuniform surface charge distribution in a nanotube. *Langmuir* 24 (10):5332–5340.

Qian, S., A. H. Wang, and J. K. Afonien. 2006. Electrophoretic motion of a spherical particle in a converging-diverging nanotube. *Journal of Colloid and Interface Science* 303 (2):579–592.

Rhee, M., and M. A. Burns. 2006. Nanopore sequencing technology: Research trends and applications. *Trends in Biotechnology* 24:580–586.

Saleh, O. A., and L. L. Sohn. 2003. An artificial nanopore for molecular sensing. *Nano Letters* 3 (1):37–38.

Schasfoort, R. B. M., S. Schlautmann, L. Hendrikse, and A. van den Berg. 1999. Field-effect flow control for microfabricated fluidic networks. *Science* 286 (5441):942–945.

Schoch, R. B., H. van Lintel, and P. Renaud. 2005. Effect of the surface charge on ion transport through nanoslits. *Physics of Fluids* 17 (10):100604.

Stein, D., M. Kruithof, and C. Dekker. 2004. Surface-charge-governed ion transport in nanofluidic channels. *Physical Review Letters* 93 (3):035901.

Storm, A. J., J. H. Chen, H. W. Zandbergen, and C. Dekker. 2005. Translocation of double-strand DNA through a silicon oxide nanopore. *Physical Review E* 71 (5):051903.

Storm, A. J., C. Storm, J. H. Chen, H. Zandbergen, J. F. Joanny, and C. Dekker. 2005. Fast DNA translocation through a solid-state nanopore. *Nano Letters* 5 (7):1193–1197.

Taniguchi, M., M. Tsutsui, K. Yokota, and T. Kawai. 2009. Fabrication of the gating nanopore device. *Applied Physics Letters* 95 (12):123701.

Trepagnier, E. H., A. Radenovic, D. Sivak, P. Geissler, and J. Liphardt. 2007. Controlling DNA capture and propagation through artificial nanopores. *Nano Letters* 7 (9):2824–2830.

Vajandar, S. K., D. Y. Xu, J. S. Sun, D. A. Markov, W. H. Hofmeister, and D. Y. Li. 2009. Field-effect control of electroosmotic pumping using porous silicon-silicon nitride membranes. *Journal of Microelectromechanical Systems* 18 (6):1173–1183.

Wanunu, M., J. Sutin, B. McNally, A. Chow, and A. Meller. 2008. DNA translocation governed by interactions with solid-state nanopores. *Biophysical Journal* 95 (10):4716–4725.

White, H. S., and A. Bund. 2008. Ion current rectification at nanopores in glass membranes. *Langmuir* 24 (5):2212–2218.

10 Electrokinetic Particle Translocation through a Nanopore Containing a Floating Electrode

In this chapter, electrokinetic particle translocation through a nanopore containing a floating electrode is studied using the mathematical model developed in Chapter 9 with modifications of the boundary conditions. The floating electrode coated on the inner surface of the nanopore gives rise to two effects: induced-charge electroosmosis (ICEO) and particle-floating electrode electrostatic interaction, which could remarkably influence the electrokinetic particle translocation process. ICEO is proportional to the square of the applied electric field, which could dramatically affect particle translocation under a relatively high electric field. Particle trapping due to the circulating ICEO near the floating electrode has been predicted, indicating a promising application for particle enrichment. The particle-floating electrode electrostatic interaction becomes important when the electrical double layers (EDLs) of the particle and the floating electrode overlap. However, the presence of the floating electrode has very limited effect on the relative deviation of the ionic current if the particle is not inside the nanopore.

10.1 INTRODUCTION

Conventional electroosmosis is linearly proportional to the applied electric field when the surface charge (zeta potential) of dielectric materials is uniform. It has been found that electroosmosis arising from the induced charge on an ideally polarizable surface shows a nonlinear relationship to the externally applied electric field. This nonlinear electrokinetic phenomenon was named induced-charge electroosmosis (ICEO) by Squires and Bazant (Bazant and Squires 2004; Squires and Bazant 2004) and has recently gained much attention in the micro-/nanofluidics community. The distribution and polarity of the induced surface charge highly depend on the surface geometry, which can easily generate circulating flows for fluid stirring and mixing (Zhao and Bau 2007; Wu and Li 2008a, 2008b; Jain, Yeung, and Nandakumar 2009). Particle enrichment and trapping have been experimentally demonstrated in a microfluidic channel with a floating electrode (Dhopeshwarkar et al. 2008; Yalcin et al. 2010, 2011). Most existing numerical simulations of ICEO focus on the fluid motion with little attention to particle motion (Wu and Li 2008a; Zhao and Yang 2009; Bazant and Squires 2010). In particular, the control of particle translocation through a nanopore is of great interest in nanopore-based sensing, discussed in Chapter 9.

In this chapter, we study the electrokinetic translocation of a DNA nanoparticle through a nanopore containing a floating electrode by numerically solving the Poisson–Nernst–Planck (PNP) equations and the modified Stokes equations. The present model is quite similar to that in Chapter 9. However, boundary conditions are modified to consider the induced charge on the ideally polarizable floating electrode. Except for ICEO arising from the floating electrode, the electrostatic interaction between the particle and the floating electrode may also come into play when their EDLs overlap, as explained in Chapter 9.

10.2   MATHEMATICAL MODEL

Let us consider a charged DNA nanoparticle translocating along the axis of a nanopore of length L_c and radius b filled with a binary KCl aqueous solution of density ρ, dynamic viscosity μ, and relative permittivity ε_f, as shown in Figure 10.1a. Either side of the nanopore connects a reservoir of length L_r and radius R_r, and the ionic concentration at both ends of the reservoirs maintains the bulk value C_0. An axisymmetric model is employed in the present study for simplicity due to the axisymmetric nature of the geometry and physical fields. The origin of a cylindrical coordinate system (r, z) is fixed at the center of the nanopore. The DNA nanoparticle is approximated as a cylinder capped with two hemispheres, as discussed in Chapter 9. The length and radius of the particle are, respectively, L_p and a. A floating electrode is coated on the inner surface of the nanopore, which is symmetric with respect to the origin of the cylindrical coordinate system. An electric field $\mathbf{E}$ is applied across the nanopore to drive the electrokinetic translocation of the particle with a surface charge density σ_p through the nanopore. The electrically conductive floating electrode is intrinsically uncharged and ideally polarizable. Therefore, the $z < 0$ portion of the floating electrode is induced to carry an opposite charge to the electrode in the reservoir on the $z < 0$ side and vice versa. Due to the electrostatic interaction between the EDL adjacent to the floating electrode and the applied electric field, ICEO is generated as shown in Figure 10.1b. Because the induced surface charges on the $z < 0$ portion and $z > 0$ portion of the floating electrode are opposite to each other, circulating flow is generated near the floating electrode, which may significantly affect particle motion.

The electric field, the ionic concentrations, and the flow field are governed by the PNP equations and the modified Stokes equations,

$$-\varepsilon_0\varepsilon_f\nabla^2\phi = F(c_1 z_1 + c_2 z_2), \tag{10.1}$$

$$\nabla \bullet \mathbf{N}_i = \nabla \bullet (\mathbf{u}c_i - D_i\nabla c_i - z_i\frac{D_i}{RT}Fc_i\nabla\phi) = 0, \; i = 1 \text{ and } 2, \tag{10.2}$$

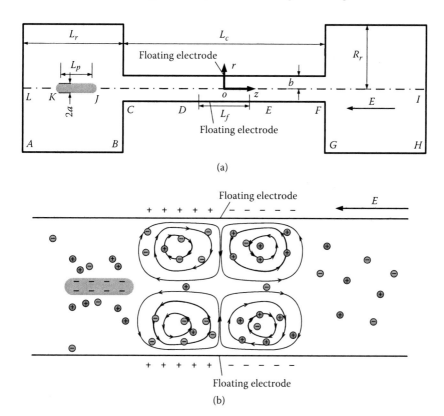

(a)

(b)

FIGURE 10.1   Schematic view of particle translocation through a cylindrical nanopore with a floating electrode (a). Charge density distribution on the floating electrode and the ICEO flow pattern in the floating electrode region (b). (From Zhang, M., Y. Ai, A. Sharma, S. W. Joo, D.-S. Kim, and S. Qian. 2011. Electrokinetic particle translocation through a nanopore containing a floating electrode. *Electrophoresis* 32:1864–1874 with permission from Wiley-VCH.)

$$-\nabla p + \mu \nabla^2 \mathbf{u} - F(z_1 c_1 + z_2 c_2)\nabla \phi = 0, \qquad (10.3)$$

$$\nabla \bullet \mathbf{u} = 0, \qquad (10.4)$$

where ε_0 is the absolute permittivity of the vacuum; ϕ is the electric potential within the fluid; F is the Faraday constant; c_1 and c_2 are, respectively, the molar concentrations of the cations (K^+) and anions (Cl^-) in the electrolyte solution; z_1 and z_2 are, respectively, the valences of cations ($z_1 = 1$ for K^+) and anions ($z_2 = -1$ for Cl^-); $\mathbf{N}_i$ is the ionic flux density of the ith ionic species; $\mathbf{u}$ is the fluid velocity; D_i is the diffusivity of the ith ionic species; R is the universal gas constant; T is the absolute temperature of the electrolyte solution; and p is the pressure.

Axial symmetric boundary conditions for all the physical fields are applied on the axis of the nanopore. The boundary conditions for the ionic concentrations are identical to those described in Chapter 9. Briefly, the ionic concentration at the two ends of the two reservoirs is the bulk concentration, $c_i(r, \pm(L_r + L_c/2)) = C_0$, $i = 1$ and 2. The normal ionic flux on the particle surface only includes the convective flux, $\mathbf{n} \bullet \mathbf{N}_i = \mathbf{n} \bullet (\mathbf{u}c_i)$, $i = 1$ and 2, where $\mathbf{n}$ is the unit normal vector directed from the particle surface into the fluid. The normal ionic fluxes on all the other boundaries are zero.

The boundary conditions for the electric potentials at the ends of the two reservoirs are, respectively, $\phi(r, -(L_r + L_c/2)) = 0$ and $\phi(r, (L_r + L_c/2)) = \phi_0$. The surface charge boundary condition on the particle surface is $-\varepsilon_0\varepsilon_f \mathbf{n} \bullet \nabla\phi = \sigma_p$. Because the floating electrode is a perfect conductor, its electric potential is a constant, $\phi = \phi_f$. In addition, the net induced charge on the floating electrode is zero (Bazant and Squires 2004; Squires and Bazant 2004, 2006),

$$\int_{-L_f/2}^{L_f/2} (\varepsilon_0\varepsilon_f \mathbf{n} \bullet \nabla\phi)\, dz = 0. \qquad (10.5)$$

All the other boundaries are assumed to be uncharged.

The boundary conditions for the flow fields are also identical to those described in Chapter 9. The fluid velocity on the particle surface only includes the particle translocation velocity U_p, which is determined based on the balance of the z component force acting on the particle using a quasi-static method (Qian, Wang, and Afonien 2006; Qian and Joo 2008; Qian et al. 2008; Hsu, Chen, and Tseng 2009; Hsu, Hsu, and Liu 2010),

$$F_E + F_H = 0. \qquad (10.6)$$

In Equation (10.6),

$$F_E = \int \varepsilon_0\varepsilon_f \left[\frac{\partial\phi}{\partial z}\frac{\partial\phi}{\partial r}n_r + \frac{1}{2}(\frac{\partial\phi}{\partial z})^2 n_z - \frac{1}{2}(\frac{\partial\phi}{\partial r})^2 n_z \right] d\Gamma \qquad (10.7)$$

is the axial electrical force obtained by integrating the Maxwell stress tensor over the particle surface, and

$$F_H = \int \left[-pn_z + 2\mu\frac{\partial u_z}{\partial z}n_z + \mu(\frac{\partial u_r}{\partial z} + \frac{\partial u_z}{\partial r})n_r \right] d\Gamma \qquad (10.8)$$

is the z component hydrodynamic force acting on the particle. Here, u_r and u_z are, respectively, the r and z components of the fluid velocity $\mathbf{u}$; n_r and n_z are, respectively, the r and z components of the unit vector $\mathbf{n}$; Γ denotes the surface of the DNA nanoparticle.

The ionic current flowing through the nanopore is

$$I = \int F(z_1\mathbf{N}_1 + z_2\mathbf{N}_2) \bullet \mathbf{n}\, dS, \qquad (10.9)$$

where S denotes the opening of either reservoir due to the current conservation. The bulk concentration C_0 as the ionic concentration scale, the particle radius a as the length scale, RT/F as the potential scale, $U_0 = \varepsilon_0 \varepsilon_f R^2 T^2 / (\mu a F^2)$ as the velocity scale, and $\mu U_0 / a$ as the pressure scale are selected to normalize all the governing equations just given:

$$-\nabla^{*2} \phi^* = \frac{(\kappa a)^2}{2} (c_1^* z_1 + c_2^* z_2), \tag{10.10}$$

$$\nabla^* \bullet \mathbf{N}_i^* = \nabla^* \bullet (\mathbf{u}^* c_i^* - D_i^* \nabla^* c_i^* - z_i D_i^* c_i^* \nabla^* \phi^*) = 0, \; i = 1 \text{ and } 2, \tag{10.11}$$

$$-\nabla^* p^* + \nabla^{*2} \mathbf{u}^* - \frac{(\kappa a)^2}{2} (z_1 c_1^* + z_2 c_2^*) \nabla^* \phi^* = 0, \tag{10.12}$$

$$\nabla^* \bullet \mathbf{u}^* = 0. \tag{10.13}$$

In these equations, $\kappa^{-1} = \sqrt{\varepsilon_0 \varepsilon_f RT / (2F^2 C_0)}$ is the Debye length. The corresponding boundary conditions are normalized as

$$c_i^* (r^*, \; \pm(L_r^* + L_c^* / 2)) = 1, \; i = 1 \text{ and } 2, \tag{10.14}$$

$$\mathbf{n} \bullet \mathbf{N}_i^* = \mathbf{n} \bullet (\mathbf{u}^* c_i^*), \; i = 1 \text{ and } 2, \tag{10.15}$$

$$\phi^* (r^*, \; -(L_r^* + L_c^* / 2)) = \phi^* (r^*, \; (L_r^* + L_c^* / 2)) - \phi_0^* = 0, \tag{10.16}$$

$$-\mathbf{n} \bullet \nabla^* \phi^* = \sigma_p^*, \tag{10.17}$$

$$\int_{-L_f^*/2}^{L_f^*/2} \frac{\partial^* \phi^*}{\partial^* \mathbf{n}} dz^* = 0, \tag{10.18}$$

where the characteristic surface charge density is $\varepsilon_0 \varepsilon_f RT / (Fa)$. The normalized z component electrical and hydrodynamic forces are, respectively,

$$F_E^* = \int \left[\frac{\partial^* \phi^*}{\partial^* z^*} \frac{\partial^* \phi^*}{\partial^* r^*} n_r + \frac{1}{2} (\frac{\partial^* \phi^*}{\partial^* z^*})^2 n_z - \frac{1}{2} (\frac{\partial^* \phi^*}{\partial^* r^*})^2 n_z \right] d\Gamma^*, \tag{10.19}$$

$$F_H^* = \int \left[-p^* n_z + 2 \frac{\partial^* u_z^*}{\partial^* z^*} n_z + (\frac{\partial^* u_r^*}{\partial^* z^*} + \frac{\partial^* u_z^*}{\partial^* r^*}) n_r \right] d\Gamma^*. \tag{10.20}$$

The dimensionless ionic current through the nanopore normalized by $FU_0 C_0 a^2$ is

$$I^* = \int (z_1 \mathbf{N}_1^* + z_2 \mathbf{N}_2^*) \bullet \mathbf{n} \, dS^*. \tag{10.21}$$

10.3   IMPLEMENTATION IN COMSOL MULTIPHYSICS

The dimensions of the geometry are $L_r = L_c = 40$ nm, $b = 4$ nm, and $R_r = 40$ nm. The radius and the length of the DNA nanoparticle are, respectively, $a = 1$ nm and $L_p = 10$ nm. Table 10.1 lists all the constants defined in COMSOL Multiphysics. The detailed implementation of the mathematical model in the graphical user interface (GUI) of COMSOL Multiphysics® is shown in Table 10.2. An example MATLAB M-file capable of geometry automation is as follows:

TABLE 10.1
Constant Table

Variable	Value or Expression	Description
D1	1.95e-9 [m^2/s]	Diffusivity of K+
D2	2.03e-9 [m^2/s]	Diffusivity of Cl−
z1	1	Valence of K+
z2	−1	Valence of Cl−
F	96485.3415 [C/mol]	Faraday constant
R0	8.314472 [J/K/mol]	Gas constant
T	300 [K]	Temperature
eps0	8.854187817e-12 [F/m]	Permittivity of vacuum
eps_r	80	Relative permittivity of fluid
eta	1.0e-3 [pa*s]	Fluid viscosity
rho	1e3 [kg/m^3]	Fluid density
V0	0.06 [V]	Applied voltage
cp	−0.01 [C/m^2]	Surface charge density of the particle
cw	0 [C/m^2]	Surface charge density of the nanopore
c0	ai^2*eps_r*eps0*R0*T/2/(F*a)^2	Bulk concentration
a	1e-9 [m]	Particle radius
lamda	sqrt(eps_r*eps0*R0*T/2/F^2/c0)	Debye length
ai	1	κa
U0	eps0*eps_r*R0^2*T^2/eta/a/F^2	Velocity scale
charges	F*a/R0/T/eps_r/eps0	Surface charge density scale
Re	rho*U0*a/eta	Reynolds number
cpd	cp*charges	Dimensionless particle's surface charge density
cwd	cw*charges	Dimensionless nanopore's surface charge density
D1d	D1*eta*F^2/ (eps0*eps_r*R0^2*T^2)	Dimensionless diffusivity of K+
D2d	D2*eta*F^2/ (eps0*eps_r*R0^2*T^2)	Dimensionless diffusivity of Cl−
V0d	V0*F/R0/T	Dimensionless applied voltage

TABLE 10.2
Model Setup in COMSOL Multiphysics 3.5a

Model Navigator	Select **Axial symmetry (2D)** and click **Multiphysics** button.
	Select **Chemical Engineering Module|Mass Transport|Nernst–Planck without Electroneutrality**. Remove the predefined variables in the **Dependent variables** and enter c1 and c2. Click **Add** button.
	Select **COMSOL Multiphysics|Electromagnetics|Electrostatics**. Click **Add** button.
	Select **COMSOL Multiphysics|Fluid Dynamics|Incompressible Navier–Stokes**. Click **Add** button.
Option Menu| **Constants**	Define variables in Table 10.1.
Physics Menu| **Subdomain Setting**	chekf mode
	Subdomain 1
	Tab c1
	D = D1d; R = 0; um = D1d/F; z = z1; u = u; v = v; V = V
	Tab c2
	D = D2d; R = 0; um = D2d/F; z = z2; u = u; v = v; V = V
	es mode
	Subdomain 1
	$\rho = 0.5*(ai)^2*(z1*c1+z2*c2)$; $\varepsilon r = 1/eps0$
	ns mode
	Subdomain 1
	Tab Physics
	$\rho = 0$; $\eta = 1$; $Fr = -0.5*ai^2*(z1*c1+z2*c2)*Vr$;
	$Fz = -0.5*ai^2*(z1*c1+z2*c2)*Vz$
	Tab Stabilization
	Deactivate Streamline diffusion and Crosswind diffusion
Physics Menu| **Boundary Setting**	chekf mode
	Axis: Axial symmetry
	Two ends of the nanopore: Concentration c1 = 1 and c2 = 1
	Particle surface: Convective flux for c1 and c2
	Walls of the nanopore and the reservoir: Insulation/Symmetry for c1 and c2
	es mode
	Axis: Axial symmetry
	Lower end of the nanopore: $V_0 = 0$
	Upper end of the nanopore: $V_0 = V0d$
	Particle surface: Surface charge $\rho s = cpd$
	Floating electrode: Electric potential $V_0 = Vf$
	Others: Zero charge/Symmetry
	ns mode
	Axis: Axial symmetry
	Two ends of the nanopore: Outlet|Pressure, no viscous stress $P_0 = 0$

continued

TABLE 10.2 (CONTINUED)
Model Setup in COMSOL Multiphysics 3.5a

	Particle surface: Inlet|Velocity $u_0 = 0$; $v_0 = vp$ where vp is the particle velocity
	Nanopore wall: Wall|No slip
	Reservoir wall: Wall|Slip
Options|Integration	Boundary selection: 5, 13, 14 (Particle surface)
Coupling	Name: Fh_z; Expression: -T_z_ns*2*pi*r; Integration order: 4
Variables|Boundary	Name: Fe_z; Expression: (-Vz*Vr*nr-0.5*Vz^2*nz+0.5*Vr^2*nz)*2*p
Variables	i*r; Integration order: 4
	Boundary selection: 7 (Floating electrode)
	Name: induced; Expression: nr*Vr*2*pi*r; Integration order: 4
Physics Menu|	Name: vp; Expression: (Fh_z+Fe_z)*1e20; Init (u): 0; Init (ut): 0
Global Equations	Name: Vf; Expression: induced*1e20; Init (u): 0; Init (ut): 0
Mesh|Free Mesh	Tab subdomain
Parameters	Subdomain 1|Maximum element size: 0.4
	Tab boundary
	Boundary 5, 6, 7, 8, 13, 14, 15, 16|Maximum element size: 0.1
Solve Menu|	Click = solve problem
Postprocessing	Result check: vp
Menu|	**Plot Parameters|Surface|Expression**: vp
	Click Apply
	Result check: Ionic current through the nanopore
	Boundary Integration|Expression:
	2*pi*r*(ntflux_c1_chekf-ntflux_c2_chekf)*nz
	Check **Compute surface integral (for axisymmetric modes)**
	Click **Apply**

```
%%%%%%%%%%%%%%%%%%%%%%%%%%%Floating electrod
e.m%%%%%%%%%%%%%%%%%%%%%%%%%%

flclear fem

% COMSOL version
clear vrsn
vrsn.name = 'COMSOL 3.5';
vrsn.ext = 'a';
vrsn.major = 0;
vrsn.build = 603;
vrsn.rcs = '$Name: $';
vrsn.date = '$Date: 2008/12/03 17:02:19 $';
fem.version = vrsn;


% NEW: Define all the particle's locations in a vector
zp=-35:1:35;
```

```
% NEW: Data storage for output
particle=zeros(length(zp),7);

% NEW: FOR loop to go through all the particle's locations
for i=1:length(zp)

% NEW: Get a particle's location in one single simulation
zp0=zp(i);

% Constants
fem.const = {'D1','1.95e-9[m^2/s]', ...
  'D2','2.03e-9[m^2/s]', ...
  'F','96485.3415[C/mol]', ...
  'R0','8.314472[J/K/mol]', ...
  'T','300[K]', ...
  'z1','1', ...
  'z2','-1', ...
  'eps_r','80', ...
  'eta','1.0e-3[Pa*s]', ...
  'v0','-0.06 [V]', ...
  'rho','1e3[kg/m^3]', ...
  'cw','0[C/m^2]', ...
  'c0','ai^2*eps_r*eps0*R0*T/2/(F*a)^2', ...
  'a','1e-9[m]', ...
  'U0','eps0*eps_r*R0^2*T^2/eta/a/F^2', ...
  'lamda','(eps_r*eps0*R0*T/(2*F^2*c0))^0.5', ...
  'ai','1', ...
  'v0d','v0*F/R0/T', ...
  'charges','F*a/R0/T/eps_r/eps0', ...
  'cwd','cw*charges', ...
  'Re','rho*U0*a/eta', ...
  'eps0','8.854187817e-12[F/m]', ...
  'D1_dless','D1*eta*F^2/(eps0*eps_r*R0^2*T^2)', ...
  'D2_dless','D2*eta*F^2/(eps0*eps_r*R0^2*T^2)', ...
  'E','v0/120/a', ...
  'cp','-0.01[C/m^2]', ...
  'cpd','cp*charges'};

% Geometry
g1=curve2([0,40],[-60,-60]);
g2=curve2([40,40],[-60,-20]);
g3=curve2([40,4],[-20,-20]);
g4=curve2([4,4],[-20,-10]);
g5=curve2([4,4],[-10,10]);
g6=curve2([4,4],[10,20]);
g7=curve2([4,40],[20,20]);
g8=curve2([40,40],[20,60]);
g9=curve2([40,0],[60,60]);
g10=curve2([0,0],[60,-60]);
g11=geomcoerce('solid',{g1,g2,g3,g4,g5,g6,g7,g8,g9,g10});
```

```
% NEW: Particle's location zp0 is used for geometry automa-
tion
g12=rect2('2','8','base','center','pos',{'0',num2str(zp0)},
   'rot','0');
g13=circ2('1','base','center','pos',{'0',num2str(zp0+4)},
   'rot','0');
g14=circ2('1','base','center','pos',{'0',num2str(zp0-4)},
   'rot','0');
g15=geomcomp({g12,g13,g14},'ns',{'R1','C1','C2'},'sf',
   'R1+C1+C2','edge','none');
g16=geomcomp({g15},'ns',{'CO2'},'sf','CO2','edge','none');
g17=geomcomp({g11,g16},'ns',{'CO1','CO3'},'sf','CO1-CO3',
   'edge','none');
g18=geomcomp({g17},'ns',{'CO2'},'sf','CO2','edge','none');
g19=fillet(g18,'radii',1.5,'point',[7,10]);

% Analyzed geometry
clear s
s.objs={g19};
s.name={'CO2'};
s.tags={'g19'};

fem.draw=struct('s',s);
fem.geom=geomcsg(fem);

% Initialize mesh
fem.mesh=meshinit(fem, ...
                  'hauto',5, ...
                  'hmax
edg',[5,0.1,6,0.1,7,0.1,8,0.1,13,0.1,14,0.1,15,0.1,16,0.1],
...
                  'hmaxsub',[1,0.4]);

% NEW: Display geometry with the sequence of the boundary
geomplot(fem, 'Labelcolor','r','Edgelabels','on','submode',
'off');

% (Default values are not included)

% Application mode 1
clear appl
appl.mode.class = 'ElectroKF_NernstPl';
appl.mode.type = 'axi';
appl.dim = {'c1','c2'};
appl.module = 'CHEM';
appl.gporder = 4;
appl.cporder = 2;
appl.assignsuffix = '_chekf';
clear prop
prop.analysis='static';
```

```
prop.equform='cons';
appl.prop = prop;
clear bnd
bnd.c0 = {0,{1;1},0,0};
bnd.type = {{'N0';'N0'},{'C';'C'},{'ax';'ax'},{'Nc';'Nc'}};
bnd.ind = [3,2,3,2,4,1,1,1,1,1,1,1,4,4,1,1];
appl.bnd = bnd;
clear equ
equ.D = {{'D1_dless';'D2_dless'}};
equ.V = {{'V';'V'}};
equ.init = 1;
equ.um = {{'D1_dless/F';'D2_dless/F'}};
equ.v = {{'v';'v'}};
equ.u = {{'u';'u'}};
equ.z = {{'z1';'z2'}};
equ.ind = [1];
appl.equ = equ;
fem.appl{1} = appl;


% Application mode 2
clear appl
appl.mode.class = 'Electrostatics';
appl.mode.type = 'axi';
appl.assignsuffix = '_es';
clear prop
clear weakconstr
weakconstr.value = 'off';
weakconstr.dim = {'lm3'};
prop.weakconstr = weakconstr;
appl.prop = prop;
clear bnd
bnd.V0 = {0,0,0,'v0d',0,0,'Vf'};
bnd.rhos = {'cpd',0,0,0,0,'cwd',0};
bnd.type = {'r','ax','V','V','nD0','r','V'};
bnd.ind = [2,4,2,3,1,6,7,6,5,5,5,5,1,1,6,6];
appl.bnd = bnd;
clear equ
equ.epsilonr = '1/eps0';
equ.rho = '0.5*(ai)^2*(z1*c1+z2*c2)';
equ.ind = [1];
appl.equ = equ;
fem.appl{2} = appl;


% Application mode 3
clear appl
appl.mode.class = 'FlNavierStokes';
appl.mode.type = 'axi';
appl.shape = {'shlag(2,''lm4'')','shlag(2,''lm5'')',
'shlag(1,''lm6'')','shlag(2,''u'')','shlag(2,''v'')',
'shlag(1,''p'')'};
```

```
appl.gporder = {30,4,2};
appl.cporder = {2,1};
appl.assignsuffix = '_ns';
clear prop
prop.analysis='static';
clear weakconstr
weakconstr.value = 'on';
weakconstr.dim = {'lm4','lm5','lm6'};
prop.weakconstr = weakconstr;
prop.constrtype='non-ideal';
appl.prop = prop;
clear bnd
bnd.symtype = {'ax','sym','sym','sym'};
bnd.v0 = {0,0,'vp',0};
bnd.type = {'sym','outlet','inlet','walltype'};
bnd.wcshape = [1;2;3];
bnd.velType = {!U0in','U0in','u0','U0in'};
bnd.ind = [1,2,1,2,3,4,4,4,4,4,4,4,3,3,4,4];
appl.bnd = bnd;
clear equ
equ.gporder = {{2;2;3}};
equ.F_z = '-0.5*ai^2*(z1*c1+z2*c2)*Vz';
equ.rho = 0;
equ.cporder = {{1;1;2}};
equ.F_r = '-0.5*ai^2*(z1*c1+z2*c2)*Vr';
equ.shape = [4;5;6];
equ.cdon = 0;
equ.sdon = 0;
equ.ind = [1];
appl.equ = equ;
fem.appl{3} = appl;
fem.sdim = {'r','z'};
fem.frame = {'ref'};
fem.border = 1;
clear units;
units.basesystem = 'SI';
fem.units = units;

% NEW: Boundary integration to calculate the hydrodynamic
force (Fh_z)
% NEW: and electrical force (Fe_z) on the particle
% NEW: Boundary integration to calculate net charge along
the floating electrode (induced)

% Coupling variable elements
clear elemcpl
% Integration coupling variables
clear elem
elem.elem = 'elcplscalar';
elem.g = {'1'};
```

```
src = cell(1,1);
clear bnd
bnd.expr = {{{},{},'nr*Vr*2*pi*r'},{{},'-T_z_
ns*2*pi*r',{}},{{}, ...
    '2*pi*r*(-Vz*Vr*nr-0.5*Vz^2*nz+0.5*Vr^2*nz)',{}}};
bnd.ipoints = {{{},{},'4'},{{},'4',{}},{{},'4',{}}};
bnd.frame = {{{},{},'ref'},{{},'ref',{}},{{},'ref',{}}};
bnd.ind = {{'1','2','3','4','6','8','9','10','11','12','15',
'16'},{'5', ...
    '13','14'},{'7'}};
src{1} = {{},bnd,{}};
elem.src = src;
geomdim = cell(1,1);
geomdim{1} = {};
elem.geomdim = geomdim;
elem.var = {'induced','Fh_z','Fe_z'};
elem.global = {'1','2','3'};
elem.maxvars = {};
elemcpl{1} = elem;
fem.elemcpl = elemcpl;

% NEW: Determine the particle velocity by balancing the
hydrodynamic force
% NEW: and electrical force on the particle
% NEW: Multiplier 1e20 is to make sure (Fh_z+Fe_z) is
extremely close to zero
% NEW: Determine the electric potential of the floating
electrode using the zero net charge

% ODE Settings
clear ode
ode.dim={'Vf','vp'};
ode.f={'induced*1e20','(Fh_z+Fe_z)*1e20'};
ode.init={'0','0'};
ode.dinit={'0','0'};
clear units;
units.basesystem = 'SI';
ode.units = units;
fem.ode=ode;

% Multiphysics
fem=multiphysics(fem);

% Extend mesh
fem.xmesh=meshextend(fem);

% Solve problem
fem.sol=femstatic(fem, ...

  'solcomp',{'lm4','v','u','V','c1','c2','lm6','lm5','p',
'Vf','vp'}, ...
```

```
'outcomp',{'lm4','v','u','V','c1','c2','lm6','p','lm5',
'Vf','vp'}, ...
'blocksize','auto');

% Save current fem structure for restart purposes
fem0=fem;

% NEW: Get velocities and displacments
data = postglobaleval(fem);
particle(i,1)=zp0;
particle(i,2)=data.y(2);

% NEW: force
% Integrate
I1=postint(fem,'T_z_ns*2*pi*r', ...
            'unit','', ...
            'recover','off', ...
            'dl',[5,13,14], ...
            'edim',1);

% Integrate
I2=postint(fem,'2*pi*r*(-Vz*Vr*nr-
0.5*Vz^2*nz+0.5*Vr^2*nz)', ...
            'unit','', ...
            'recover','off', ...
            'dl',[5,13,14], ...
            'edim',1);

particle(i,3)=I1;
particle(i,4)=I2;

% NEW: Current
% NEW: Anode
% Integrate
I3=postint(fem,'2*pi*r*(ntflux_c1_chekf-ntflux_c2_
chekf)*nz', ...
            'unit','mol/s', ...
            'recover','off', ...
            'dl',4, ...
            'edim',1);

particle(i,5)=I3;

% NEW: Current
% NEW: Cathode
% Integrate
I4=postint(fem,'2*pi*r*(ntflux_c1_chekf-ntflux_c2_
chekf)*nz', ...
            'unit','mol/s', ...
            'recover','off', ...
```

```
'dl',2,  ...
'edim',1);

particle(i,6)=I4;

particle(i,7)=(abs(I3)-abs(I4))/abs(I3)*100;

% NEW: Write the data into a file
dlmwrite('E500kv.dat',particle,',');

% NEW: Output the COMSOL Multiphysics GUI .mph file if
necessary
flsave(strcat('E500kv_zp=',num2str(zp0),'.mph'),fem);
end
```

10.4   RESULTS AND DISCUSSION

Usually, the nanopore is inherently charged. However, the effect of the nanopore's surface charge is assumed to be zero in the present study to highlight the effect due to the induced charge arising from the floating electrode. Three main factors—the applied electric field strength, the ratio of the particle radius to the Debye length, and the length of the floating electrode—on DNA translocation through a nanopore are investigated in this section. The electric field strength used in the following is obtained by dividing the electric potential difference between the two reservoirs by the total length of the computational domain. All the following results are presented in dimensionless form.

10.4.1   EFFECT OF ELECTRIC FIELD

The externally applied electric field plays an important role on the induced surface charge on the floating electrode, which in turn affects DNA translocation through the nanopore. Figure 10.2 shows the induced surface charge density along the floating electrode under different electric fields when $\kappa a = 1$ (dashed line, $E_z^* = 3.87 \times 10^{-4}$; circles, $E_z^* = 1.93 \times 10^{-2}$) and $\kappa a = 4$ (solid line, $E_z^* = 3.87 \times 10^{-4}$; squares, $E_z^* = 1.93 \times 10^{-2}$). Obviously, the $z^* < 0$ portion of the floating electrode carries a positive charge, which is opposite to the induced charge on the $z^* > 0$ portion. In addition, the induced surface charge shows a linear relationship to its local position except for the two ends of the floating electrode. The induced surface charge for $E_z^* = 1.93 \times 10^{-2}$ is divided by a factor of 50 to obtain a clear comparison with the case for $E_z^* = 3.87 \times 10^{-4}$. The perfect coincidence of symbols and lines indicates that the induced surface charge is linearly proportional to the applied electric field intensity. Furthermore, the induced surface charge increases as κa increases under a specific external electric field.

Figure 10.3 depicts the effect of the applied electric field on the particle mobility when $\kappa a = 4$, $L_f = L_c/2$ (a) and $L_f = L_c$ (b). The particle mobility is defined as the ratio of the particle velocity to the applied electric field, $\eta_p^* = u_p^*/E_z^*$. When the

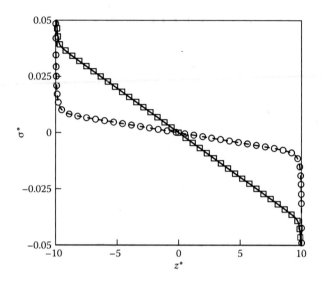

FIGURE 10.2 Surface charge density induced on the float electrode when $\kappa a = 1$ (dashed line, $E_z^* = 3.87 \times 10^{-4}$; circles, $E_z^* = 1.93 \times 10^{-2}$) and $\kappa a = 4$ (solid line, $E_z^* = 3.87 \times 10^{-4}$; squares, $E_z^* = 1.93 \times 10^{-2}$). Symbols are divided by 50 for a clear comparison. (From Zhang, M., Y. Ai, A. Sharma, S. W. Joo, D.-S. Kim, and S. Qian. 2011. Electrokinetic particle translocation through a nanopore containing a floating electrode. *Electrophoresis* 32:1864–1874 with permission from Wiley-VCH.)

floating electrode is not coated on the inner surface of the nanopore, the particle mobility is almost a constant inside the nanopore and independent of the applied electric field, which is regarded as the reference mobility. The particle mobility is increased inside the nanopore because of the enhanced electric field inside the nanopore compared to that in the reservoir.

When the floating electrode is coated on the inner surface of the nanopore and has a length of $L_f = L_c/2$, particle mobility becomes highly dependent on the applied electric field. When the applied electric field is relatively low, $E_z^* = 3.87 \times 10^{-4}$, the predicted particle mobility is nearly identical to the reference mobility. As the applied electric field is increased to $E_z^* = 3.87 \times 10^{-3}$, particle mobility is the same as the reference mobility in the reservoir region. When the particle approaches the floating electrode, the particle mobility decreases at the beginning and minimizes around $z_p^* = -5$. Subsequently, particle mobility starts to increase and matches the reference mobility at $z_p^* = 0$. When the particle moves away from the floating electrode, particle mobility increases and then maximizes around $z_p^* = 5$. After that, particle mobility begins to decrease and eventually matches the reference mobility when it exits the nanopore. If the applied electric field is further increased to $E_z^* = 1.93 \times 10^{-2}$, the variation tendency of the particle mobility is very similar to the case for $E_z^* = 3.87 \times 10^{-3}$. However, the deviation from the reference mobility becomes even more significant. The particle mobility becomes

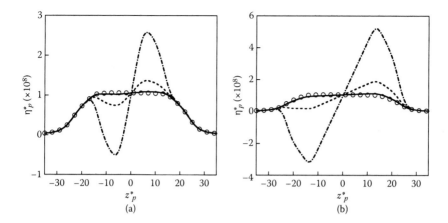

FIGURE 10.3 Variation of particle mobility along the axis of the nanopore when $\kappa a = 4$ and $L_f = L_c/2$ (a) and $L_f = L_c$ (b). Solid line, dashed line, and dash-dotted line represent, respectively, $E_z^* = 3.87 \times 10^{-4}$, $E_z^* = 3.87 \times 10^{-3}$, and $E_z^* = 1.93 \times 10^{-2}$ with floating electrode, while circles represent the reference mobility for a dielectric nanopore (i.e., $L_f = 0$). (From Zhang, M., Y. Ai, A. Sharma, S. W. Joo, D.-S. Kim, and S. Qian. 2011. Electrokinetic particle translocation through a nanopore containing a floating electrode. *Electrophoresis* 32:1864–1874 with permission from Wiley-VCH.)

negative in the $z^* < 0$ region, which implies that the particle is trapped inside the nanopore. Particle trapping inside a microchannel containing a floating electrode has been experimentally observed (Yalcin et al. 2010, 2011). If the floating electrode covers the entire inner surface of the nanopore, the applied electric field shows a more significant effect on particle mobility. When the applied electric field is $E_z^* = 1.93 \times 10^{-2}$, the particle cannot even enter the nanopore.

To understand the effect of the floating electrode on particle translocation through the nanopore, Figures 10.4a–10.4d show the flow fields at four different particle locations when $\kappa a = 4$, $E_z^* = 1.93 \times 10^{-2}$, and $L_f = L_c/2$. The induced surface charge on the floating electrode leads to the formation of induced EDL, which interacts with the applied electric field and thus generates ICEO inside the nanopore. As indicated in Figure 10.3, the $z^* < 0$ and $z^* > 0$ portions carry opposite surface charges. Therefore, the induced ICEOs next to the $z^* < 0$ and $z^* > 0$ portions oppositely face each other, which leads to a pair of circulating flows near the floating electrode, as shown in Figure 10.4a. When the particle is far away from the floating electrode, $z_p^* = -35$, the circulating flow is symmetric with respect to $z^* = 0$. In all these cases, the particle translocates from the $z^* < 0$ region to the $z^* > 0$ region.

Obviously, the induced ICEO along the axis in the $z^* < 0$ region is opposite to the particle translocation, indicating a retardation effect. However, the induced ICEO along the axis in the $z^* > 0$ region tends to enhance particle translocation. When the particle is located at $z_p^* = -12$, its motion is mainly affected by ICEO next to the $z^* < 0$ portion of the floating electrode. Accordingly, particle mobility

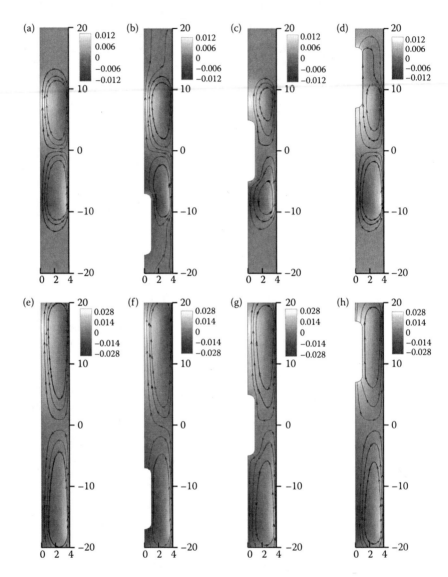

FIGURE 10.4 Flow field near the floating electrode when $\kappa a = 4$, $E_z^* = 1.93 \times 10^{-2}$, $z_p^* = -35$ (a, e), $z_p^* = -12$ (b, f), $z_p^* = 0$ (c, g), and $z_p^* = 12$ (d, h). The lengths of the floating electrode for (a–d) and (e–h) are, respectively, $L_f = L_c/2$ and $L_f = L_c$. Gradient levels denote the fluid velocity in the z direction, and streamlines with arrows denote the fluid velocity vector. The floating electrode locates at $-10 \le z^* \le 10$. (From Zhang, M., Y. Ai, A. Sharma, S. W. Joo, D.-S. Kim, and S. Qian. 2011. Electrokinetic particle translocation through a nanopore containing a floating electrode. *Electrophoresis* 32:1864–1874 with permission from Wiley-VCH.)

is reduced because of the opposite ICEO, as shown in Figure 10.3a. When the particle is located at $z_p^* = 0$, particle motion is affected by both ICEOs arising from the $z^* < 0$ and $z^* > 0$ portions of the floating electrode. Owing to the symmetric nature of the circulating flow, the retardation and enhancement effects cancel each other. Thus, the particle mobility matches the reference mobility, as revealed in Figure 10.3a. When the particle is located at $z_p^* = 12$, the particle motion is mainly affected by ICEO in the $z^* > 0$ portion of the floating electrode. As a result, the particle mobility is enhanced in this region, as shown in Figure 10.3a.

When the floating electrode covers the entire inner surface of the nanopore, $L_f = L_c$, the ICEO circulating flow takes effect along the entire nanopore. The flow patterns shown in Figures 10.4e–104h are very similar to Figures 10.4a–10.4d, respectively. As stated in Figure 10.2, the induced charge is linearly proportional to its local position. As a result, a longer floating electrode leads to a higher magnitude surface charge. It is also shown that the maximum flow velocity in Figures 10.4e–10.4h is larger than that in Figures 10.4a–10.4d. Therefore, a longer floating electrode induces a more significant effect on particle translocation through the nanopore.

Figure 10.5 depicts the effect of the floating electrode on the ionic current when $\kappa a = 4$, $L_f = L_c/2$ (Figure 10.5a), and $L_f = L_c$ (Figure 10.5b). When the applied electric field is relatively low, $E_z^* = 3.87 \times 10^{-4}$, the effect of the floating electrode on the ionic current is negligible. When the applied electric field is relatively high, $E_z^* = 1.93 \times 10^{-2}$, the induced surface charge is significantly increased. As a result, a large number of ions are attracted to the nanopore, which accordingly increases the ionic conductance of the nanopore. Therefore, the ionic current through the nanopore is increased under a relatively high electric field. The current deviation, defined as $\chi^* = (I^* - I_b^*)/I_b^*$, is a useful parameter for characterizing the current change due to the translocating particle. Figures 10.5c and 10.5d show, respectively, the effect of the floating electrode on the ionic current deviation when $\kappa a = 4$, $L_f = L_c/2$, and $L_f = L_c$. As the EDL is relatively thin, current blockade is predicted, which qualitatively agrees with the experimental observations obtained from the literature (Meller, Nixon, and Branton 2001; Li et al. 2003; Storm, Storm, et al. 2005; Storm, Chen, et al. 2005). However, the presence of the floating electrode has limited effect on the current deviation.

10.4.2   Effect of Ratio of Particle Radius to Debye Length

In the previous studies, the EDLs of the particle and the floating electrode do not overlap. Therefore, ICEO is the only factor that is responsible for the alternation of the particle mobility. In this section, we investigate the effect of EDL overlapping on particle translocation through the nanopore. Figure 10.6 shows the particle mobility as a function of the particle's location when $\kappa a = 1$, $L_f = L_c/2$ (a), and $L_f = L_c$ (b) under three different electric fields: $E_z^* = 3.87 \times 10^{-4}$ (solid lines), $E_z^* = 3.87 \times 10^{-3}$ (dashed lines), and $E_z^* = 1.93 \times 10^{-2}$ (dash-dotted lines).

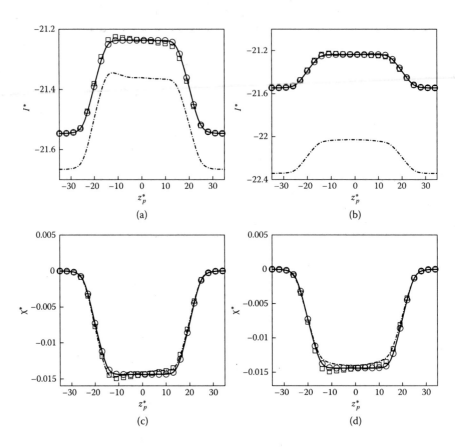

FIGURE 10.5 Ionic current (a, b) and ionic current deviation (c, d) through the nanopore as a function of the particle position when $\kappa a = 4$ and $L_f = L_c/2$ (a, c) and $L_f = L_c$ (b, d). Solid line (circles) and dash-dotted line (squares) represent, respectively, $E_z^* = 3.87 \times 10^{-4}$ and $E_z^* = 1.93 \times 10^{-2}$ with (without) floating electrode. The ionic current for $E_z^* = 3.87 \times 10^{-4}$ is multiplied by 50 for comparison. (From Zhang, M., Y. Ai, A. Sharma, S. W. Joo, D.-S. Kim, and S. Qian. 2011. Electrokinetic particle translocation through a nanopore containing a floating electrode. *Electrophoresis* 32:1864–1874 with permission from Wiley-VCH.)

The particle mobility in the absence of the floating electrode is included as the reference mobility.

Obviously, the variation of particle mobility under EDL overlapping is totally different from the cases shown in Figure 10.3. Take $L_f = L_c/2$ as an example; when the applied electric field is relatively low, $E_z^* = 3.87 \times 10^{-4}$, particle mobility is identical to the reference mobility when the particle is far away from the floating electrode. As the particle moves toward the floating electrode, the particle mobility begins to increase and then maximizes around $z_p^* = -12$. After that, it decreases below the reference mobility and minimizes at $z_p^* = 0$. When the particle moves out of the floating electrode, particle mobility starts to increase and

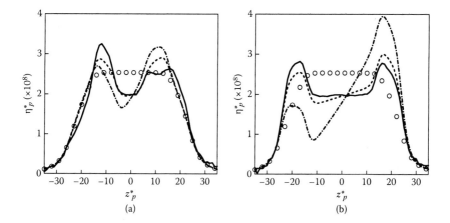

FIGURE 10.6 Variation of particle mobility along the axis of the nanopore when $\kappa a = 1$ and $L_f = L_c/2$ (a) and $L_f = L_c$ (b). Solid line, dashed line, and dash-dotted line represent, respectively, $E_z^* = 3.87 \times 10^{-4}$, $E_z^* = 3.87 \times 10^{-3}$, and $E_z^* = 1.93 \times 10^{-2}$ with a floating electrode, while circles represent the reference mobility without floating electrode. (From Zhang, M., Y. Ai, A. Sharma, S. W. Joo, D.-S. Kim, and S. Qian. 2011. Electrokinetic particle translocation through a nanopore containing a floating electrode. *Electrophoresis* 32:1864–1874 with permission from Wiley-VCH.)

maximizes around $z_p^* = 12$. Once the particle totally exits the nanopore, particle mobility matches the reference mobility again. When the applied electric field increases to $E_z^* = 3.87 \times 10^{-3}$, the variation tendency of the particle mobility is similar to the case for $E_z^* = 3.87 \times 10^{-4}$. However, the enhancement of particle mobility is reduced in the $z^* < 0$ region and is increased in the $z^* > 0$ region. When the applied electric field further increases, the decrease (increase) in the enhancement of the particle mobility in the $z^* < 0$ ($z^* > 0$) region becomes more obvious. The effect of a longer floating electrode ($L_f = L_c$) on particle mobility, as shown in Figure 10.6b, is similar to that shown in Figure 10.6a and thus is not discussed here.

The flow fields inside the nanopore at four different particle locations when $\kappa a = 1$, $L_f = L_c/2$, $E_z^* = 3.87 \times 10^{-4}$ (a–d), and $E_z^* = 1.93 \times 10^{-2}$ (e–h) are shown in Figure 10.7. As explained previously, a pair of ICEO circulating flows is generated in the floating electrode region when the particle is far away, as shown in Figures 10.7a and 10.7e. When the applied electric field is relatively low, $E_z^* = 3.87 \times 10^{-4}$, the induced charge is very small. As a result, the flow field inside the nanopore is dominated by the particle velocity, as shown in Figures 10.7b–10.7d. When the applied electric field is increased 50 times, ICEO is increased about 2,500 times. It is confirmed that ICEO is proportional to the square of the applied electric field. Therefore, ICEO under a relatively high electric field could retard the particle translocation in the $z^* < 0$ region and enhance the particle translocation in the $z^* > 0$ region, as shown in Figures 10.7f and 10.7h.

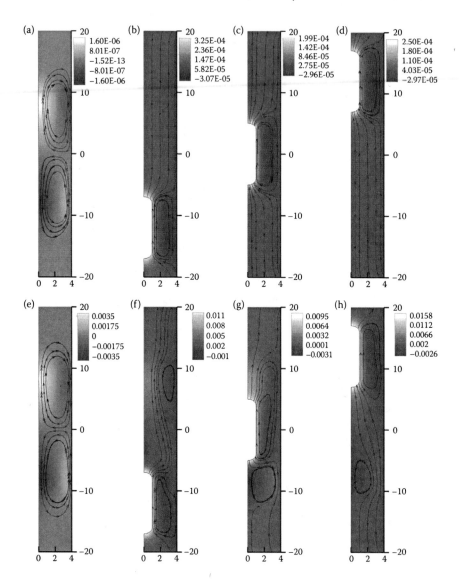

FIGURE 10.7 Flow field near the floating electrode when $\kappa a = 1$, $L_f = L_c/2$, $z_p^* = -35$ (a, e), $z_p^* = -12$ (b, f), $z_p^* = 0$ (c, g), and $z_p^* = 12$ (d, h). The electric fields for (a–d) and (e–h) are, respectively, $E_z^* = 3.87 \times 10^{-4}$ and $E_z^* = 1.93 \times 10^{-2}$. Gradient levels denote the fluid velocity in the z direction, and streamlines with arrows denote the fluid velocity vector. The floating electrode locates at $-10 \le z^* \le 10$. (From Zhang, M., Y. Ai, A. Sharma, S. W. Joo, D.-S. Kim, and S. Qian. 2011. Electrokinetic particle translocation through a nanopore containing a floating electrode. *Electrophoresis* 32:1864–1874 with permission from Wiley-VCH.)

In Chapter 9, it was revealed that the electrostatic interaction between the particle and the charged nanopore becomes important when their EDLs overlap. Figure 10.8 shows the net charge described by $c_1^* - c_2^*$ near the floating electrode at four different particle locations when $\kappa a = 1$, $L_f = L_c/2$, and $E_z^* = 1.93 \times 10^{-2}$. The ion distribution for $E_z^* = 3.87 \times 10^{-4}$ is similar to Figure 10.8. Positive and negative net charges are, respectively, induced on the $z^* > 0$ and $z^* < 0$ portions of the floating electrode. As a result, more negative (positive) ions are attracted to the $z^* < 0$ ($z^* > 0$) portions of the floating electrode, as shown in Figure 10.8a. When the particle is located at $z_p^* = -12$, the particle is mainly affected by the $z^* < 0$ portion of the floating electrode, which carries an opposite charge to the particle. Therefore, an attractive electrostatic interaction facilitates the particle translocation, as shown in Figure 10.6a. When the particle is located at $z_p^* = 0$, the particle is affected by both $z^* < 0$ and $z^* > 0$ portions of the floating electrode. The attractive electrostatic interaction between the particle and the $z^* < 0$ portion of the floating electrode retards particle translocation. Meanwhile, the electrostatic interaction between the particle and the $z^* > 0$ portion of the floating electrode is repulsive, which also retards particle translocation. Therefore, particle mobility is minimized at $z_p^* = 0$, as shown in Figure 10.6a. When the particle is located at $z_p^* = 12$, the particle is mainly affected by the $z^* > 0$ portion of the floating

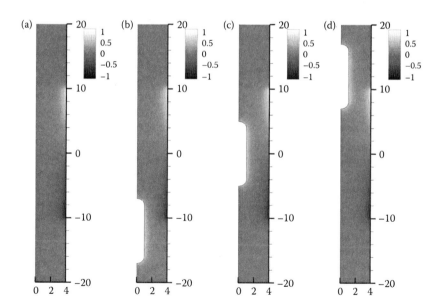

FIGURE 10.8 Net concentration distribution $c_1^* - c_2^*$ near the floating electrode when $\kappa a = 1$, $L_f = L_c/2$, $E_z^* = 1.93 \times 10^{-2}$, $z_p^* = -35$ (a), $z_p^* = -12$ (b), $z_p^* = 0$ (c), and $z_p^* = 12$ (d). The floating electrode locates at $-10 \le z^* \le 10$. (From Zhang, M., Y. Ai, A. Sharma, S. W. Joo, D.-S. Kim, and S. Qian. 2011. Electrokinetic particle translocation through a nanopore containing a floating electrode. *Electrophoresis* 32:1864–1874 with permission from Wiley-VCH.)

electrode, which exerts a repulsive force on the particle. Thus, particle mobility is enhanced. The particle-floating electrode electrostatic interaction is the main factor that alters the particle mobility when the applied electric field is relatively low and the EDLs overlap.

When the applied electric field is relatively high, $E_z^* = 1.93 \times 10^{-2}$, particle translocation is affected by both the particle-floating electrode electrostatic interaction and the ICEO. When the particle is located at $z_p^* = -12$, the particle-floating electrode electrostatic interaction enhances particle translocation. However, ICEO retards particle translocation. In addition, the particle-floating electrode electrostatic interaction is proportional to the applied electric field, and ICEO is proportional to the square of the applied electric field. Apparently, ICEO increases faster than the particle-floating electrode electrostatic interaction. Therefore, the enhancement of the particle mobility decreases in the $z^* < 0$ region as the applied electric field increases. When the particle is located at $z_p^* = 0$, ICEO arising from the $z^* < 0$ and $z^* > 0$ regions cancel each other. Consequently, the particle mobility at $z_p^* = 0$ matches the mobility under a relatively low electric field. When the particle is located at $z_p^* = 12$, particle mobility is enhanced by both the particle-floating electrode electrostatic interaction and ICEO. As a result, the enhancement of particle mobility increases in the $z^* > 0$ region as the applied electric field increases. The effect of the floating electrode on the ionic current when $\kappa a = 1$ is similar to that shown in Figure 10.5 and thus is not discussed here. In addition, the presence of the floating electrode exhibits a limited effect on the current deviation.

10.5   CONCLUDING REMARKS

Electrokinetic particle translocation through a nanopore containing a floating electrode has been studied using a continuum model composed of the PNP equations and the modified Stokes equations. A nonuniform surface charge is induced along the floating electrode, which shows a linear dependence on the applied electric field. Two effects arising from the floating electrode, ICEO and particle-floating electrode electrostatic interaction, could significantly affect particle translocation. As ICEO is proportional to the square of the applied electric field, it becomes negligible under a relatively low electric field. When the applied electric field is relatively high, ICEO could retard particle translocation when it approaches the floating electrode and enhance the particle translocation when it moves away from the floating electrode. When ICEO dominates particle electrophoresis, it could trap the particle near the floating electrode region, which shows a potential application for particle enrichment. The particle-floating electrode electrostatic interaction becomes significant only when the EDLs of the particle and the floating electrode overlap. An attractive or a repulsive electrostatic force acts on the particle, depending on the polarities of surface charges on the particle and the floating electrode. Generally, the particle-floating electrode electrostatic interaction facilitates particle translocation at the two ends of the floating electrode and retards particle translocation in the middle region of the floating electrode. In summary, the floating electrode could be used to control particle translocation through a nanopore. However, the presence of the floating electrode had limited effect on the ionic current deviation.

REFERENCES

Bazant, M. Z., and T. M. Squires. 2004. Induced-charge electrokinetic phenomena: Theory and microfluidic applications. *Physical Review Letters* 92 (6):066101.

Bazant, M. Z., and T. M. Squires. 2010. Induced-charge electrokinetic phenomena. *Current Opinion in Colloid and Interface Science* 15 (3):203–213.

Dhopeshwarkar, R., D. Hlushkou, M. Nguyen, U. Tallarek, and R. M. Crooks. 2008. Electrokinetics in microfluidic channels containing a floating electrode. *Journal of the American Chemical Society* 130 (32):10480–10481.

Hsu, J. P., Z. S. Chen, and S. Tseng. 2009. Effect of electroosmotic flow on the electrophoresis of a membrane-coated sphere along the axis of a cylindrical pore. *Journal of Physical Chemistry B* 113 (21):7701–7708.

Hsu, J. P., W. L. Hsu, and K. L. Liu. 2010. Diffusiophoresis of a charge-regulated sphere along the axis of an uncharged cylindrical pore. *Langmuir* 26 (11):8648–8658.

Jain, M., A. Yeung, and K. Nandakumar. 2009. Efficient micromixing using induced-charge electroosmosis. *Journal of Microelectromechanical Systems* 18 (2):376–384.

Li, J. L., M. Gershow, D. Stein, E. Brandin, and J. A. Golovchenko. 2003. DNA molecules and configurations in a solid-state nanopore microscope. *Nature Materials* 2 (9):611–615.

Meller, A., L. Nivon, and D. Branton. 2001. Voltage-driven DNA translocations through a nanopore. *Physical Review Letters* 86 (15):3435–3438.

Qian, S., and S. W. Joo. 2008. Analysis of self-electrophoretic motion of a spherical particle in a nanotube: Effect of nonuniform surface charge density. *Langmuir* 24 (9):4778–4784.

Qian, S., S. W. Joo, W. S. Hou, and X. X. Zhao. 2008. Electrophoretic motion of a spherical particle with a symmetric, nonuniform surface charge distribution in a nanotube. *Langmuir* 24 (10):5332–5340.

Qian, S., A. H. Wang, and J. K. Afonien. 2006. Electrophoretic motion of a spherical particle in a converging-diverging nanotube. *Journal of Colloid and Interface Science* 303 (2):579–592.

Squires, T. M., and M. Z. Bazant. 2004. Induced-charge electro-osmosis. *Journal of Fluid Mechanics* 509:217–252.

Squires, T. M., and M. Z. Bazant. 2006. Breaking symmetries in induced-charge electroosmosis and electrophoresis. *Journal of Fluid Mechanics* 560:65–101.

Storm, A. J., J. H. Chen, H. W. Zandbergen, and C. Dekker. 2005. Translocation of double-strand DNA through a silicon oxide nanopore. *Physical Review E* 71 (5):051903.

Storm, A. J., C. Storm, J. H. Chen, H. Zandbergen, J. F. Joanny, and C. Dekker. 2005. Fast DNA translocation through a solid-state nanopore. *Nano Letters* 5 (7):1193–1197.

Wu, Z. M., and D. Q. Li. 2008a. Micromixing using induced-charge electrokinetic flow. *Electrochimica Acta* 53 (19):5827–5835.

Wu, Z. M., and D. Q. Li. 2008b. Mixing and flow regulating by induced-charge electrokinetic flow in a microchannel with a pair of conducting triangle hurdles. *Microfluidics and Nanofluidics* 5 (1):65–76.

Yalcin, S. E., A. Sharma, S. Qian, S. W. Joo, and O. Baysal. 2010. Manipulating particles in microfluidics by floating electrodes. *Electrophoresis* 31 (22):3711–3718.

Yalcin, S. E., A. Sharma, S. Qian, S. W. Joo, and O. Baysal. 2011. On-demand particle enrichment in a microfluidic channel by a locally controlled floating electrode. *Sensors and Actuators B: Chemical* 153 (1):277–283.

Zhang, M., Y. Ai, A. Sharma, S. W. Joo, D.-S. Kim, and S. Qian. 2011. Electrokinetic particle translocation through a nanopore containing a floating electrode. *Electrophoresis* 32:1864–1874.

Zhao, C. L., and C. Yang. 2009. Analysis of induced-charge electro-osmotic flow in a microchannel embedded with polarizable dielectric blocks. *Physical Review E* 80 (4):046312.

Zhao, H., and H. H. Bau. 2007. Microfluidic chaotic stirrer utilizing induced-charge electro-osmosis. *Physical Review E* 75 (6):066217.

Index